OXFORD PAP

A Dictionar

Space Exploration

Dr Dasch currently manages the NASA National Space Grant Program and EPSCoR (Experimental Program to Stimulate Competitive Research). His more than 150 publications are in Marine Geochemistry, Geochronology (terrestrial and lunar rocks and meteorites), and science education. In addition to a Fulbright Fellowship Award, Dr Dasch has served as chairman of the Geological Society of America (Cordilleran Section), President of the Oregon Academy of Science, and is a recipient of the Burlington Northern Faculty Fellowship for Teaching. He recently received the International Astronautical Federation Frank J. Malina Medal for Space Education in Beijing, China. He is a Distinguished Alumnus of Sul Ross State University.

A Dictionary of

Space Exploration

Edited by

DR E. JULIUS DASCH

UNIVERSITY PRESS

OXFORD
UNIVERSITY PRESS

Great Clarendon Street, Oxford OX2 6DP

Oxford University Press is a department of the University of Oxford.
It furthers the University's objective of excellence in research, scholarship,
and education by publishing worldwide in

Oxford New York

Auckland Cape Town Dar es Salaam Hong Kong Karachi
Kuala Lumpur Madrid Melbourne Mexico City Nairobi
New Delhi Shanghai Taipei Toronto

With offices in

Argentina Austria Brazil Chile Czech Republic France Greece
Guatemala Hungary Italy Japan Poland Portugal Singapore
South Korea Switzerland Thailand Turkey Ukraine Vietnam

Published in the United States
by Oxford University Press Inc., New York

First published 2005

British Library Cataloguing in Publication Data
Data available

Library of Congress Cataloging in Publication Data
Data available

Typeset by SPI Publisher Services, Pondicherry, India
Printed in Great Britain by
Clays Ltd, St Ives plc

ISBN 0-19-280631-9 978-0-19-280631-4

Contents

List of Appendix Tables

AAS Abbreviation for either *American Astronautical Society or *American Astronomical Society.

AATSR Abbreviation for *advanced along track scanning radiometer.

ablation shield The heat shield on a spacecraft to protect it during re-entry. The intense heat generated by friction during the high-speed entry of the craft into the Earth's atmosphere burns away ablative (evaporating) materials on the heat shield, absorbing the heat and protecting the craft. The space shuttle uses a different kind of non-ablative protection made up of thousands of heat-absorbing tiles.

ABMA Abbreviation for *Army Ballistic Missile Agency.

abort A premature end of a space flight due to danger to the crew, the mission, or the environment, such as an accident or systems failure. This can occur during launch or during a mission. The *Apollo 13* mission was aborted on its third day after an oxygen tank exploded in the service module, and the crew was saved by an innovative new flight plan. NASA's space shuttle official launch aborts are: return to launch site (RTLS), trans-Atlantic landing (TAL), and abort to orbit (ATO). Only ATO has had to be used, on one mission in 1985. There are also several unofficial contingency aborts that could save the crew in the event of an emergency during a launch, but there is no guarantee of saving the craft or the crew. An abort is impossible during a shuttle descent, such as when the *Columbia* broke apart on 1 February 2003, killing its crew of seven.

abort sensing and implementation system (ASIS) The automatic detection system added to the *Atlas rocket in 1959 to create near-perfect reliability for crewed space flights. ASIS could sense and signal an impending catastrophic failure and then activate the crew's escape plan prior to the failure. Before this addition, the Atlas rocket had only launched weapons.

acceleration, secular A continuous and non-periodic change in orbital velocity of one body around another, or the axial rotation period of a body.

An example is the axial rotation of the Earth. This is gradually slowing down owing to the gravitational effects of the Moon and the resulting production of tides, which have a frictional effect on the Earth. However, the angular momentum of the Earth–Moon system is maintained, because the momentum lost by the Earth is passed to the Moon. This results in an increase in the Moon's orbital period and a consequential moving away from the Earth. The overall effect is that the Earth's axial rotation period is increasing by about 15 millionths of a second per year, and the Moon is receding from the Earth at about 4 cm per year.

a

accretion disc A flattened ring of gas and dust orbiting an object in space, such as a star or *black hole. The orbiting material is accreted (gathered in) from a neighbouring object such as another star. Giant accretion discs are thought to exist at the centres of some galaxies and *quasars.

If the central object of the accretion disc has a strong gravitational field, as with a neutron star or a black hole, gas falling onto the accretion disc releases energy, which heats the gas to extreme temperatures and emits short-wavelength radiation, notably X-rays.

ACE Acronym for **Advanced Composition Explorer*.

ace The human link between the NASA flight team and other teams on the ground when the *deep-space network tracks a spacecraft. The ace analyses real-time data from the spacecraft, watches for problems, and initiates the appropriate responses.

achondrite A stony *meteorite representing differentiated planetary material. Because differentiation is an igneous process, these are igneous rocks or breccias (broken fragments) of igneous rocks. They comprise about 15% of all meteorites and lack the *chondrules (silicate spherules) found in *chondrites.

acquisition of signal (AOS) The reception of information from spacecraft instruments (*telemetry) or locating spacecraft position by radar.

active galaxy A type of galaxy that emits vast quantities of energy from a small region at its centre, the active galactic nucleus (AGN). Active galaxies are subdivided into *radio galaxies, *Seyfert galaxies, BL Lacertae objects, and *quasars.

Active galaxies are thought to contain black holes with a mass 10^8 times that of the Sun, drawing stars and interstellar gas towards them in a process of *accretion. The gravitational energy released by the in-falling material is the power source for the AGN. Some of the energy may appear as a pair of opposed jets emerging from the nucleus. The orientation of the jets to the line of sight and their interaction with surrounding material determines the type of active galaxy that is seen by observers. *See also* STAR BURST GALAXY.

active microwave instrument (AMI) A scientific instrument on the *European Remote-Sensing Satellites, *ERS-1* and *ERS-2*. It combines a microwave imager and a *synthetic aperture radar to produce images of the Earth, as well as a *scatterometer to measure wind in the Earth's atmosphere.

ACTS Acronym for **Advanced Communications Technology Satellite*.

Adamson, James Craig (1946–) A US astronaut employed 1981–92 at mission control in the Johnson Space Center. He flew on two space shuttle missions: an 80-orbit flight in *Columbia* in August 1989 and a 142-orbit mission aboard *Atlantis* in August 1991.

Adamson was selected as an astronaut in 1984 and remained at NASA until 1992, after which he continued as a management consultant and member of the NASA Advisory Council. In 1994 he became an executive vice-president of the Lockheed company (later *Lockheed Martin).

ADEOS Abbreviation for **Advanced Earth-Observing Satellite.*

ADF Abbreviation for *Astrophysics Data Facility.

advanced along-track scanning radiometer (AATSR) A radiometer that measures Earth's surface temperatures. An upgrade of the *along-track scanning radiometer, it was carried aboard the European Space Agency's **Envisat.*

Advanced Communications Technology Satellite (ACTS) A satellite that tested new technology for communications satellites. Developed and operated by NASA's *Glenn Research Center, it was launched in September 1993 and conducted 103 experiments in 74 000 hours before ending its work in May 2000. *ACTS* utilized spot-beam antennae and on-board switching and processing systems for the tests.

http://acts.grc.nasa.gov/ Project information on the *ACTS* project, run by NASA's Space Communications Program at Cleveland, Ohio. The site includes detailed descriptions of *ACTS* spacecraft, operations, and experiments, as well as the latest news and events.

Advanced Composition Explorer (ACE) A NASA satellite launched on 25 August 1997 carrying six high-resolution spectrometers to measure the composition and abundances of atomic particles of solar, interplanetary, interstellar, and galactic origins, spanning the energy range from a few hundred eV (solar wind ions) up to 600 MeV (galactic cosmic ray nuclei). *ACE* also carried three instruments to monitor energetic electrons, hydrogen and helium ions, and a magnetometer. The spacecraft occupies a halo orbit about the L1 Earth–Sun *Lagrangian point from where it provides advance warning (about one hour) of geomagnetic storms that can overload power grids, disrupt communications on Earth, and endanger astronauts.

Advanced Earth-Observing Satellite (ADEOS) A Japanese remote-sensing satellite launched on 17 August 1996. It gathered data on climate change, the environment, and land and ocean processes. On 30 June 1997, ADEOS suffered damage to its solar array that made it go out of control and led to its subsequent failure.

There were eight instruments on board *ADEOS*, including the ocean colour and temperature scanner (OCTS), the interferometric monitor for greenhouse gases (IMGG), and the total ozone mapping spectrometer (TOMS). A second satellite, *ADEOS-II*, was launched on 14 December 2002. Its two main instruments are the advanced microwave scanning radiometer (AMSR), to observe the Earth's water cycle, and the global imager (GLI), for precise observations of the land, oceans, and clouds.

Advanced Land Observing Satellite (ALOS) A Japanese remote-sensing satellite that will map land areas, survey Earth resources, and monitor disaster sites using techniques developed from its predecessors such as the **Advanced Earth-Observing Satellite* (*ADEOS*). *ALOS* has three sensors: a stereo mapping camera to measure land elevation, a visible and near-infrared radiometer to observe the covering of land surfaces, and a synthetic aperture radar that enables day-and-night and all-weather land observation. *ALOS* was scheduled to be launched from the Tanegashima Space Centre in 2005.

Advanced Life-Support Program A NASA research programme to develop systems of long-term life support for humans in space, considered essential to the exploration of the Moon and Mars. Located at the *Johnson Space Center, the research team investigates ways to produce food, purify water and air, and regenerate oxygen. Experiments involve growing crop plants for food and oxygen regeneration, and researching biological and physico-chemical processes to turn waste into usable resources.

advanced microwave sounder unit (AMSU) A system of three separate instruments used on Earth-observation satellites, such as *NOAA-15* launched by NASA in 1998 for the *National Oceanic and Atmospheric Administration (NOAA). The microwave radiometer AMSU-B was supplied by Britain's Meteorological Office to provide global atmospheric humidity profiles.

advanced multi-mission operations system (AMMOS) The upgraded *telemetry system to monitor data from spacecraft in Earth orbit or deep space. Introduced in 1996, AMMOS has helped track probes such as the **Mars Global Surveyor* and **Cassini*. Its software is the multi-mission ground data system.

Advanced Radio Interferometry between Space and Earth (ARISE) A future NASA mission to launch one, or possibly two, 25-m radio telescopes to observe space in conjunction with numerous ground radio telescopes using *very long-baseline interferometry. The satellite for the ARISE mission, tentatively scheduled for 2008, will have a highly elliptical Earth orbit. It will look at super-massive *black holes and galactic nuclei to study how galaxies evolve and how matter reacts under great pressures. Scientists believe ARISE will produce images of black holes that are 5 000 times better than those from the **Hubble Space Telescope*.

advanced radioisotope power system (ARPS; or radioisotope thermal generator (RTG)) A small spacecraft generator producing electricity from heat generated by radioactive decay. Although public protests have occurred over the danger of radioactive fallout in the event of a rocket explosion, such a device is currently required for operating power in deep space, away from the Sun. ARPS has powered many Russian and US artificial satellites and space probes, including NASA's **Voyager* and **Galileo*.

Advanced Research in Telecommunications Systems (ARTES) A European Space Agency research programme concerning telecommunications and global navigation. It includes definitions of telecommunications standards, feasibility studies for new programmes, development and testing of equipment, and creation of infrastructures and partnerships.

***Advanced Satellite for Cosmology and Astrophysics* (*ASCA*)** Japan's fourth X-ray satellite, a joint project with NASA launched in February 1993. Its original name was *ASTRO-D*. Carrying four large-collection-area telescopes to scan X-rays in space, its missions included the verification of black holes and of particle acceleration. *ASCA* was destroyed when re-entering the atmosphere in March 2001, after observing more than 2 000 subjects.

advanced synthetic aperture radar (ASAR) An upgraded version of *synthetic aperture radar. It can operate continuously for 30 minutes to observe land and sea topography and other characteristics under all weather conditions.

advanced very high-resolution radiometer (AVHRR) A meteorological imaging system to detect radiation in order to determine the Earth's cloud cover and surface temperatures of its clouds, land, and water. AVHRR-1, with a four-channel radiometer, was launched in October 1988 on *Tiros-N*. The five-channel AVHRR-2 was launched in June 1981 on *NOAA-7*, and the six-channel AVHRR-3 in May 1998 on *NOAA-15*.

Aelita A Russian biomedical device to study the vascular systems and brains of cosmonauts, introduced on the *Salyut 7* space station launched in July 1982.

Aerobee rocket A US military *sounding rocket. The most successful sounding rocket of the space age, it could lift 45 kg of scientific instruments to an altitude of 121 km. Aerobee was developed by the Office of Naval Research and the Applied Physics Laboratory of Johns Hopkins University, and produced by Aerojet General for the US military. The many versions included the Aerobee-Hi and Aerobee 350. When the programme ended in 1985, 1058 rockets had been launched. The second-stage motor of the **Vanguard* rocket was based on the Aerobee-Hi.

aerobraking (or aerodynamic braking) The slowing of a space vehicle by using a planet's atmosphere. This is often done with parachutes or speed-brake flaps, such as the rudder of the space shuttle.

aerodynamic drag The atmospheric resistance that slows down a moving body; *see* DRAG.

aeronautics The science of travel through the Earth's atmosphere, including aerodynamics, aircraft structures, jet and rocket propulsion, and aerial navigation. It is distinguished from *astronautics, which is the science of travel through space.

In **subsonic aeronautics** (below the speed of sound), aerodynamic forces increase at the rate of the square of the speed.

Trans-sonic aeronautics covers the speed range from just below to just above the speed of sound and is crucial to aircraft design. Ordinary sound waves move at about 1 225 kph at sea level, and air in front of an aircraft moving slower than this is 'warned' by the waves so that it can move aside. However, as the flying speed approaches that of the sound waves, the warning is too late for the air to escape, and the aircraft pushes the air aside, creating shock waves, which absorb much power and create design problems. On the ground the shock waves give rise to a sonic boom. It was once thought that the speed of sound was a speed limit to aircraft, and the term sound barrier came into use.

a

Supersonic aeronautics is concerned with speeds above that of sound and in one sense may be considered a much older study than aeronautics itself, since the study of the flight of bullets (ballistics) was undertaken soon after the introduction of firearms.

Hypersonics is the study of airflows and forces at speeds above five times that of sound (Mach 5); for example, for guided missiles, space rockets, and advanced concepts such as *HOTOL (horizontal takeoff and landing). For all flight speeds, streamlining is necessary to reduce the effects of air resistance.

Astronavigation (navigation by reference to the stars) is used in aircraft as well as in ships and is a part of aeronautics.

http://www.aiaa.org/ Access to the *AIAA Bulletin*. This site also includes details of the Institute's research departments, recent conferences, technical activities, and project updates. If the extensive front page doesn't have what you need, the site is also fully searchable.

http://www.grc.nasa.gov/WWW/K-12/airplane/ NASA's introduction to aeronautics. Tutorials, practical experiments (build your own wind tunnel!), and information about aerodynamics, rockets, and space flight. An index of aerodynamics makes browsing the documents very easy.

aeroshell The *ablation shield that protects a spacecraft during entry into a planetary atmosphere.

Aerospace Directing and Controlling Centre The mission control centre for Chinese spacecraft. The facility, opened in 1999, was built in Beijing to handle the **Shen Zhou* spacecraft programme whose first launch was that year. The centre was in charge of China's first crewed spacecraft launched on 15 October 2003.

Aerospike The test engine developed for NASA's X-33 spacecraft, a small-scale prototype of a proposed reusable launch vehicle (RLV) that was to replace the space shuttle. The X-33, however, was cancelled in 2001. The Aerospike engine had a variable nozzle system to compensate for different altitudes. Rocketdyne Propulsion and Power, part of the *Boeing Company, developed and assembled it.

AFTE Acronym for *autogenic-feedback training exercise.

Agena rocket A NASA upper-stage rocket also used as an uncrewed docking target. The Lockheed company (later *Lockheed Martin) developed the rocket, 7.6 m long and 1.5 m in diameter, for the US Air Force. A series of Agena rockets were launched in 1966 to provide astronauts with rendezvous and docking experience. That year *Gemini 8*, commanded by Neil *Armstrong, was the first to attempt and accomplish this.

Agenzia Spaziale Italiana (ASI; English: Italian Space Agency) A member of the *European Space Agency, established by the Italian government in 1988. It has actively participated in many projects, including **Hermes* and **Spacelab*. ASI developed *Leonardo*, the multi-purpose logistics module that space shuttle astronauts attach to the **International Space Station*, to load and unload supplies and equipment.

http://www.asi.it/ Outlines the organization and objectives of the ASI (Italian Space Agency). The site profiles national and collaborative missions and has biographies of Italian astronauts. There are links to Italian Space Agency sites with more in-depth information on current space projects.

AIAA Abbreviation for *American Institute of Aeronautics and Astronautics.

aimpoint The point in a planet's plane of rotation about the Sun at which astronauts aim their spacecraft to enter an orbit around that planet, or for a fly-by.

airlock A cylindrical module in a space shuttle or space station used by astronauts to exit to space and return. When beginning *extravehicular activity (EVA), astronauts enter the airlock and close the inside hatch. After checking that the spacesuits operate perfectly, they undergo *prebreathing and then vent air out of the airlock. They then open the outer hatch and each pulls in a safety *tether from outside and attaches it before entering space. After an EVA, the astronauts float into the airlock, remove the tethers and hook them outside, close the outer hatch, repressurize the airlock, and open the inner hatch to return to the main spacecraft.

In the airlock is an **airlock adapter plate**, a framed fixture used to store spacesuits. Astronauts normally use the area in front of the fixture to put on and remove their spacesuits. The floor of the airlock is a **foot restraint platform**, which has a grid design into which a triangular metal piece on the sole of each shoe can be inserted.

air-sampling system A NASA term for bottles kept aboard the space shuttle to monitor air purity. They are kept in a *modular locker and removed for air sampling during a mission.

Akers, Thomas Dale (1951–) US astronaut. He clocked more than 800 hours in space during four space shuttle flights, including more than 29 hours of extravehicular activity. He was responsible for the space probe *Ulysses* during the October 1990 flight of *Discovery* that deployed it, flew on the first flight of *Endeavour* in May 1992, helped repair the *Hubble Space Telescope* from *Endeavour* in December 1993, and was the flight engineer when *Atlantis* docked with the Russian space station *Mir* in September 1996.

Akers was selected as an astronaut in 1987, later becoming the assistant director (technical) of the Johnson Space Center. He left NASA in 1997.

albedo The fraction of the incoming light reflected by a body such as a planet. A body with a high albedo (near 1) is very bright, while a body with a low albedo (near 0) is dark. The Moon has an average albedo of 0.12, Venus 0.76, and Earth 0.37.

alcohol An organic chemical compound used as an early rocket fuel. Wernher *von Braun's World War II German team developed alcohol-burning rockets. In the influential 1946 *Project RAND report, *Preliminary Design of an Experimental World-Circling Spaceship*, US engineers considered a four-stage rocket burning alcohol and liquid oxygen, but recommended a two-stage vehicle using liquid hydrogen and liquid oxygen.

a

Aldrin, Buzz (born Edwin Eugene Aldrin) (1930–) A US astronaut who, with Neil *Armstrong, landed on the Moon on 20 July 1969 during the *Apollo 11* mission, becoming the second person to set foot on the Moon.

During the *Gemini 12* flight with James *Lovell in 1966, Aldrin spent 5.5 hours in outer space without any ill effects. His 'walk' in space set a record for *extravehicular activity and proved that people could work outside an orbiting vehicle. He resigned from NASA in 1971, having spent 289 hours and 53 minutes in space.

Born at Montclair, New Jersey, Aldrin graduated from the US Military Academy at West Point, New York State, and flew for the US Air Force during the Korean War (1950–3) and later in West Germany. He received a Doctor of Science degree (ScD) from the Massachusetts Institute of Technology in 1963. In 1972 he resigned from the air force to found Research and Engineering Consultants. He also founded ShareSpace, a non-profit organization

(Image © NASA)

Buzz Aldrin photographed by Neil Armstrong during the historic *Apollo 11* Moon walk. Visible to the lower right of the picture is the foot of the *Eagle* lunar module. The astronaut's crisp footprints reveal the fine-grained, talc-like nature of the topmost lunar soil.

supporting space travel for the public. His books include (with Wayne Warga) *Return to Earth* (1975) and the science-fiction novel (with John Barnes) *Encounter with Tiber* (1996).

ALEXIS A small US satellite, known in full as *Array of Low-Energy X-ray Imaging Sensors*, which carried six wide-angle telescopes to survey the sky at ultrasoft X-ray and extreme ultraviolet wavelengths. It also contained a broadband VHF ionospheric survey experiment called Blackbeard. *ALEXIS* was launched by a Pegasus booster on 25 April 1993.

ALH 84001 A meteorite thought to be from Mars that was found at Allan Hills, Antarctica, in 1984. It is 4.5 billion years old, with minor, secondary veins about a billion years younger. NASA scientists at the Johnson Space Center reported in 1996 that fossil-like structures in ALH 84001, along with other evidence, indicated life on Mars. In particular, tiny magnetite crystals inside the meteorite have crystal forms like those produced inside bacteria on Earth. Whether or not this evidence proves that the rock, or Mars, once harboured life is still the subject of scientific debate.

http://curator.jsc.nasa.gov/curator/antmet/marsmets/alh84001/sample.htm NASA meteorite pages at Johnson Space Center include detailed information about the analysis of the meteorite from Mars, ALH84001. There are lots of pictures, plus articles about the search for past life on Mars.

Allen, Andrew Michael (1955–) US astronaut. He was aboard the space shuttle *Atlantis* in July and August 1992 to deploy the **European Retrievable Carrier* and demonstrate the *tethered satellite system (TSS). In March 1994, and February–March 1996, he was a member of the *Columbia* crews that conducted *microgravity tests. He flew a second TSS mission, STS-75, in 1996. The TSS was lost on the latter flight. He was selected as an astronaut in 1987 and logged over 900 hours in space before retiring from NASA in 1997.

Allen, Joseph Percival, IV (1937–) US scientist-astronaut. He was a mission specialist on *Columbia* in November 1982, the first space shuttle flight to carry a four-person crew, and again on *Discovery* in November 1984, when the crew successfully performed the first space salvage, retrieving two communications satellites. He was selected as a scientist-astronaut in 1967, logging 314 hours in space before leaving NASA in 1985.

all-sky monitor (ASM) An artificial-satellite unit made up of three wide-angle cameras to monitor bright X-ray sources. It rotates to scan 80% of the X-ray sky every 90 minutes, giving scientists an immediate image of any new phenomenon. Built by the Center for Space Research at the Massachusetts Institute of Technology, ASM was aboard NASA's **Rossi X-Ray Timing Explorer*, launched in December 1995.

along-track scanning radiometer (ATSR) An Earth-observation instrument of the European Space Agency that produces infrared images at a resolution of 1 km. Launched on **European Remote-Sensing Satellites*, the instruments scan the land, oceans, and atmosphere, contributing to the knowledge of climate studies and climate change. ATSR-1 was aboard *ERS-1* (launched in 1991) and ATSR-2 on *ERS-2* (launched in 1995). The *advanced along-track scanning radiometer (AATSR) was developed for the **Envisat*.

ALOS Abbreviation for **Advanced Land-Observing Satellite*.

Alouette Either of two Canadian satellites designed to study the ionosphere, tracking solar disturbances that affect radio communications. *Alouette 1* was Canada's first satellite and the first to be built by a country other than the USA or USSR. It was launched in September 1962 and set a longevity record of ten years. *Alouette 2* was launched in November 1965.

alpha proton X-ray spectrometer (APXS) An instrument used to identify and measure the concentrations of chemical elements on other planets. It exposes dust and rocks to a radioactive source that produces measurable alpha particles, protons, and X-rays, which in turn create measurable radioactive isotopes of the elements to be studied. NASA, which derived the spectrometer from a similar Russian instrument, attached it to the *Sojourner rover on the *Mars Pathfinder* mission which landed on Mars in July 1997, and to the Mars rovers, Spirit and Opportunity, which landed in January 2004.

ALSA Acronym for *astronaut life-support assembly.

ALSEP Acronym for **Apollo* lunar surface experiments package.

altimetry The process of measuring altitude, or elevation. Satellite altimetry involves using an instrument—commonly a laser—to measure the distance between the satellite and the ground.

altitude (or elevation) The angular distance of an object above the horizon, ranging from 0° on the horizon to 90° at the zenith. Together with the *azimuth, it forms a system of horizontal coordinates for specifying the positions of celestial bodies.

Altman, Scott Douglas (1959–) US astronaut. He was one of the seven-person crew that conducted *microgravity research aboard the space shuttle *Columbia* during a 16-day flight in April–May 1998. He was also in the crew on *Atlantis* that prepared the **International Space Station* in February 2000 for the first resident crew, and flew on **Columbia* in 2002. He was selected as an astronaut in 1994 and logged more than 38 days in space. He is now chief of the Shuttle Branch for the Astronaut Office.

aluminium (chemical symbol Al) A lightweight, silver-white, ductile, malleable, metallic element, used in the construction of many spacecraft. The frame and body of the space shuttle are made of aluminium, and an *ablation shield protects the aluminium body from re-entry heat.

American Astronautical Society (AAS) An independent US scientific and technical organization dedicated to the advancement of space science and exploration. Established in 1954, it had about 1 500 members in 2004, most of them engineers and scientists. Its headquarters are in Springfield, Virginia.

American Astronomical Society (AAS) An independent US organization promoting astronomy and astronautics. Established in 1899, it had nearly 7 000 members in 2004. The Division of Planetary Sciences, formed in 1968, is now its largest division, with more than 1 200 planetary scientists and astronomers. The AAS's headquarters are in Washington, DC.

http://www.aas.org/ Detailed information about the organization and activities of the society including their journals, meetings, and membership. Among the resources is an interactive database that allows visitors to search for Web sites and materials that have been reviewed by AAS members.

American Institute of Aeronautics and Astronautics (AIAA) A US society for aerospace engineers and scientists. It was formed in 1963 by the merger of the American Rocket Society (1930) and the Institute of Aerospace Sciences (1932). The AIAA calls itself the 'world's largest professional society devoted to the progress of engineering and science in aviation, space, and defence'. It had more than 31 000 members in 2004. Its headquarters are in Reston, Virginia.

Ames Research Center The NASA space-research field centre at Moffett Field, California, for the study of aeronautics and life sciences. It was established in 1939 as the Ames Aeronautical Laboratory, part of the *National Advisory Committee for Aeronautics, and became part of NASA in 1958. Ames managed the **Pioneer* probe series, designed the **Galileo* probe, and is involved in the search for extraterrestrial life. Its passive dosimeter system (PDS) monitors radiation on the **International Space Station*. Plans have been made to add a 200-hectacre (500-acre) NASA Research Park, including a laboratory and education facility for scientists and students.

http://www.arc.nasa.gov/ Latest news of Ames's space-technology and astrobiology research. These pages outline the purpose of astrobiology, future missions, and techniques. The 'Ask an Astrobiologist' feature allows registered users to put questions to top astrobiologists and participate in online discussions. The site is also a useful gateway to other astrobiology resources.

AMMOS Acronym for *advanced multi-mission operations system.

AMSU Acronym for *advanced microwave sounder unit.

AMU Acronym for *astronaut manoeuvring unit.

analyser of space plasma and energetic atoms (ASPERA) An artificial-satellite imager with four sensors to detect interplanetary plasma and electromagnetic fields. ASPERA-3 was launched on the European Space Agency's **Mars Express* on 2 June 2003 to study the interaction between the solar wind and the Martian atmosphere, determining how strongly interplanetary plasma and the electromagnetic field affect the atmosphere.

Anders, William Alison (1933–) US astronaut. He flew on *Apollo 8*, the first crewed flight to the Moon, orbiting it in December 1968 with Frank *Borman and James *Lovell. He was selected as an astronaut in 1963, and left NASA in 1969 to serve as executive secretary of the National Aeronautics and Space Council.

Anderson, Clayton Conrad (1959–) US astronaut. He joined the Johnson Space Center in 1983 and held several positions before being selected as an astronaut in 1998. He headed the trajectory design team as flight design manager for the **Galileo* mission, launched in 1989, and the **Gamma-Ray Observatory*, launched by the space shuttle *Atlantis* in 1991. He became chief of the flight design branch in 1993 and manager of the Emergency Operations Center in 1996.

Anderson, Michael Philip (1959–2003) US astronaut. He flew on the space shuttle *Endeavour* in January 1998 when it docked with the Russian space station *Mir* for the transfer of more than 4 080 kg of equipment, supplies, and water. He was selected as an astronaut in 1994. He died when the space shuttle *Columbia* broke up on re-entry in February 2003 after a 16-day mission to the *International Space Station.*

Anik A series of Canadian communications satellites, including *Anik D2*, which was deployed in *geosynchronous orbit from the space shuttle *Discovery* in November 1984.

announcement of opportunity (AO) A call for proposals for a satellite mission. AO is generally used to solicit proposals for unique, high-cost research investigation opportunities that typically involve flying experimental hardware provided by the proposer on one of NASA's Earth-orbiting or free-flying space flight missions. Proposals selected through AOs can be long-term projects, involve budgets of many millions of dollars for the largest programmes, and are usually awarded through contracts, even for non-profit organizations, although occasionally grants are used.

anomaly NASA term for an unexpected occurrence during a space flight.

ANS Abbreviation for **Astronomical Netherlands Satellite*.

Ansari X Prize A $10 million prize offered in 1996 to the first team to build and launch a spacecraft capable of taking three people to 100 km altitude and repeat the feat within a fortnight, before 1 January 2005. The offer was made by the privately funded X Prize Foundation of St Louis, Missouri. In May 2004, the prize was renamed the Ansari X Prize following a donation from entrepreneurs Anousheh and Amir Ansari. Two dozen teams from seven countries registered for the competition which was won on 4 October 2004 by Mojave Aerospace Ventures of Mojave, California, with **SpaceShipOne*.

anti-g garment An early name for a *g-suit.

antisuffocation valve A safety valve on the right backside of the space shuttle *launch/entry suit, to be released in case of an oxygen emergency. It was part of a design implemented after the *Challenger* disaster in 1986.

AO Abbreviation for *announcement of opportunity.

A-OK 'Very OK' or 'Great'. John 'Shorty' Powers, NASA's public-relations officer, introduced the term by misquoting astronaut Alan *Shepard during the first sub-orbital flight in May 1961. Powers soon discovered that Shepard had only voiced an enthusiastic 'OK!', but the press had already popularized the expression.

AOS Abbreviation for *acquisition of signal.

aphelion The point at which an object, travelling in an elliptical orbit around the Sun, is at its farthest from the Sun. This is a solar-orbit *apoapsis. The Earth is at its aphelion on 5 July.

apoapsis During an orbit of a spacecraft or other body, the furthermost point from the planet or body orbited. Specific apoapsis terms include **apogee** for an Earth orbit, **aphelion** for the Sun, **apolune** for the Moon, and **apojove** for Jupiter.

apogee The point at which a spacecraft, or other object, travelling in an elliptical orbit around the Earth is at its farthest from the Earth. This is an Earth-orbit *apoapsis.

Apollo asteroid A member of a group of *asteroids whose orbits cross that of the Earth. They are named after the first of their kind, Apollo, discovered in 1932 by German astronomer Karl Reinmuth and then lost until 1973. Apollo asteroids are so small and faint that they are difficult to see except when close to the Earth (Apollo is about 2 km across).

Apollo asteroids can collide with the Earth from time to time. In December 1994 the Apollo asteroid 1994 XM1 passed within 100 000 km of the Earth, the closest observed approach of any asteroid. A collision with an Apollo asteroid 65 million years ago may have been one of the causes of the extinction of the dinosaurs. A closely related group, the Amor asteroids, come close to Earth but do not cross its orbit.

***Apollo* lunar surface experiments package (ALSEP)** The overall name for the scientific instruments first deployed on the Moon by *Apollo 12* astronauts Pete *Conrad and Alan *Bean in November 1969. The package, designed to reveal the Moon's internal structure, included an ion detector, seismometer, magnetometer, laser ranging reflector, and solar wind spectrometer. Later ALSEP packages were deployed by *Apollo 14*, *Apollo 15*, *Apollo 16*, and *Apollo 17*.

http://www.lpi.usra.edu/expmoon/Apollo_Experiments.html Describes the *Apollo* lunar surface experiments package (ALSEP). For each of the six *Apollo* missions that landed on the Moon, there are details of the experiments carried out on the lunar surface by the crew, with supporting photographs.

***Apollo* project** The US space project to land a person on the Moon, achieved on 20 July 1969, when Neil *Armstrong was the first to set foot there. He was accompanied on the Moon's surface by Buzz *Aldrin; Michael *Collins remained in the orbiting command module.

The programme was announced in 1961 by US president John F Kennedy. The world's most powerful rocket, Saturn V (*see* SATURN ROCKET), was built to launch the *Apollo* spacecraft, which carried three astronauts. When the spacecraft was in orbit around the Moon, two astronauts would descend to the surface in the lunar module to take samples of rock and soil and set up experiments that would send data back to Earth. After four preparatory flights, *Apollo 11* made the first lunar landing. Five more crewed landings followed, the last in 1972. The total cost of the programme was over US$24 billion.

The *Apollo*–Saturn rocket complex stood 111 m tall. Saturn's first stage separated and second stage fired at 72 km; the third stage ignited at 177 km for extra power to put *Apollo* into Earth orbit at 28 000 kph, and later fired to send *Apollo* towards the Moon.

Apollo 1 During a preliminary check on the ground the three crew were killed by a fire on 27 January 1967. After this, NASA conducted five uncrewed test flights.

Apollo 7 The first successful *Apollo* mission to carry a crew, *Apollo 7* was a test flight sent into orbit around the Earth on 11 October 1968.

Apollo 8 Launched on 21 December 1968, this was the first mission to take a crew around the Moon.

Apollo 9 Launched on 3 March 1969, this mission tested the lunar module in orbit around the Earth.

Apollo 10 Launched on 18 May 1969, this mission successfully tested the lunar module 14.5 km above the surface of the Moon.

Apollo 11 After a launch on 16 July 1969, Armstrong and Aldrin landed the lunar module (named *Eagle*) in an area called the Sea of Tranquillity on the Moon's surface on 20 July 1969. Armstrong had to land manually because the automatic navigation system was heading for a field of boulders. On landing, Armstrong announced, 'Tranquillity base here. The *Eagle* has landed.' The module remained on the Moon for 22 hours, during which time the astronauts collected rocks, set up experiments, and mounted a US flag. Apart from a slight wobble when rejoining the command module, the return flight went without a hitch. After splashdown, the astronauts were quarantined as a precaution against unknown illnesses from the Moon.

Apollo 12 Launched on 14 November 1969, this mission achieved another successful Moon landing, in spite of twice being struck by lightning.

Apollo 13 Intended to be the third Moon landing, *Apollo 13* was launched on 11 April 1970 with the crew of John Swigert, Fred *Haise, and James *Lovell. On the third day of the mission Swigert reported to Houston, 'We've had a problem here.' An electrical fault had caused an explosion in one of the oxygen tanks in the service module, cutting off supplies of power and oxygen to the command module. The planned landing was abandoned and the rocket was sent round the Moon before heading back to Earth. The crew used the lunar module *Aquarius* as a 'lifeboat', though they had to endure near-freezing temperatures to save power, making sleep almost impossible. Attempting re-entry in the crippled ship almost led to disaster but the crew splashed down safely on 17 April.

Apollo 14 Launched on 31 January 1971, this mission reached the Moon on 5 February and returned to Earth on 8 February with samples of lunar rock.

Apollo 15 Launched on 26 July 1971, this mission used the first surface vehicle on the Moon, the *lunar roving vehicle.

Apollo 16 Launched on 16 April 1972, this mission gathered lunar soil and rock during 71 hours 2 minutes on the Moon.

Apollo 17 Launched on 7 December 1972, this was the last of the *Apollo* Moon landings. Detailed geological studies were carried out during a record 74 hours on the Moon, and large amounts of rock and soil were brought back.

http://www.hq.nasa.gov/alsj/a11/a11.html NASA page reliving the excitement of the *Apollo 11* mission, with recollections from the participating astronauts, images, audio clips, access to key White House documents, and a bibliography.

http://www.hq.nasa.gov/office/pao/History/apollo/welcome.html Simple site giving an overview of the *Apollo* programme, with information about each *Apollo* and *Skylab* mission, statistics, and a brief summary. There are links to other NASA *Apollo*-related pages.

(Image © NASA)

In a life raft following splashdown, *Apollo 11* astronauts Neil Armstrong, Michael Collins, and Buzz Aldrin wear biological isolation garments put on within their spacecraft. They are undergoing disinfection by US Navy personnel.

***Apollo–Soyuz* Test Project (ASTP)** A joint US-Soviet mission from 15 to 24 July 1975, in which the *Apollo 18* and *Soyuz 19* spacecrafts docked (17–19 July). They were connected by a special airlock/docking module so the two crews—three astronauts and two cosmonauts—could visit one another. The module was required because the atmosphere of the NASA vehicle was pure oxygen at less than sea-level pressure, while the Soviets used oxygen-nitrogen at sea-level pressure.

***Apollo Telescope Mount* (*ATM*)** A module of **Skylab* that contained five instruments for solar research. Crew members had to engage in extravehicular activity to retrieve the exposed film that recorded the Sun's spectrum. The module had four solar-panel wings for additional power. Its name was inherited from a cancelled *Apollo* mission.

apolune (or apselene) An *apoapsis in the orbit of a spacecraft or other body around the Moon.

***Applications Technology Satellite* (*ATS*)** A series of six NASA satellites launched to test new instruments and technology for communications, meteorological, and navigation satellites, especially those in *geosynchronous orbit. *ATS-1* was launched in December 1966 and contributed to communications and advance weather forecasting. *ATS-6*, launched in May 1974, sent educational and public-health telecasts to rural areas in the USA and India.

apsis (plural: apsides) Either of two points in an orbit, one being at maximum and the other at minimum distance from the controlling central body. The *apogee and *perigee of the Moon are apsides, as are the *aphelion and *perihelion of a planet.

APT Acronym for *automatic picture transmission.

Apt, Jay (born Jerome Apt III) (1949–) US astronaut. He was on four space shuttle flights, during which he made two space walks and logged 847 hours in space. He was a member of the crew of *Atlantis* in April 1991 that deployed the **Gamma-Ray Observatory*, flight engineer of *Endeavour* in September 1992 on a cooperative mission with Japan, commander of *Endeavour* in April 1994, mapping Earth's air pollution, and served aboard *Atlantis* in September 1996, docking with Russia's *Mir* space station. He joined NASA in 1980 and was a flight controller for shuttle payload operations when selected as an astronaut in 1985. He left NASA in 2001.

APU Acronym for *auxiliary power unit.

APXS Abbreviation for *alpha proton x-ray spectrometer.

Aqua A NASA satellite launched on 4 May 2002 to gather information on the Earth's water cycle. *Aqua* measures evaporation from the oceans, water vapour in the atmosphere, clouds, precipitation, soil moisture, sea ice, land ice, and snow cover. It is one of three spacecraft in NASA's Earth Observing System (EOS), the other two being **Aura* and **Terra*.

Arabsat A series of telecommunications satellites developed for the Arab nations, first launched in February 1985 by the Ariane 3 rocket.

Archambault, Lee Joseph (1960–) US astronaut selected in 1998. While awaiting his first flight, he worked on flight instrument upgrades for the space shuttle. He has also served as an Astronaut Support Person for launch and landing operations at the Kennedy Space Center.

arc minute (symbol ′) A unit for measuring small angles, used in geometry, surveying, map-making, and astronomy. An arc minute is one-sixtieth of a degree and is divided into 60 arc seconds (symbol ″). Small distances in the sky, as between two close stars or the apparent width of a planet's disc, are expressed in minutes and seconds of arc.

argument The angular distance of an orbiting spacecraft or other body from the object being orbited. For example, the **argument of periapsis** is the angle between the *periapsis (closest point of the orbiting body to the object it orbits) and the ascending *node (point where the orbit intersects the plane of the *ecliptic, moving from south to north of the ecliptic). The **argument of perihelion** is the angle between the ascending node and the perihelion (closest point between the Sun and an orbiting body).

Ariane rocket A series of launch vehicles built by the *European Space Agency, mainly to put telecommunications satellites into orbit two at a time. The first flight was in 1979. The launch site is the *Centre Spatial Guyanais, at Kourou in French Guiana. Ariane is a three-stage rocket using liquid fuels. Small solid-fuel and liquid-fuel boosters can be attached to its first stage to increase carrying power. Since 1984 it has been operated commercially by *Arianespace.

A more powerful version of the rocket, Ariane 5, was launched on 4 June 1996. However, it pitched over and disintegrated. A fault in the software controlling the take-off trajectory was to blame. In October 1998, another Ariane 5 rocket completed a successful mission, launching a dummy satellite into orbit. The model has had further successes, including the launch of India's *Insat 3B* satellite in March 2000, and the European Space Agency's *Rosetta* spacecraft launched on 2 March 2004.

Plans to launch the more powerful Ariane 5 EC-B were cancelled in 2005.

Arianespace A private company that has commercially operated the *Ariane rocket since 1984 for private and government customers. Arianespace is financed by European banks and aerospace industries. It holds more than 50% of the world market for placing satellites in geostationary transfer orbits.

Ariel A series of six UK satellites launched by NASA between 1962 and 1979. The most significant was *Ariel 5* in 1974, which made a pioneering survey of the sky's X-ray wavelengths.

ARISE Acronym for *Advanced Radio Interferometry between Space and Earth.

Armstrong, Neil Alden (1930–) US astronaut. On 20 July 1969, he became the first person to set foot on the Moon and made his now famous remark, 'That's one small step for man, one giant leap for mankind.' The Moon landing was part of the **Apollo* project.

He joined the US National Aerospace Program in 1962, and commanded *Gemini 8* in March 1966, linking with an uncrewed *Agena rocket. With Buzz *Aldrin and Michael *Collins in *Apollo 11* on 16 July 1969, he lifted off from the Kennedy Space Center in Florida, to land four days later on the Moon. The mission collected soil samples, explored, and set up scientific instruments during 22 hours on the lunar surface.

Born in Wapakoneta, Ohio, Armstrong gained his pilot's licence at 16, studied aeronautics at Purdue University, Indiana, and served as a US naval pilot in Korea in 1949–52 before joining NASA as a test pilot. Armstrong was professor of aeronautical engineering at the University of Cincinnati, Ohio, between 1971 and 1981, and then became chair of the aerospace company CTA, Inc.

http://www.jsc.nasa.gov/Bios/htmlbios/armstrong-na.html Biographical notes from NASA of the very first Moon walker. Details include personal information and education, as well as his work at NASA.

Army Ballistic Missile Agency (ABMA) A US Army missile and space facility at Huntsville, Alabama. In the 1950s, German rocket engineer Wernher *von Braun led a team of more than 2 000 at his Guided Missile Development Division, to develop large launching vehicles. Most of the ABMA staff were transferred to create NASA's *Marshall Space Flight Center in 1960.

ARPS Acronym for *advanced radioisotope power system.

Artemis The European Space Agency (ESA) telecommunications satellite launched on 12 July 2001 from Kourou, French Guiana, by the Ariane 5 launcher as part of the ESA's *Data-Relay and Technology Mission. Its rocket suffered a malfunction, and the satellite went into an abnormally low transfer orbit. After 18 months, however, this was corrected by on-board technology. The satellite carries three communication payloads (data relay, navigation enhancement, and extension of mobile communication via satellite) and an *ion engine thruster system to correct its position in geostationary orbit. The ion propulsion system uses only 10% of the fuel compared to a conventional system using chemical fuel. However, due to partial launch failure, chemical fuel was used to raise the orbit from 17 000 km to a circular orbit of 31 000 km altitude. The remaining 5 000 km to the geostationary orbit would be achieved by means of the ion thruster system. *Artemis* reached its geostationary orbit in January 2003, and has an operational lifetime of five to seven years. For their work in rescuing the satellite, its team received the 2003 Space Operation and Support Award from the *American Institute of Aeronautics and Astronautics.

ARTES Contraction of *Advanced Research in Telecommunications Systems.

Aryabhata India's first artificial satellite, launched by a Soviet rocket in April 1975. It was built by the Indian Space Research Organization and named after a 5th-century Indian mathematician. A power failure on its fourth day in orbit ended the intended experiments in X-ray astronomy and other areas, and all signals were lost the next day. *Aryabhata* burned up on re-entry in February 1992.

ASAR Acronym for *advanced synthetic aperture radar.

ASCA Acronym for **Advanced Satellite for Cosmology and Astrophysics.*

ascan Informal NASA shortened term for 'astronaut candidate'.

Ashby, Jeffrey Shears (1954–) US astronaut. Serving as a space shuttle pilot, he flew on *Columbia* during its mission in July 1999 to deploy the **Chandra X-Ray Observatory*, on the *Endeavour* flight in April 2001 to the *International Space Station* (*ISS*) to deliver and install the *Canadarm and resupply the logistics module **Raffaello*, and on the *Atlantis* flight in 2002 to the *ISS* to install the Integrated Truss Segment. He has logged 660 hours in space and 436 orbits. Ashby was selected as an astronaut in 1994.

ASI Abbreviation for *Agenzia Spaziale Italiana.

***AsiaSat 3* (later *HGS-1*)** A communications satellite that failed to achieve proper orbit but was salvaged by a return trip to the Moon. Intended for geostationary orbit over Asia, it was launched as *AsiaSat 3* in December 1997 on a Proton rocket, but went too low into orbit when the fourth stage fired late. The owner, Asia Satellite Telecommunications of Hong Kong, abandoned the satellite, which was assumed by Hughes Global Services Inc. Using an untried manoeuvre, Hughes fired an on-board motor in May 1998 to send *AsiaSat 3* on a nine-day trip to the Moon, using the Moon's gravity to return it to the correct Earth orbit. It was renamed *HGS-1* and now performs normally in geostationary orbit over the Americas, the Atlantic, and part of west Africa.

ASI/MET Contraction of *atmospheric structure instrument/meteorology package.

ASIS Acronym for *abort sensing and implementation system.

ASM Abbreviation for *all-sky monitor.

ASPERA Contraction of *analyser of space plasma and energetic atoms.

assembly Part of a *subsystem of a spacecraft. An attitude control subsystem, for example, may contain three or four reaction-wheel assemblies.

assembly, test, and launch operations (ATLO) The sequence of tightly scheduled events from the assembly of a spacecraft until its launch. NASA assembles a spacecraft in a *clean room, and tests it with computer programs, including one that simulates launch vibrations. The spacecraft is then transported to the launch site for additional tests, connected to the rocket's upper stage, and placed atop the *launch vehicle. The ATLO ends with the countdown and launch.

Association of Specialist Technical Organizations for Space (ASTOS) An association established in 1988 in the UK to support small and medium-sized companies involved in space-related projects. It had 19 member companies in 2004. Its services include gathering and exchanging information, creating working groups for specific subjects, and providing a voice for its members on the strategic direction of the UK's space policy.

asteroid Any of many thousands of small bodies, made of rock and minerals, that orbit the Sun. Most lie in a region called the **asteroid belt** between the orbits of Mars and Jupiter, and are thought to be fragments left over from the formation of the Solar System. About 100 000 asteroids may exist, but their total mass is only a few hundredths of the mass of the Moon. These rocky fragments range in size from 1 km to 900 km in diameter.

The largest asteroids are sometimes called minor planets; these include *Ceres (the largest asteroid, 940 km in diameter) and Vesta (which has a light-coloured surface and is the brightest as seen from Earth). Some asteroids are in orbits that bring them close to Earth and some, such as the *Apollo asteroids, which include *Eros and *Icarus, even cross Earth's orbit. They may be remnants of former comets.

The first asteroid was discovered by the Italian astronomer Giuseppe Piazzi at the Palermo Observatory, Sicily, on 1 January 1801. The first asteroid moon was observed, by the space probe **Galileo* in 1993, orbiting asteroid Ida.

Recent research

NASA's **NEAR Shoemaker* was launched in 1996, and from February 2000 orbited the asteroid Eros to collect data on asteroid composition. In 2001 it survived a deliberate crash landing to become the first spacecraft to land on an asteroid.

The NASA-funded near-earth asteroid tracking (NEAT) system at the Maui Space Surveillance Site in Hawaii, detected more than 18 000 asteroids between 1995 and 2001. In 2003 astronomers of the Lowell Observatory Near-Earth Object Search Programme in Arizona recorded the closest approach of a natural object to Earth—an asteroid, designated SQ222, roughly 10 m across, passed within 88 000 km of Earth, and was only observed as it was moving away from the planet.

There are an estimated 1 000–2 000 near-Earth asteroids (NEAs) with a diameter greater than 1 km, according to research published by US astronomers in 2000. In December 2001, US astronomers calculated that the chances of Earth being struck by such an asteroid were 1 in 5 000. This was considerably lower than the previous estimate of 1 in 1 500.

http://www.solarviews.com/eng/asteroid.htm What is the difference between an asteroid and a meteorite? Find out at this site, which contains a table of statistics about asteroids, a chronology of asteroid exploration, and images of selected asteroids.

http://nssdc.gsfc.nasa.gov/planetary/planets/asteroidpage.html Documents spacecraft encounters with asteroids and comets, and outlines future missions. The site contains a wealth of useful background information, an image gallery, and links to many important asteroid resources on the Internet.

ASTOS Acronym for *Association of Specialist Technical Organizations for Space.

ASTP Abbreviation for **Apollo–Soyuz* Test Project.

astrobiology The study of life in the universe. It is especially concerned with the origin, evolution, distribution, and future of life. Astrobiology combines biology, astronomy, geology, chemistry, and other sciences. NASA has an Astrobiology Institute at its *Ames Research Center, which works with academic and other research organizations to conduct research, promote the subject, and train researchers.

http://www.astrobiology.com/ Superb gateway to astrobiology on the Internet. With news of astrobiology articles appearing in scientific literature, the site features comprehensive links to astrobiology resources, from spacesuit design and life-support systems, to exobiology and terraforming.

ASTRO-F A Japanese infrared astronomy observatory due for launch in 2006. *ASTRO-F* will carry a 0.7-m telescope to make an all-sky survey with better sensitivity and resolution than ESA's **Infrared Astronomical Satellite* (*IRAS*) in 1983. Among *ASTRO-F*'s aims will be to help our understanding of how galaxies, stars, and planetary systems form and evolve.

astrogeology The study of geology in the universe, especially the Solar System. Many spacecraft concentrate on the geology of the Earth and other planets, as well as asteroids and comets. The *United States Geological Survey has an Astrogeology Branch, founded by US astronomer Eugene Shoemaker.

astrometry The measurement of the precise positions of stars, planets, and other bodies in space. Such information is needed for practical purposes including accurate timekeeping, surveying and navigation, and calculating orbits and measuring distances in space. Astrometry is not concerned with the surface features or the physical nature of the body under study.

Before telescopes, astronomical observations were simple astrometry. Precise astrometry has shown that stars are not fixed in position but have a *proper motion caused as they and the Sun orbit the Milky Way galaxy. The nearest stars also show *parallax (apparent change in position), from which their distances can be calculated. Above the distorting effects of the atmosphere, satellites such as **Hipparcos*, launched by the European Space Agency in 1989, can make even more precise measurements than ground telescopes, so refining the distance scale of space.

astronaut A person trained to make flights into space; the general term is often used to include cosmonauts (the Russian equivalent).

By 2001 NASA had employed more than 350 astronauts. Potential astronauts apply to the **Astronaut Candidate Program** and receive appointments from the **Astronaut Selection Office**. European Space Agency astronauts train at the European Astronauts Centre in Cologne, Germany.

NASA selections are usually made every two years for space shuttle *pilots and *mission specialists. The finalists undergo a week of personal interviews and medical evaluations. Selected astronauts are assigned to the Astronaut Office at the *Johnson Space Center for a one- to two-year training and evaluation programme, involving classes, training in *shuttle mission simulators, and then with flight controllers in the *mission control centre.

http://www.spacefacts.de/ Fascinating directory of space travellers. The site offers portraits and biographies of US, Russian, and international astronauts, in an easy-to-navigate database. The entries include personal data, missions, and career information.

Astronaut Hall of Fame US museum honouring astronauts. Located in Titusville, Florida, near the *Kennedy Space Center, it contains tributes to astronauts inducted into the hall (52 by 2004), and the world's largest collection of astronaut mementos. It was created in 1984 by the non-profit Mercury Seven Foundation (now the Astronaut Scholarship Foundation), established by the *'Original Seven' *Mercury* project astronauts. Since 1987 it has operated in conjunction with the US Space Camp on the same site. In 2003, Sally Ride became the first woman astronaut inducted into the Hall of Fame.

astronautics The science of space travel. It was formerly subsumed under *aeronautics.

Both subjects include **hypersonics**, the study of airflows and forces at speeds above five times that of sound (Mach 5), as experienced by space rockets and guided missiles.

http://www.jpl.nasa.gov/basics/ Illustrated tutorials covering essential aspects of space flight, including the space environment, flight projects, and operations. The site has a useful glossary of technical terms and abbreviations.

astronaut life-support assembly (ALSA) An *extravehicular activity pack introduced for **Skylab* astronauts. It was worn on the chest and connected to the spacecraft by an umbilical cord device. The ALSA regulated an astronaut's oxygen, water, and electrical power received from the spacecraft.

astronaut manoeuvring unit (AMU) A large *extravehicular activity unit created for **Gemini* project astronauts, with control arms on either side for steering. Extra layers were added to the legs of the *Gemini* spacesuit to protect the astronaut from the firing of the hydrogen peroxide thrusters. The AMU flew on *Gemini 9*, but was never operated.

astronaut wings A special patch awarded by the US Air Force and US Navy to their NASA astronauts. It is worn above the large name tag on the flight suit.

***Astronomical Netherlands Satellite* (*ANS*)** A joint Dutch–NASA orbiting astronomical observatory, launched on 30 August 1974, which observed celestial objects at ultraviolet and X-ray wavelengths. During its observing lifetime of 20 months, from September 1974 to June 1976, *ANS* measured the positions, spectra, and time variations of galactic and extragalactic X-ray sources in the energy range 2 to 15 keV, and observed about 400 objects in the ultraviolet.

astronomical unit (AU) A unit equal to the mean distance of the Earth from the Sun: 149.6 million km. It is used to describe planetary distances. Light travels this distance in approximately 8.3 minutes.

astronomy The science of the celestial bodies: the Sun, the Moon, and the planets; the stars and galaxies; and all other objects in the universe. It is concerned with their positions, motions, distances, and physical conditions and with their origins and evolution. Astronomy thus divides into fields such as astrophysics, celestial mechanics, and *cosmology. *See also* GAMMA-RAY ASTRONOMY, INFRARED ASTRONOMY, RADIO ASTRONOMY, ULTRAVIOLET ASTRONOMY, and X-RAY ASTRONOMY.

The powers of astronomy to explore the universe have been greatly extended by the use of rockets, satellites, space stations, and space probes. Even the range and accuracy of the conventional telescope may be greatly improved free from the Earth's atmosphere. When the USA launched the *Hubble Space Telescope* into permanent orbit in 1990, it was the most powerful optical telescope yet constructed, with a 2.4-m in mirror. It detects celestial phenomena seven times more distant (up to 14 billion light years) than any Earth-based telescope.

Astro Observatory One of two NASA orbiting ultraviolet observatories flown on **Spacelab* missions aboard the space shuttle in 1990 and 1995. Each observatory measured ultraviolet radiation from celestial objects, such as planets, stars, quasars, and nebulae (clouds of gas and dust). They each carried three instruments: the Hopkins Ultraviolet Telescope (HUT), the Ultraviolet Imaging Telescope (UIT), and the Wisconsin ultraviolet photo-polarimetry experiment (WUPPE). Astro 1 was launched on 2 December 1990 on *Columbia* and made 231 observations of 130 celestial objects during its ten-day flight. Astro 2 was launched on 2 March 1995 on *Endeavour* and made 385 observations of 260 celestial objects during 16.5 days, producing five times as much data as the first mission.

astrophysics The study of the physical nature of stars, galaxies, and the universe. It began with the development of spectroscopy in the 19th century, which allowed astronomers to analyse the composition of stars from their light. Astrophysicists view the universe as a vast natural laboratory in which they can study matter under conditions of temperature, pressure, and density that are unattainable on Earth.

Astrophysics Data Facility (ADF) NASA's facility that supports the processing, management, archiving, and distribution of mission data concerning astrophysics. Located at the *Goddard Space Flight Center, the facility designs and operates astrophysics data systems. It also works with Goddard's Laboratory for High-Energy Astrophysics (LHEA) and Laboratory for Astronomy and Solar Physics (LASP) to manage data for specific missions. The ADF is part of the Space Science Data Operations Office (SSDOO).

ASTROSAT An Indian multi-wavelength astronomy satellite planned for launch in 2006 or 2007. It will take spectra and measure brightness variations of galactic and extra-galactic sources at optical, ultraviolet, and X-ray wavelengths simultaneously.

Athena rocket US rocket that has two versions: the three-stage Athena 1 that can launch up to 788 kg, and the four-stage Athena 2 capable of launching up to 1 958 kg. Athena rockets have flown six times, with Athena 1 first launched in 1997 and Athena 2 in 1998. Built by Lockheed Martin, the design was selected by NASA in March 2001 as one of three (the others are the Delta and Atlas) for its launch services for payloads that weigh 1 485 kg or more.

Athena rockets launched NASA's *Lewis* remote-sensing satellite in 1997 and the **Lunar Prospector* to the Moon in 1998. In September 2001, the Athena launched several small experimental satellites for NASA and the US Air Force from Kodiak Island in Alaska.

Atlantis The fourth of the US *space shuttles. It made its first flight on 3 October 1985 with a crew of five.

In 1989 *Atlantis* launched the **Galileo* spacecraft on a six-year journey to the planet Jupiter and in 1995 undertook the first of seven missions to dock with the Russian space station **Mir*, exchanging crew members. After a refit in 1997–8, it resumed service with a series of launches from mid-1999 in support of the **International Space Station* (*ISS*), and on 14 February 2001 successfully delivered the US science laboratory module *Destiny* to the *ISS*. After the *Columbia* shuttle tragedy in 2003, when the shuttle broke up on re-entry, *Atlantis* was one of the three remaining shuttles, along with *Discovery* and *Endeavour*. It was named after a research vessel at the Woods Hole Oceanographic Institute in Massachusetts active from 1930 to 1966.

ATLAS Contraction of *Atmospheric Laboratory for Applications and Science.

(Image © NASA)

The upgraded interior of the space shuttle *Atlantis*. Nicknamed the 'glass cockpit', the new multifunction electronic display subsystem (MEDS) gives clear graphical displays of spacecraft attitude and flight properties.

a

Atlas rocket US rocket, originally designed and built in 1946 as an intercontinental missile for the US Air Force but subsequently adapted for space use.

The rocket used a 'balloon tank' construction with thin stainless steel sections that had to be pressurized by helium gas when not fuelled to prevent their collapse. Atlas rockets, using a liquid oxygen fuel, launched US astronauts into orbit during the **Mercury* project (1961–3).

Upper stages added at a later date created the Atlas–Agena and Atlas–Centaur rockets. The addition of an *Agena rocket as the upper stage on an Atlas rocket was used by the US Air Force to launch spy satellites. The Atlas–Agena rocket launched the **Mariner* spacecraft, and in 1965 and 1966 the Atlas rocket placed the Agena rocket in orbit as a docking target for *Gemini* spacecraft.

The addition of a *Centaur rocket to the Atlas rocket created a more powerful launch vehicle, which could lift 4 670 kg. The combination sent the **Surveyor* spacecraft to the Moon and was also often used by the US Air Force, forming the basis of a commercial Atlas fleet.

ATLO Acronym for *assembly, test, and launch operations.

ATM Abbreviation for *Apollo Telescope Mount.

atmospheric entry probe A space probe designed to collect and transmit data about a planet's atmosphere, usually including its composition, pressure, density, and temperature. Such a mission is relatively short, and the spacecraft often has no propulsion, requiring only batteries and equipment to measure and communicate. Such probes are often destroyed by the pressures of the alien environments.

An atmospheric entry probe is usually deployed by the larger spacecraft that transports it to the planet, such as the Jupiter probe launched by **Galileo* in 1995 and the *Huygens* probe being carried to Saturn's moon Titan by **Cassini* for a 2004 arrival. A planet's atmosphere can also be measured by a *lander that carries instruments such as the *atmospheric structure instrument/meteorology package.

Atmospheric Laboratory for Applications and Science (ATLAS) A series of three laboratories flown on **Spacelab* missions aboard the space shuttle to investigate the interactions of the Sun on the Earth's atmosphere.

ATLAS-1 was launched on 24 March 1992 on *Atlantis* and returned on 2 April. It had 12 instruments (from the USA, France, Germany, Belgium, Switzerland, the Netherlands, and Japan) to study ultraviolet astronomy, atmospheric chemistry, solar radiation, and space plasma physics.

ATLAS-2 was flown on *Discovery*, launched on 8 April 1993 and returned 17 April. It collected data on the relationship between the Sun's energy output and the Earth's middle atmosphere, focusing on the effects of this on the ozone layer.

ATLAS-3 was launched on *Atlantis* on 3 November 1994 and landed on 14 November. It measured the northern hemisphere's middle atmosphere in late autumn, when the Antarctic ozone hole diminishes, to study the effects of the hole on the mid latitudes.

atmospheric revitalization rack NASA term for the space shuttle rack that houses the carbon dioxide removal system.

atmospheric structure instrument/meteorology package (ASI/MET) An instrument package used by spacecraft to provide data on a planet's atmosphere. Its sensors measure atmospheric density, temperature, pressure, and wind from altitudes in excess of 100 km down to the planet's surface. An ASI/MET subsystem was carried aboard NASA's **Mars Pathfinder* in 1997.

ATS Abbreviation for *Applications Technology Satellite.

ATSR Abbreviation for *along-track scanning radiometer.

attitude control The control of the orientation of a spacecraft. The position of stars is used to calculate this, and attitude is maintained by the **attitude and orbit control subsystem** (AOCS). This employs control moment gyroscopes or other stabilizers to measure a spacecraft's attitude, and uses a computer program to correct problems via the thrusters, turning the spacecraft through a roll, pitch, or yaw angle. The AOCS also controls and maintains the spacecraft's orbit by the use of thrusters.

ATV Abbreviation for *Automated Transfer Vehicle.

A-type rocket (Russian name: R-7 rocket) A name used in the West for a series of Soviet rocket boosters. The A-1 launched **Sputnik* and the cosmonaut Yuri *Gagarin, the first person in space. The A-2, with upper stages added, was the launch vehicle for the **Soyuz* spacecraft.

Augustine Advisory Committee A committee formed in 1990 by the *National Space Council to advise on the future of the US space programme. Norman Augustine, chair and chief executive of the Martin Marietta Corporation (later merged with Lockheed to form *Lockheed Martin), headed the 12-member committee. Its report, released in December 1990, recommended an increase in satellites and space probes for scientific purposes, including environmental research. It did not encourage crewed missions to the Moon or Mars, advising NASA to concentrate on the space shuttle and space station programmes.

Aura A pollution-monitoring satellite launched on 15 July 2004, the third and final part of NASA's Earth Observing System (EOS). *Aura*'s purpose is to help establish whether the ozone layer is recovering, whether air quality is getting worse, and whether the Earth's climate is changing. *Aura* will continue the observations made by NASA's **Upper Atmosphere Research Satellite* (UARS) and *Total Ozone Mapping Spectrometer (TOMS). In addition, *Aura* will probe the Earth's troposphere, the region of the atmosphere from the ground to about 10 km. *Aura* complements the two other Earth Observing System satellites, **Terra* and **Aqua*.

Aurora The long-term plan initiated by the European Space Agency for the robotic and human exploration of the Solar System, notably Mars, the Moon and the asteroids. Among its aims is to investigate the possibility of life in other parts of the Solar System. The Aurora programme will involve a series of major Flagship missions preceded by smaller technology demonstrations known as Arrow missions.

The first two Flagship missions are *ExoMars, which will land an automatic rover on the planet, and a Mars sample return mission. In this latter mission a spacecraft will be placed into orbit around Mars from which a descent module will be released to land on the surface. Once samples of Martian soil have been collected an ascent vehicle will carry them into orbit around Mars to rendezvous with a re-entry vehicle which will return them to Earth. In preparation, two Arrow missions will be undertaken: a re-entry vehicle to simulate the conditions that will be experienced by a sample capsule returning from Mars and a mission to demonstrate the use of aerocapture, a technique that saves on fuel by using friction with the planet's upper atmosphere to brake a spacecraft into orbit.

Further steps may include a robotic Mars base and a human landing on the Moon. The final goal is to send European astronauts to Mars as part of an international endeavour in the 2030s.

http://www.esa.int/SPECIALS/Aurora/index.html

aurora The coloured light in the night sky near the Earth's magnetic poles, called **aurora borealis** ('northern lights') in the northern hemisphere and **aurora australis** ('southern lights') in the southern hemisphere. Although aurorae are usually restricted to the polar skies, fluctuations in the *solar wind occasionally cause them to be visible at lower latitudes. An aurora is usually in the form of a luminous arch with its apex towards the magnetic pole, followed by arcs, bands, rays, curtains, and coronae, usually green but often showing shades of blue and red, and sometimes yellow or white. Aurorae are caused at heights of over 100 km by a fast stream of charged particles from solar flares and low-density 'holes' in the Sun's corona. These are guided by the Earth's magnetic field towards the north and south magnetic poles, where they enter the upper atmosphere and bombard the gases in the atmosphere, causing them to emit visible light.

http://www.exploratorium.edu/learning_studio/auroras/ Take a tour of the aurorae. This page provides an in-depth explanation for the aurora phenomenon, includes a number of images taken from Earth and space, and sheds light on where they are best viewed. You can view a video clip of the aurora borealis and listen to audio clips of David Stern of NASA.

Aurora 7 A **Mercury* project spacecraft that flew the second orbital mission in May 1962, with astronaut Scott *Carpenter. *Aurora 7* lost fuel when two of its systems were operated simultaneously. Attitude control became difficult to maintain, and Carpenter had to fire his thrusters and retrorockets manually, landing 402 km from the proposed landing site in the Pacific Ocean. Some on-board experiments also failed during the flight. See illus. on p. 28.

a

(Image © NASA)
Engineers of the McDonnell Aircraft Company check the *Aurora 7* spacecraft prior to astronaut Scott Carpenter's triple-orbit mission round Earth. The *Mercury* project flight suffered technical difficulties and very nearly ended in disaster.

AUSSAT An organization formed in 1981 by the federal government of Australia and Telecom Australia to own and operate Australia's domestic satellite system. It has now been renamed the Optus Corporation. The first satellite, *AUSSAT-1*, was launched by the US space shuttle *Discovery* in 1985, and the third satellite was launched from French Guiana in 1987 by the European Space Agency's Ariane rocket launcher. The *AUSSAT* satellite system enables people in remote outback areas of Australia to receive television broadcasts.

auto An automatic pilot control system in the space shuttle. The other two control options are *control-stick steering and *direct control.

autogenic-feedback training exercise (AFTE) A NASA procedure to improve an astronaut's tolerance to *space sickness. It combines biofeedback and autogenic therapy, which is a learned self-regulation technique. AFTE can control space sickness symptoms involving the autonomous nervous system, such as nausea and vomiting. The exercise was developed at the Ames Research Center and was first used on the third *Spacelab* mission in 1985.

Automated Transfer Vehicle (ATV) A crewless supply ship for the *International Space Station* (ISS), built by the European Space Agency. The cylindrical ATV, 10.3 m long and 4.5 m in diameter, with X-shaped solar arrays, can ferry 7.5 tonnes of cargo at a time to the *ISS* from its launch site at Kourou, French Guiana. Astronauts aboard the *ISS* will be able to enter the pressurized cargo section of the ATV, while its engines will be used to adjust the station's orbit. Each ATV will remain docked with the *ISS* for up to six months before undocking and burning up in the Earth's atmosphere, taking with it up to 6.5 tonnes of *Station* waste. The ATV is due to enter service in 2006. At least seven ATVs will be built, making annual trips to the *ISS*.

automatic interplanetary station The Russian name for an uncrewed planetary space probe. These have included *Luna 3* to the Moon and *Zond 2* to Mars. The spacecraft are fully automated, operating without ground-control signals, using automatic systems for tasks ranging from temperature control to film developing.

automatic picture transmission (APT) A system that provides real-time data from meteorological satellites, such as the European Space Agency's **Meteosat* series. The transmission, in analogue video format, can be displayed as an image on a computer screen when a satellite is within range of a ground station. APT is derived from the *advanced very high-resolution radiometer (AVHRR) system.

APT was used on the **Tiros* series of satellites launched in 1966 by NASA and the Environmental Science Services Administration.

auxiliary power unit (APU) A unit on the space shuttle that is an auxiliary power source to gimbal (turn) the main engines. It is used during the shuttle's launch (activated by the pilot five minutes before blast-off) and re-entry.

Avco Corporation US company involved in the *Apollo* programme. It worked on the re-entry considerations in the design of the spacecraft and also developed Chartek fire-retardant material for the *Apollo* spacesuit. Chartek delayed the suit's temperature build-up during a fire and could swell to six times its thickness to create a protective barrier.

Located in Wilmington, Massachusetts, Avco Manufacturing was the third-largest US producer of World War II materials. It became the Avco Corporation in 1959 and was merged with Textron Inc., of Providence, Rhode Island, in 1985.

Avdeyev, Sergei Vasilyevich (1956–) Russian cosmonaut. Selected as a cosmonaut in 1985, Avdeyev holds the world record of 748 days of space flight time, during three missions aboard the *Mir* space station and *Soyuz* TM-15, TM-22, and TM-28 in 1992, 1995, and 1998. He also made ten space walks totalling 42 hours. He retired in 2003.

AVHRR Abbreviation for *advanced very high-resolution radiometer.

avionics Any electronic subsystem used in a spacecraft for its navigation, guidance, and communications. These critical areas are handled by computer and sensing systems and data transmission circuits located in the **avionics bay**. In the space shuttle, this is located in the mid-deck, below and forward of the flight deck. Avionics systems are also used for aircraft and missiles.

azimuth The angular distance of an object eastwards along the horizon, measured from due north, between the astronomical meridian (the vertical circle passing through the centre of the sky and the north and south points on the horizon) and the vertical circle containing the celestial body whose position is to be measured.

Babakin Science and Research Centre A Russian facility in Moscow that develops and manufactures spacecraft for Rosaviakosmos's planetary missions. It is part of the Lavochkin Association which develops spacecraft. Established in the early 1980s, the centre helped develop the two *_Vega_ probes launched in 1984 and the two *_Phobos_ probes launched in 1988, among others. In 2002, the centre joined the EADS Space Transportation Company of Bremen, Germany, to establish the Return and Rescue Space Systems Company.

backpack An informal term for as astronaut's *life-support system or *manned manoeuvring unit, worn on the back. The backpack and spacesuit were put on separately by _Apollo_ astronauts, but by the time of the space shuttle flights they had become one piece.

back-up An auxiliary system in a spacecraft that can quickly replace a primary system that fails. A back-up can also be used for additional power: one main computer and two back-ups are needed to operate the robot arm on the _International Space Station._

back-up An astronaut or cosmonaut serving as a reserve crew member on a mission. Back-ups replace scheduled crew members who become ill or develop other problems.

Bagian, James Philipp (1952–) US astronaut and physician. On board the space shuttle _Discovery_ in March 1989, he became the first person to inject the drug Phenergan to treat *space sickness, a treatment now used by NASA. His second shuttle mission, in June 1991, was aboard the first dedicated space and life-science mission, _Spacelab_ Life Sciences (SLS-1), to monitor physical reactions to *microgravity. Bagian became an astronaut in 1980.

Baikonur Cosmodrome (or Baykonur Cosmodrome) A launch site for spacecraft, located near Baikonur (formerly Tyuratam and Leninsk), Kazakhstan, near the Aral Sea. It was built in 1955. The first Soviet satellites and all Soviet (later Russian) space probes, as well as crewed *_Soyuz_ missions, were launched from here. In July 2000 the _Zvezda_ module for the *_International Space Station (ISS)_ was sent into orbit from Baikonur. Following NASA's suspension of shuttle flights after the 2003 _Columbia_ space shuttle disaster, later flights to the _ISS_ have been launched from Baikonur. The cosmodrome, which Russia now rents from Kazakhstan, is the world's largest space launching facility. It covers an area of 12 200 sq km, much larger than its US equivalent, the *Kennedy Space Center in Florida. This includes dozens of launching pads, five tracking control centres, nine tracking stations, and a rocket test range. On 11 May 2002, a hanger used to assemble and test space vehicles collapsed, killing eight people.

The cosmodrome is some 322 km southwest of the mining town of Baikonur, but the Soviets informed the International Aeronautical Federation that Baikonur was Gagarin's launch site, and the name remained. However, in 1996 Russian president Boris Yeltsin changed the name of the town of Tyuratam, where the cosmodrome is situated, to Baikonur.

http://www.russianspaceweb.com/baikonur.html In-depth and illustrated accounts of the history of the Baikonur launch centre, with detailed information on the Energiya-*Buran* and *Soyuz* facilities, launch sites, and control station at Baikonur. There is also information on, as well as photographs of, the town of Baikonur (formerly Tyuratam).

Baker, Ellen Louise Shulman (1953–) US astronaut and physician. In October 1989, she flew aboard the space shuttle *Atlantis* that deployed the space probe **Galileo*. In June 1993, she flew aboard *Columbia*, on the first flight of the **Microgravity Science Laboratory*. In June 1995, she flew on *Atlantis* for the first docking of a shuttle with the Russian space station *Mir*. Baker joined NASA as a medical officer in 1981 and was selected as an astronaut in 1984.

Baker, Michael Allen (1953–) US astronaut. He is NASA's manager of International and Crew Operations for the *International Space Station*. He has undertaken four space shuttle flights, including an *Endeavour* mission in September 1994 that marked the second flight of the *Space Radar Laboratory, and the January 1997 launch of *Atlantis* to the Russian space station *Mir*. Baker was selected as an astronaut in 1985.

ballistic descent The direct falling descent of a spacecraft. Such a re-entry does not have the curved path of a *ballistic trajectory flight path.

ballistic trajectory The curved flight path of a spacecraft after the launch vehicle's power has ended. The launching of early spacecraft had to be configured carefully to follow the correct ballistic trajectory.

balloon A craft consisting of a gasbag, filled with gas lighter than the surrounding air, from which a basket or instruments can be suspended. Balloons are now used as a means of meteorological, infrared, gamma-ray, and ultraviolet observation.

NASA has developed an *Ultra-Long Duration Balloon to circle the Earth at the edge of space. Balloon packages are also used as space probes, for example, by the Soviet **Vega* mission that placed atmospheric balloons in Venus's atmosphere in 1985.

Barringer Crater (or Arizona Mete or Crater or Coon Butte) An impact crater near Winslow in Arizona, caused by the impact of a 50-m iron *meteorite about 50 000 years ago. It is 1.2 km in diameter and 200 m deep, and the walls are raised 50–60 m above the surrounding desert.

It is named after the US mining engineer Daniel Barringer who proposed in 1902 that it was an impact crater rather than a volcanic feature, an idea confirmed in the 1960s by US astronomer Eugene Shoemaker, who studied the crater for his doctoral research at the California Institute of Technology.

Barry, Daniel Thomas (1953–) US astronaut and physician. He flew with the space shuttle *Endeavour* in January 1996, when he undertook a six-hour *space walk to evaluate techniques to be used to assemble the **International Space Station* (*ISS*). He was also aboard the space shuttle *Discovery*, launched in May 1999 to deliver supplies to the *ISS* before the arrival of the first resident crew, and on the *Discovery* flight in August 2001 to the *ISS* to exchange crews and to attach the *Leonardo* module. Barry was selected as an astronaut in 1992.

BARSC Abbreviation for *British Association of Remote-Sensing Companies.

Bassett, Charles Arthur, II (1931–66) US astronaut. Selected as pilot of the spacecraft *Gemini 9*, he was killed when a T-38 jet aircraft crashed on 28 February 1966, before the craft's launch, scheduled for June 1966. Bassett was selected as an astronaut in 1963. His earlier work at NASA had involved him in training and flight simulation.

beacon A downlink from a spacecraft that immediately shows that the spacecraft is in one of several possible states. An on-board system assesses the craft's overall health. It then adds one of four tones to its download carrier to indicate how urgently it needs to contact the *deep-space station (NASA ground tracking and communications station) with this data.

Beagle 2 A spacecraft *lander of the European Space Agency. *Beagle 2*, a UK-led project, was scheduled to analyse soil on Mars in search of life after being carried there by **Mars Express*, which was launched on 2 June 2003. Using an entry, descent, and landing system (EDLS), *Beagle 2* was programmed to land on a flat region, Isidis Planitia. It descended there on 25 December 2003 but the transmission was lost and further contact proved impossible. Scientists hope to send another *Beagle* lander to Mars in 2009 as part of the **Aurora* programme.

Bean, Alan LaVern (1932–) US astronaut. A member of the *Apollo 12* mission (*see* APOLLO PROJECT), on 19 November 1969 he became the fourth person to walk on the Moon. With Pete *Conrad, he made two long exploratory lunar walks, totalling some eight hours, and collected 27 kg of lunar rocks and soil for scientific study. He was commander of the second crew aboard the **Skylab* mission for two months from July 1973.

Bean left the navy in 1975, remaining as NASA's head of the Astronaut Candidate Operations and Training Group until he retired in 1981.

Bean was born in Wheeler, Texas, and in 1955 received a Bachelor of Science (BSc) degree in aeronautical engineering from the University of Texas. He served in the US Navy as a pilot and test pilot before entering the space programme in October 1963. During his retirement Bean concentrated on art, producing pictures of the *Apollo* mission and other space events.

Belka and Strelka The two dogs launched aboard the Soviet spacecraft *Korabl Sputnik 2* (also known as *Sputnik 5*) on 19 August 1960 for 17 orbits to test life-support systems. Other animal passengers included 40 mice, two rats, and hundreds of insects, as well as plant specimens. The capsule was an uncrewed version of the *Vostok* spacecraft that carried the first person, Yuri *Gagarin, into space a year later.

Bell Aerosystems US company whose space work included development of the lunar landing research vehicle (LLRV) and the ascent engine for the *lunar module (LM). The LLRV was used to train *Apollo* astronauts to fly the LM. Bell also produced the Bell XS-1 experimental aircraft (later known as the X-1), the first to fly faster than the speed of sound, and the engine for the *Gemini*-Agena Target Vehicle (GATV) used for the 1966 rendezvous of *Gemini 10* with an Agena target. The company, with headquarters in Buffalo, New York, later became a division of Textron Inc., of Providence, Rhode Island.

Bell Bomi An early 1951 concept for a reusable spaceplane whose principles became part of the *space shuttle design two decades later. Envisioned by Bell Aircraft Corporation, it was a two-stage vehicle having a large booster with five rocket engines and a small spaceplane with three. The two vehicles would be separated 2 minutes and 2 seconds after lift-off, with the winged booster gliding back to Earth as the spaceplane continued its flight. Bell predicted speeds of up to 8 450 kph.

Belyayev, Pavel (1925–70) Soviet cosmonaut. His one flight, on *Voskhod 2* in March 1965, was made up of a series of emergencies. During the mission, the other cosmonaut, Aleksei *Leonov, made the first *space walk. When he returned, the hatch door would not seal properly. On re-entry, the spacecraft's retrorockets failed and had to be manually fired. There were also problems with the service module, which began to gyrate when it had a late separation from its re-entry capsule. Belyayev was selected as a cosmonaut in 1960, and left the space programme in 1970.

bends (or compressed-air sickness or caisson disease) A syndrome arising from too rapid a release of nitrogen from solution in the blood, as a result of decompression. The problem is generally associated with deep-sea divers, but astronauts must be protected from the bends by sufficient cabin pressure to prevent nitrogen from forming painful gas bubbles.

During the **Apollo–Soyuz* Test Project, astronauts transferring from the space shuttle to the docking module had to increase the module's atmospheric pressure slowly to about two-thirds of sea-level pressure to avoid the bends.

BepiColombo A joint mission to Mercury by the European Space Agency and Japan, planned for launch in 2012. It will consist of two separate orbiters: the *Mercury Planetary Orbiter* (*MPO*), supplied by ESA, and the *Mercury Magnetospheric Orbiter* (*MMO*) supplied by the Japan Aerospace Exploration Agency, JAXA. *MPO* will study the surface and internal composition of the planet, while the *MMO* will study Mercury's magnetosphere, the region of space around the planet that is dominated by its magnetic field. Each spacecraft will observe the planet from polar orbit for one Earth year. On the way to Mercury, both craft will be powered by solar-electric propulsion, as demonstrated by ESA's **SMART-1* mission to the Moon. BepiColombo is named after the Italian scientist, mathematician and engineer Giuseppe ('Bepi') Colombo.

BeppoSAX An Italian X-ray satellite with participation from the Netherlands. 'Beppo' was the nickname of the Italian astronomer Giuseppe Occhialini, whose memory the satellite honoured, and SAX is from 'Satellite per Astronomia X'. Launched on 30 April 1996, *BeppoSAX* made spectroscopic studies of X-ray sources and looked for gamma-ray bursts, discovering 50 during its 6-year operational lifetime. It re-entered the atmosphere in April 2003.

Berezovoi, Anatoly (1942–) Soviet cosmonaut. Along with Valentin Lebedev, he set a record of 211 days in space in the space station *Soyuz T-5*, arriving in May 1982 and returning in December. They conducted numerous experiments, including ones in biology, agriculture, and geology, and launched an amateur-radio satellite designed by Russian students. Berezovoi also took a space walk of nearly three hours. He was selected as a cosmonaut in 1970 and left the space programme in 1992.

berthing An astronaut's action in attaching modules by using a robot arm. This is in contrast to *docking.

Beta cloth A woven glass-fibre cloth used for the outer layer of NASA *spacesuits. It replaced nylon after a fire in January 1967 killed three astronauts in their spacecraft on the launch pad. Coated with Teflon, Beta cloth can withstand temperatures of up to 2 760 °C. The material was also used for the inner layers of the *Apollo* lunar boot and is used for space shuttle sleeping bags.

B F Goodrich Company US company involved in developing and manufacturing NASA's *spacesuits, including the early *Mercury* and *Gemini* spacesuits.

BIG Acronym for *biological isolation garment.

Big Bang The hypothetical but widely accepted 'explosive' event that marked the origin of the universe as we know it. At the time of the Big Bang, the entire universe was squeezed into an infinitely small, hot, superdense state. The Big Bang explosion threw this compact material outwards, producing the expanding universe seen today (*see* REDSHIFT). The cause of the Big Bang is unknown; observations of the current rate of expansion of the universe suggest that it took place about 13–14 billion years ago. The Big Bang theory began modern *cosmology.

According to a modified version of the Big Bang theory, called the **inflationary theory**, the universe underwent a rapid period of expansion shortly after the Big Bang, which accounts for its current large size and uniform nature. The inflationary theory is supported by the most recent observations of *cosmic background radiation.

Big Crunch In cosmology, possible fate of the universe in which it ultimately collapses to a point following the halting and reversal of the present expansion. *See also* BIG BANG and CRITICAL DENSITY.

binary star One of a pair of stars moving in orbit around their common centre of mass. Observations show that most stars are binary, or even multiple; an example is the nearest star system to the Sun, Rigil Kent (Alpha

Centauri). A *spectroscopic binary is a binary in which two stars are so close together that they cannot be seen separately, but their separate light spectra can be distinguished by a spectroscope. Another type is the *eclipsing binary, a double star in which the two stars periodically pass in front of each other as seen from Earth. When one star crosses in front of the other, the total light received on Earth from the two stars declines. The first eclipsing binary to be noticed was Algol, in 1670, by Italian astronomer Geminiano Montanari.

The binary system Rigil Kent, for example, consists of a star almost identical to the Sun with another star about a third as bright closer to it than Neptune is to the Sun (that is, closer than 4.4 billion km). Each of these stars appears to describe an ellipse about the other in about 80 years. A third, much fainter star, Proxima Centauri, is too far away to disturb their mutual orbit appreciably.

The study of binary star systems has provided the only reliable information about the masses of stars. For a few stars it has also yielded direct measures of their dimensions, shapes, and effective temperatures.

biological isolation garment (BIG) An early biological protective suit with a faceplate, given to astronauts returning from space. The suits had no ventilation and were uncomfortably hot. After the splashdown of the *Apollo 11* astronauts in the Pacific Ocean in July 1969, US Navy swimmers threw into the spacecraft the garments to be donned before the astronauts went to the mobile quarantine facility on the deck of the USS *Hornet*.

biomedical monitoring sensor An aeromedical sensor introduced by NASA for its **Mercury* project flights. There were four sensors taped to the astronaut's chest for electrocardiograph readings, a thermistor in the helmet to check respiration rate and depth, an arm cuff to monitor blood pressure, and a rectal probe for body temperature. The data was transmitted to flight surgeons on the ground.

BIO-Plex Contraction of *Bioregenerative Planetary Life-Support Systems Test Complex.

Bioregenerative Planetary Life-Support Systems Test Complex (BIO-Plex) A project conducted for NASA's *Advanced Life-Support Program. It involves a complex of five enclosed chambers that provide all the air and water and the majority of food for four astronauts by using biological and physiochemical life-support technologies. The testing, at the *Johnson Space Center, began in 2001 and is scheduled to finish in 2006, with humans living in the chambers between 15 and 180 days.

Biosatellite *Cosmos* A Soviet satellite launched in September 1987 to investigate the effects of *microgravity on rhesus monkeys. The week-long flight extended research done on **Spacelab* and involved scientists from the USA and other nations. Some of the research data was flawed, however, when one monkey escaped his constraints and tampered with electrodes.

bipropellant Rocket *propellant, the liquid fuel and liquid oxidizer of which are stored in separate tanks, from where they are pumped into the combustion chamber.

Black Arrow rocket Britain's first space launch vehicle, designated R3, and the only one to launch a satellite. Although it put the first British satellite, **Prospero*, into orbit in October 1971, the Black Arrow programme was cancelled in July of that year. Only four were launched, all from Woomera, Australia, with two failing. One of the rockets is now on display in the Science Museum in London, England.

black hole An object in space whose gravity is so great that nothing can escape from it, not even light. It is thought to form when a massive star shrinks at the end of its life. A black hole sucks in more matter, including other stars, from the space around it. Matter that falls into a black hole is squeezed to infinite density at the centre of the hole. Black holes can be detected because gas falling towards them becomes so hot that it emits X-rays.

Black holes containing the mass of millions of stars are thought to lie at the centres of *quasars. Satellites have detected X-rays from a number of objects that may be black holes, but only a small number of likely black holes have been identified in our Galaxy.

Cygnus X-1, first discovered in 1964 by astronomers at the US Naval Research Laboratory, is an X-ray source in the constellation of Cygnus. A0620–00, in the constellation of Monoceros, is one of the best black-hole candidates in the Galaxy, discovered in the 1980s by US astronomers Jeffrey McClintock of the Harvard-Smithsonian Center for Astrophysics and Ronald Remillard of the Massachusetts Institute of Technology. V404 Cygni, close to Cygnus X-1, is a possible black hole discovered in 1992. Nova Muscae, identified as a black hole in 1992 by McClintock, Remillard, and US astronomer Charles Bailyn of Yale University, New Haven, Connecticut, lies approximately 18 000 light years from Earth. The *Hubble Space Telescope* discovered in 1997 evidence of a black hole 300 million times the mass of the Sun. It is located in the middle of galaxy M84 about 50 million light years from Earth. In March 2001, NASA scientists, using images from the orbiting **Chandra X-Ray Observatory* of X-ray emissions from space objects as they may have appeared 12 billion years ago, concluded that there may have been about 300 billion black holes when the universe was young.

http://archive.ncsa.uiuc.edu/Cyberia/NumRel/BlackHoles.html
Well-written guide to black holes. There is a good balance of text and images, with clear explanatory diagrams. Video and audio clips of scientists explain some of the current theories of the evolution and behaviour of black holes.

Blaha, John Elmer (1942–) US astronaut. His five space shuttle flights include a record 14-day life-science research mission on the space shuttle *Columbia* (October–November 1993), regarded by NASA as its most successful and efficient *Spacelab* flight. During the mission, the seven crew members performed medical experiments on themselves. In September 1996, the space shuttle *Atlantis* took Blaha to the *Mir* space station, where he stayed for four months. He was selected as an astronaut in 1980 and left NASA in 1997.

blast-off The launch of a spacecraft, rocket, or missile.

b

Bloomfield, Michael John (1959–) US astronaut. He has flown on three space shuttle missions, aboard *Atlantis* (September 1997) to dock with the Russian space station *Mir*, with *Endeavour* (November 2000) to deliver supplies to the first resident crew of the *International Space Station* (*ISS*), and on *Atlantis* (April 2002) to install the SO truss, the backbone of the *ISS*. He had logged 753 hours in space by 2004. Bloomfield was selected as an astronaut in 1995.

blueshift A manifestation of the *Doppler effect in which an object appears bluer when it is moving towards the observer or the observer is moving towards it (blue light is of a higher frequency than other colours in the spectrum). The blue shift is the opposite of the *redshift.

Bluford, Guion Stewart, Jr (1942–) US astronaut. A mission specialist, his four space shuttle flights included two pioneering missions aboard *Challenger*: on the first flight with a night-time launch and landing (30 August–5 September 1983); and in October 1985 to carry eight crew members on the first **Spacelab* mission under German control. Bluford became an astronaut in 1979 and left NASA in 1993.

BNSC Abbreviation for *British National Space Centre, the UK government agency that coordinates the British space programme.

Bobko, Karol Joseph (1937–) US astronaut. His three space shuttle flights included the first *Challenger* mission, which he piloted in April 1983 and from which the first extravehicular activity from a shuttle was conducted. He was also the commander of the first flight of *Atlantis* in October 1985. Bobko was assigned to the **Manned Orbital Laboratory* (*MOL*) programme in 1966. He became a NASA astronaut in 1969 when the *MOL* was terminated, and left NASA in 1989.

body flap A part of a space shuttle's aft fuselage, below the cones of the main engines. The movable, horizontal body flap is adjusted to balance the shuttle's attitude when it re-enters the Earth's atmosphere. It also protects the engine nozzles from re-entry heat. After landing, the device has a ground clearance of 3.68 m.

Boeing Company US military and commercial aircraft manufacturer that has produced many components for NASA. These include the *Delta rocket series and the *Inertial Upper-Stage rockets for such probes as **Galileo*, **Magellan*, and **Ulysses*. Boeing is also the prime contractor for all of the US components of the **International Space Station*, including the *Destiny* science laboratory. Boeing's Delta II rocket launched the twin rovers to Mars in 2003, and its subsidiary, Spectrolab, made the high-efficiency solar cells to power the rovers. These solar cells have powered more than 500 satellites and interplanetary missions in the last 40 years.

boilerplate The model of a spacecraft used for testing. It normally has no working parts and is constructed of heavier metal than the final functional vehicle.

Bolden, Charles Frank, Jr (1946–) US astronaut. His four space shuttle flights included the *Discovery* mission launched in April 1990 to deploy the **Hubble Space Telescope*. He also commanded the *Discovery* mission launched in February 1994, the first flight to include a Russian cosmonaut as a mission specialist. Bolden was selected as an astronaut in 1980.

bolometric magnitude A measure of the brightness of a star over all wavelengths. Bolometric magnitude is related to the total radiation output of the star. *See* MAGNITUDE.

Bondarenko, Valentin (1937–61) Soviet cosmonaut. The youngest member of the original group of cosmonauts, Bondarenko died from shock eight hours after accidentally starting a fire in a pressurized pure oxygen chamber by tossing cotton wool soaked in alcohol onto an electric hotplate, during a ten-day series of tests in the chamber. He was selected as a cosmonaut in 1960.

boom A long metal beam extending from a spacecraft and serving as a structure subsystem. The *Voyager* spacecraft, for example, had a boom appendage to mount its sensitive magnetometer 6–13 m from the spacecraft's electric currents. The boom is collapsed into a protective canister for the launch and then deployed in flight.

booster A first-stage rocket of a launching vehicle, or an additional rocket strapped to the main rocket to assist lift-off.

The US *Delta rocket, for example, has a cluster of nine strap-on boosters that fire on lift-off. Europe's Ariane 3 rocket uses twin strap-on boosters, as does the US space shuttle.

Borman, Frank (1928–) US astronaut. He commanded the *Apollo 8* mission (*see* APOLLO PROJECT) that made the first crewed flight around the Moon, in 1968. He previously commanded, in 1965, a *Gemini 7* endurance flight, piloted by James *Lovell. Borman spent 14 days in orbit and made the first space rendezvous, with *Gemini 6* astronauts. He was selected as an astronaut in 1962.

Borman was subsequently appointed deputy director of NASA's Flight Crew Operations Branch. He left NASA in 1969.

Bowersox, Kenneth Dwane (1956–) US astronaut. He flew two space shuttle missions to repair and service the **Hubble Space Telescope*: aboard *Endeavour* in December 1993 and, as commander, on *Discovery* in February 1997. He also made two shuttle flights aboard *Columbia* with the **Microgravity Science Laboratory*, in June 1992 and, as commander, in October 1995. He was a member of Expedition 6 on the *International Space Station*, being launched in *Endeavour* in November 2002 for five and a half months. He is currently the director of flight crew operations at the Johnson Space Center. Bowersox was selected as an astronaut in 1987.

bow shock crossing The phenomenon recorded by a space probe that leaves the solar wind and flies into a planet's magnetosphere. This happens during the end of the vehicle's *far encounter phase involving the planet or at the beginning of the *near encounter phase. The bow shock crossing is

identified from data collected by a magnetometer, a plasma instrument, and a plasma-wave instrument. When the solar-wind magnetosphere boundary swings back and forth (often over millions of kilometres), the spacecraft will encounter many bow shock crossings.

Brady, Charles Eldon, Jr (1951–) US astronaut. He flew on the longest space shuttle mission, logging more than 405 hours on **Columbia* from 20 June to 7 July 1996 with the **Microgravity Science Laboratory*. The seven-member crew included nationals from France, Canada, Italy, and Spain. Brady was selected as an astronaut in 1992.

Brand, Vance DeVoe (1931–) US astronaut. He piloted the command module on the **Apollo–Soyuz* Test Project launched in July 1975. He was commander of three space shuttle missions: aboard *Columbia* in November 1982 (the first fully operational shuttle flight), on *Challenger* in February 1984, and on *Columbia* again in December 1990. Brand was selected as an astronaut in 1966 and logged 746 hours in spacecraft. He resigned as an astronaut in 1992 and was appointed deputy director for Aerospace Projects at NASA's *Dryden Flight Research Center.

Brandenstein, Daniel Charles (1943–) US astronaut. He piloted the first space shuttle to have a night launch and landing (*Challenger*, 30 August and 3 September 1983), and commanded the crews of three further flights. These included the longest shuttle mission at that time (the 10-day, 21-hour flight of *Columbia* in January 1990) and the first flight of *Endeavour* in May 1992. Brandenstein was selected as an astronuat in 1978 and retired from NASA in 1992. He became president of the *National Space Society in 1998.

Brasilsat Several satellites launched by Brazil to create a telecommunications network. *Brasilsat 1* was launched by an Ariane 3 rocket in February 1985 and *Brasilsat 2* in May 1986. Other South American nations can lease or buy television channels on the network.

Bridges, Roy Dunbard, Jr (1943–) US astronaut, director of NASA's Kennedy Space Center, and director of Langley Research Center from 2003. His one space flight was aboard the shuttle *Challenger* in July–August 1985. Bridges was selected as an astronaut in 1980.

British Association of Remote-Sensing Companies (BARSC) A British trade association involved with *remote-sensing spacecraft. Established in 1985, its members' activities include designing and developing satellite payloads, processing and interpreting images from spacecraft, and developing data transmitters and receiver stations. The BARSC works to ensure that its members' interests are represented on national, international, and governmental committees. It had 19 member companies in 2004.

British Interplanetary Society (BIS) A British space organization established in 1933 with charitable status. The society, located in London, calls itself 'the world's longest established organization devoted solely to supporting and promoting the exploration of space and astronautics'. The society publishes *Spaceflight* magazine and, since 2004, *Voyage* magazine for students aged 10–14.

It has worldwide membership and is noted for its visionary concepts. In 1948, one of its study groups envisioned a delta-winged reusable space orbiter, and it was at a BIS meeting in August 1968 that George *Mueller, the associate administrator for Manned Space Flight at NASA, publicly revealed plans for an orbiter 'shuttling between Earth and the installations which will be operating in space'.

http://www.bis-spaceflight.com/ Home page of the British Interplanetary Society (BIS), the world's longest established organization devoted solely to supporting and promoting the exploration of space and astronautics. The site offers *Spaceflight* magazine, the society journal, and membership of the society.

British National Space Centre (BNSC) Britain's space agency, established in 1985. It acts on behalf of the UK government and research councils, who spend some £160 million each year on civil space, with about 60% of this channelled through the *European Space Agency. The BNSC has been involved in numerous space missions, including Europe's **Giotto* spacecraft, the **Mars Express* launched in 2003, including its **Beagle 2* rover, and the **Rosetta* spacecraft and lander launched in 2004. It also helps develop innovative space technology and satellite payloads, and assists the UK space industry in maximizing business opportunities, especially concerning telecommunications and navigation satellites.

http://www.bnsc.gov.uk/ Regularly updated with news of space missions and space technology research with UK involvement. The site's resource centre has a glossary, Frequently Asked Questions about British space activities, and a useful links page, with the URLs of national space agencies.

Brown, Curtis Lee, Jr (1956–) US astronaut. He logged more than 1 383 hours in space on six *space shuttle flights, including the *Endeavour* mission for Japanese **Spacelab* experiments in September 1992, and the *Discovery* flight in December 1999 to upgrade the **Hubble Space Telescope*. He has since left NASA. Brown was selected as an astronaut in 1987.

Brown, David McDowell (1956–2003) US astronaut and physician. A mission specialist, he was assigned to a space shuttle launch in 2001, having previously supported *payload development for the **International Space Station*. Brown was selected as an astronaut in 1996; he died when the space shuttle *Columbia* broke up on re-entry in February 2003.

Brown, Mark Neil (1951–) US astronaut. His two space shuttle flights were as a mission specialist aboard *Columbia* (August 1989) and on the *Discovery* flight that deployed the *Upper-Atmosphere Research Satellite (September 1991). Brown became an astronaut in 1985, and left NASA in 1993 to head the Space Division of the General Research Corporation.

brown dwarf An object less massive than a star but denser than a planet. Brown dwarfs do not have enough mass to ignite nuclear reactions at their centres, but shine by heat released during their contraction from a gas cloud. Groups of brown dwarfs have been discovered, and some astronomers believe that vast numbers of them exist throughout the Galaxy.

b

The first brown dwarf to be positively identified was Gliese 229B (GI229B), in the constellation Lepus, by US astronomers using images from the Hubble Space Telescope. It is about 50 times as massive as Jupiter but emits only 1% of the radiation of the smallest known star and has a surface temperature of 650 °C.

B-type rocket Western designation for the smallest Russian rockets, used to launch small satellites. An example of this two-stage rocket is the *Cosmos B-1*, first launched in 1962, which was based on a Soviet intermediate-range ballistic missile (IRBM), the SS-5 Sandal. The present *Cosmos* version was developed from another IRBM, the SS-5 Skean.

Buchli, James Frederick (1945–) US astronaut. He made four space shuttle flights, including the German **Spacelab* mission aboard *Challenger*, which was launched in October 1985 with a crew of eight, the largest team launched into space. Buchli became an astronaut in 1979. He retired as an astronaut in 1992 and joined the Boeing Defense and Space Group.

Budarin, Nikolai Mikhailovich (1953–) Russian cosmonaut. Twice a resident of the space station **Mir*, he first flew there aboard NASA's space shuttle *Atlantis*, launched in June 1995, and remained on *Mir* until September, when he returned on the **Soyuz* transport vehicle. Budarin subsequently flew to *Mir* on *Soyuz* in January 1998 as a board engineer, remaining on the space station until August. He later logged over five months on the *International Space Station* as the flight engineer of Expedition Six, launched on *Endeavour* on 23 November 2002 and remaining until 3 May 2003. Budarin joined the *Energiya Rocket and Space Complex in 1976 as an engineer and was selected as a cosmonaut in 1989.

built-in hold During a countdown to a spacecraft launch, a delay purposely added in advance. Two-day countdowns can have up to seven built-in holds to give the inspection crews a rest. These can last from 10 minutes to about 26 hours. Built-in holds are also added to assess the weather. A launch of the space shuttle *Discovery* on 7 November 1984 was postponed during a built-in hold due to rough upper-atmosphere conditions.

Bull, John Sumter (1934–) US astronaut. He was selected as an astronaut in 1966, but resigned two years later for health reasons. He worked at NASA's *Ames Research Center 1973–85, conducting simulation and flight-test research on advanced flight systems for aircraft. In 1986 he began managing NASA's autonomous systems technology for space applications. He retired from NASA in 1989 and became a consultant in aerospace research and technology.

***Buran* (Russian: snowstorm)** Soviet space shuttle. It had only one uncrewed orbital flight, on 15 November 1988, before being officially cancelled in 1993. The programme was authorized in 1976 to compete with NASA's shuttle. *Buran* was launched on an *Energiya rocket and was limited to two orbits because of its small computer memory. Funding was then cut, and work on two additional shuttles was halted. In June 2001, *Buran* was prepared for relaunch. Russia planned to use the shuttle to transport space tourists.

Burbank, Daniel Christopher (1961–) US astronaut. A mission specialist, he flew on the space shuttle *Atlantis* in September 2000 to prepare the *International Space Station* for the arrival of the first resident crew. He was scheduled for another flight in 2004 which was postponed following the 2003 *Columbia* space shuttle disaster. Burbank was selected as an astronaut in 1996.

burn The single firing of a rocket engine or thruster, particularly to adjust a spacecraft's position or to begin its re-entry.

burnout The moment during a space flight when a rocket's fuel or oxidizer propellant has been entirely consumed, and so the engine stops firing. The rocket is then jettisoned from a spacecraft or goes into free flight. The **burnout velocity** is the maximum velocity reached by a rocket or spacecraft when burnout occurs.

Burnout weight is the weight of a rocket after burnout has occurred. This is contrasted with its *initial weight.

burp firing An early NASA testing procedure before a scheduled launch, consisting of firing the rocket or spacecraft thrusters on the launch pad for a short time without moving the craft. It was discontinued in 1968, beginning with the *Apollo 8* mission, to keep the fuel and oxidizer separate in case the launch was delayed, creating greater launch-time flexibility.

Bursch, Daniel Wheeler (1957–) US astronaut. He was the flight engineer of the fourth resident crew on the *International Space Station*, launched on *Endeavour* on 5 December 2001 and, with Carl Walz, he set the US space flight endurance record of 196 days. His previous three space shuttle flights include the *Discovery* mission in September 1993 that deployed the **Advanced Communications Technology Satellite*. In January 2003, he was assigned as an instructor for two years at the Naval Post Graduate School in Monterey, California. Bursch was selected as an astronaut in 1990.

Bykovsky, Valery Fyodorovich (1934–) A Soviet cosmonaut who made the longest-ever solo space flight, orbiting the Earth 81 times over a period of five days in June 1963. During the last three days, Valentina *Tereshkova was in orbit simultaneously aboard *Vostok 6*. He made two subsequent flights, aboard *Soyuz 22* in 1976 and *Soyuz 31* in 1978.

Cabana, Robert Donald (1949–) US astronaut. His four space shuttle flights include the *Discovery* mission in October 1990 to deploy the **Ulysses* spacecraft. He also flew with the first *International Space Station* mission in December 1998, on board *Endeavour*, to connect the *Unity* and *Zarya* modules. Cabana was selected as an astronaut in 1985.

cabin In a spacecraft, the enclosed area occupied by the crew. Because the crew occupy the cabin, maintaining **cabin pressure** is important. The space shuttles maintain a cabin pressure equal to atmospheric pressure at sea level, using the **atmosphere revitalization pressure control system (ARPS)**.

California Institute of Technology (or Caltech) A university in Pasadena, California, that is the home of NASA's *Jet Propulsion Laboratory. It was established in 1891 and is noted for its many Nobel prizewinners. The aeronautical laboratory was already developing liquid and solid propellants for rockets by 1939 and worked closely with the US Army on missiles during World War II. In 1987, Caltech and Russia's Space Research Institute signed an agreement to collaborate on research.

http://www.astro.caltech.edu/ Information about the research activities of Caltech's astronomy department, with links to observatories operated by Caltech, such as Palomar and Owens Valley Radio Observatory (OVRO).

Callisto The second-largest moon of the planet Jupiter, 4 800 km in diameter, orbiting every 16.7 days at a distance of 1.9 million km. Its surface is covered with large craters.

The US space probe *Galileo* detected oxygen on Callisto in 1997, suggesting that its surface is made of water ice. In 1998 *Galileo*'s magnetometer measured electrical currents near Callisto's surface that are consistent with the existence of a buried salty ocean beneath the icy crust.

Camarda, Charles (1952–) US astronaut experienced in thermal structures and materials engineering. His first flight was on the *Discovery* space shuttle in July–August 2005. Camarda was selected for astronaut training in 1996.

Cameron, Kenneth Donald (1949–) US astronaut. He was the pilot of the *Discovery* shuttle mission in April 1991 that deployed the **Gamma-Ray Observatory*. He commanded the flights of *Discovery* in April 1993 and *Atlantis* in November 1995, the latter being NASA's second shuttle to dock with the Russian space station *Mir*. He was selected as an astronaut in 1984 and left NASA in 1996.

Canadarm (or remote manipulator system (RMS)) A system used by the space shuttle and **International Space Station* (*ISS*) to deploy or recover satellites and other payloads. It is also used to move astronauts when they stand on its mobile *foot restraint during extravehicular activity. The original 15-m version, always mounted in the cargo bay of the space shuttle, was first used by *Columbia* on a flight in March 1982. The two arms are each 6.7 m long, and the *end effector, or hand, is used to grasp payloads and move them in or out of the cargo bay.

Canadarm was designed and constructed by SparAerospace Ltd of Toronto, Canada. It can lift loads as large as a single-decker bus. A specially trained mission specialist uses two hand controls to operate the device, which has electric motors at its three joints. The operator receives television pictures from cameras located at these joints.

Canadarm 2 is longer (17.6 m), stronger, and more flexible than the first version. It was delivered to and installed on the *ISS* in April 2001, and immediately handed its 1 350-kg packing crate back to the Canadarm on *Endeavour*, the first robot-to-robot transfer in space.

(Image © NASA)

US astronaut James Voss attached to space shuttle *Atlantis*'s Canadarm during a 6 hour 44 minute EVA (extravehicular activity) on STS-101. Voss and colleague Jeffrey Williams made repairs and fitted new parts to the *International Space Station*.

Canadian Space Agency (CSA) The organization responsible for the Canadian space programme. The CSA has contributed instruments and payloads for many space missions. Its high-flux telescope on **Ulysses* (launched in 1990) was the first Canadian instrument flown in deep space. Its robot arm, *Canadarm, is a crucial part of the **International Space Station* (*ISS*). **RADARSAT 1*, launched in November 1995, was the first Canadian Earth-observation satellite. The CSA implemented a new Canadian space programme in 1999 for 'space knowledge, applications, and industry development.' By 2004, eight Canadian astronauts had flown in space and six were eligible for future flights. The agency is scheduled to send one astronaut to the *ISS* every three years for a period of three months each.

Canada became the third nation to have a satellite in space with **Alouette 1* in September 1962, and selected its first six astronauts in 1983. The CSA was established in March 1989 from the Space Division of the National Research Council Canada and other federal departments. It is located in the John H Chapman Space Centre at Saint-Hubert, Quebec.

http://www.space.gc.ca/home/index.asp Impressive site presenting Canada's space programme. There are details of the CSA's *RADARSAT* series, space technology research, and Canada's contribution to the *International Space Station*. Among the many resources are a news archive, image gallery, and activities for younger students.

CapCom Contraction of *capsule communicator.

Cape Canaveral (formerly Cape Kennedy (1963–73)) A rocket launch site used by *NASA, a promontory on the Atlantic coast of Florida, 367 km north of Miami. The *Kennedy Space Center is nearby.

By March 2001, around 3200 missiles and rockets had been launched from the Cape Canaveral site. These include lift-offs to the Moon during the **Apollo* project, for which a new launch complex was constructed on 324 sq km of land, doubling the size of the Cape Canaveral site.

To celebrate 50 years since the first launch at the site, 50 model rockets were launched on 24 July 2000, close to the original launch pad.

CAPS Acronym for *crew altitude protection system.

capsule The original name for a crewed spacecraft. It was more appropriate at first because of the thimble shape and the limited room inside a spacecraft. US astronaut Michael *Collins said that by the time he joined the space programme in 1963, astronauts detested the word and 'spacecraft' was always used: 'A capsule was something you swallowed.'

capsule communicator (CapCom) A ground-based NASA communicator who maintains contact with astronauts during space missions. There were four capsule communicators for the *Apollo 11* mission that landed on the Moon in 1969: the landing CapCom, the post-landing and goodnight Capcom, the EVA CapCom (for extravehicular activity), and the lunar module launch CapCom. The abbreviated form is now generally considered to stand for 'spacecraft communicator'.

captured rotation (or synchronous rotation) A circumstance in which one body in orbit around another, such as the moon of a planet, rotates on its axis in the same time as it takes to complete one orbit. As a result, the orbiting body keeps one face permanently turned towards the body about which it is orbiting. An example is the rotation of our own *Moon, which arises because of the tidal effects of the Earth over a long period of time.

carbon-carbon An extremely hard compound used for the heat shield on a space shuttle's nose and to cover the leading edges of its wings. It reaches a temperature of 1 370°C during re-entry to the Earth's atmosphere.

carbon cycle A sequence of nuclear fusion reactions in which carbon atoms act as a catalyst to convert four hydrogen atoms into one helium atom with the release of energy. The carbon cycle is the dominant energy source for ordinary stars of mass greater than about 1.5 times the mass of the Sun.

Nitrogen and oxygen are also involved in the sequence so it is sometimes known as the **carbon–nitrogen–oxygen cycle** or **CNO cycle**.

carbon dioxide removal system A spacecraft system to absorb carbon dioxide, which is exhaled by astronauts. NASA, which uses 100% oxygen in its spacecraft, has the cabin gases pass through a filter of lithium hydroxide. It combines with the carbon dioxide to produce lithium carbonate, with water as a by-product. The system is stored in the *atmospheric revitalization rack.

carbon-fibre reinforced plastic (CFRP) A lightweight but extremely stiff plastic material that expands very little when heated. It has been used to construct spacecraft instruments, such as the *microwave radiometer (MWR) and the global ozone monitoring by occulation of stars (GOMOS) instrument.

Carey, Duane Gene (1957–) US astronaut. He was the pilot of *Columbia* on its mission on 1–12 March 2002 to service the *Hubble Space Telescope*. Prior to his assignment to a space mission, he was involved in the technical aspects of spacecraft systems and operations. Carey was selected as an astronaut in 1996.

cargo The general name for equipment and supplies carried aboard a spacecraft, as in the *cargo bay of the space shuttle. The *Leonardo* multi-purpose logistic module (MPLM) is called a 'cargo carrier' because it transports laboratory racks with experiments, supplies, and equipment to the **International Space Station*.

cargo bay (or payload bay) An area of a spacecraft used to store satellites, *payloads, and other equipment. On the space shuttle, the area can be viewed from the *crew station through two large windows, and is 18 m in length and 4.6 m in diameter, taking up most of the shuttle's fuselage. It is large enough to accommodate five satellites or major parts of the **Spacelab* laboratory.

The cargo bay has two large doors running along its length. When they are open, giant cooling radiators on their interior sides dissipate heat generated by the shuttle's electrical systems.

Carpenter, (Malcolm) Scott (1925–) US astronaut. He was one of NASA's *'Original Seven' selected in 1959, and piloted *Aurora 7*, the second US crewed orbital flight, in May 1962. He was seconded from NASA in 1965 to participate in the US Navy's SEALAB-II programme. Returning to NASA, he was involved in the design of the *lunar module for the *Apollo* project. He left NASA in 1967.

Carr, Gerald Paul (1932–) US astronaut. He was commander of the third and final mission to **Skylab* and achieved a record duration time for crewed flights in space of 84 days, 1 hour, and 15 minutes, from 16 November 1973 to 8 February 1974. Carr was selected as an astronaut in 1966 and left NASA in 1977.

Carter, Manley Lanier, Jr (1947–91) US astronaut and physician. He was a mission specialist on the *Discovery* space shuttle flight of November 1989, carrying military and other payloads. He was subsequently assigned to the first *International Microgravity Laboratory* flight. Carter was selected as an astronaut in 1984. He died in the crash of a commercial aircraft in April 1991.

Casper, John Howard (1943–) US astronaut and manager of the Space Shuttle Management Integration and Planning Office at the Johnson Space Center. Prior to this, he was director of Safety, Reliability, and Quality Assurance there. He has commanded three of his four space shuttle flights, including the *Endeavour* mission of January 1993 to deploy the *Tracking and Data-Relay Satellite. He logged over 825 hours in space. Casper was selected as an astronaut in 1984.

Cassini–Huygens A space probe developed jointly by NASA and the European Space Agency to explore the planet *Saturn. *Cassini* was launched on 15 October 1997 to study the planet's atmosphere, rings, and moons, and to drop off a sub-probe, *Huygens*, to land on Saturn's largest moon, *Titan. The spacecraft entered orbit around Saturn on 1 July 2004 and *Huygens* was released on 25 December, reaching the surface of Titan on 14 January 2005.

The *Cassini–Huygens* mission was launched on a Titan 4 rocket, with its electricity supplied by 32 kg of plutonium. This was the largest amount of plutonium ever to be sent into space, and provoked fears of contamination should *Cassini*, or its rocket, malfunction.

http://saturn.jpl.nasa.gov/cassini/english/ Well-organized site documenting the *Cassini–Huygens* mission to Saturn. Computer simulations show where the probe is currently. There is an impressive gallery featuring the latest pictures returned by the spacecraft, time-lapse movies, and animations. There are also interesting details of what is already known about Saturn's moons.

catwalk A long steel-grated walkway used by astronauts just before a launch to go from the service-tower lift to the sterile *white room next to the spacecraft's open hatch. The catwalk swings away from the shuttle seven minutes before launch.

CCD Abbreviation for *charge-coupled device.

CCLRC Abbreviation for *Council for the Central Laboratory of the Research Councils.

CDA Abbreviation for *cosmic dust analyser.

CDP Abbreviation for *Crustal Dynamics Project.

CDR Abbreviation for *commander.

celestial mechanics The branch of astronomy that calculates the orbits of celestial bodies and their gravitational attractions, and the orbits of artificial satellites and space probes. A spacecraft approaching a planet can use radio science to measure its own acceleration caused by the planet's gravity. This can be translated into a measurement of the planet's mass and, if images are available, its density.

celestial sphere An imaginary sphere surrounding the Earth, on which the celestial bodies seem to lie. The positions of bodies such as stars, planets, and galaxies are specified by their coordinates on the celestial sphere. The equivalents of latitude and longitude on the celestial sphere are called *declination and *right ascension (which is measured in hours from 0 to 24).

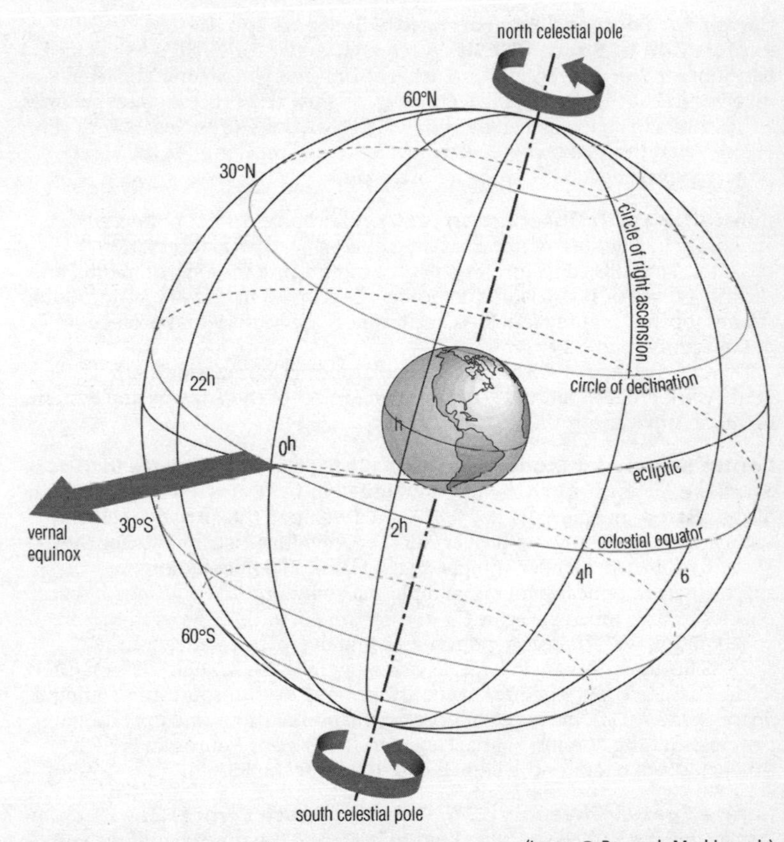

(Image © Research Machines plc)
The main features of the celestial sphere. Declination runs from 0° at the celestial equator to 90° at the celestial poles. Right ascension is measured in hours eastwards from the vernal equinox, one hour corresponding to 15° of longitude.

The **celestial poles** lie directly above the Earth's poles, and the **celestial equator** lies over the Earth's Equator. The celestial sphere appears to rotate once around the Earth each day, actually a result of the rotation of the Earth on its axis.

Centaur NASA's upper-stage rocket used to lift heavy satellites and space probes in combination with the *Atlas rocket and the *Titan rocket. The US Air Force developed the liquid-hydrogen and liquid-oxygen engine for the first Centaur in 1958. The rocket's deployment of deep-space probes has included the two **Voyager* probes in 1977, when the Centaur's sophisticated computer flew each spacecraft into low orbit, calculated the flight path and velocity, then shut off the rocket engines and separated as *Voyager* headed for Jupiter.

Center for Food and Environmental Systems for Human Exploration of Space (CFESH) A research centre for NASA's *Advanced Life-Support Program. Located at Tuskegee University, Alabama, the team was involved in 2001 in developing technology to grow and process sweet potatoes and peanuts in controlled environments. CFESH also works closely with the *Bioregenerative Planetary Life-Support Systems Test Complex (BIO-Plex) project to provide food in space for astronauts.

Centre for Earth Observation (CEO) A European Union project to encourage a wider use of information gathered by *Earth-observation satellites. The CEO concentrates on space monitoring for environmental and security purposes. It also works on product development and marketing, doing studies for the shipping industry, agribusiness, environmental protection organizations, and other groups.

The CEO began in 1995 as part of the European Commission's 4th Framework Programme and is now a main project of the Strategy and Systems of Space Applications Unit.

Centre National d'Etudes Spatiales (CNES; National Centre for Space Studies) A French space agency, established in 1961. It is a leading member of the *European Space Agency (ESA) and developed the *Ariane rocket. It takes a major part in space probes, such as providing instruments for the **Rosetta* orbiter and lander launched on 2 March 2004, including spectrometers, optical cameras, and plasma analysers. CNES proposes space policies to the French government and carries out various programmes in collaboration with industry and research and defence organizations.

CNES headquarters are in Paris. Divisions include the Launch Division (DLA), in the Parisian suburb of Eury, responsible for space transport; the Toulouse Space Centre (CST), in charge of programme preparations and operational systems; and the *Centre Spatial Guyanais (CSG), near Kourou in French Guiana, which operates the launch base there for the ESA.

Centre Spatial Guyanais (CSG; Guianese Space Centre) The European Space Agency's launch site, near Kourou in French Guiana. It handles about 12 launches a year. The location on the northeast coast of South America is ideal for launching geostationary satellites, since they must orbit above the Equator. CSG is operated by the *Centre National d'Etudes Spatiales, which began launching French rockets there in 1969. The facility has a launch complex, ELA 2, comprising a launch preparation zone, a launch zone, and a launch control centre, to launch *Ariane 4 rockets, an ELA 3 complex for Ariane 5, and a payload preparation complex.

C

centrifugal force The apparent force arising for an observer moving with a rotating system. For an object of mass m moving with a velocity v in a circle of radius r, the centrifugal force F equals mv^2/r (outwards).

centrifuge A machine that creates centrifugal force by spinning around a centre. NASA uses one, located at Brooks Air Force Base, Texas, to simulate *g-forces encountered by astronauts during a launch. The US Air Force uses it to train pilots, and has trained more than 100 astronauts since 1991. The 'g qualification ride' involves two nine-minute sessions in the centrifuge, which reaches a threshold of 3 g, that is, a gravitational force three times a person's own body weight.

The **International Space Station* will also have a centrifuge accommodation module to simulate Earth's gravity for comparative studies, and it will be able to simulate the gravity of the Moon and Mars for future space missions.

centripetal force The force that acts radially inwards on an object moving in a curved path. For example, with a weight whirled in a circle at the end of a length of string, the centripetal force is the tension in the string. For an object of mass m moving with a velocity v in a circle of radius r, the centripetal force F equals mv^2/r (inwards). The reaction to this force is the *centrifugal force.

CEO Abbreviation for *Centre for Earth Observation.

CEOS Abbreviation for *Committee on Earth-Observation Satellites.

Ceres The largest asteroid, 940 km in diameter, and the first to be discovered (by Italian astronomer Giuseppe Piazzi in 1801). Ceres orbits the Sun every 4.6 years at an average distance of 414 million km. Its mass is about 0.014 of that of the Earth's Moon.

Cernan, Eugene Andrew (1934–) US astronaut. He piloted *Gemini 9* in June 1966, piloted *Apollo 10*, orbiting the Moon in May 1969, and commanded *Apollo 17* in December 1972. He was the second US astronaut to walk in space and the last person to walk on the Moon, having been one of two astronauts to fly there twice (the other being James *Lovell).

Cernan and Jack *Schmitt stayed on the lunar surface for three days, setting a record for the longest lunar surface extravehicular activity (22 hours and 6 minutes). Cernan logged more than 566 hours in space, including more than 73 hours on the Moon. He was selected as an astronaut in 1963 and resigned from NASA in 1976.

CFESH Abbreviation for *Center for Food and Environmental Systems for Human Exploration of Space.

CFRP Abbreviation for *carbon-fibre reinforced plastic.

Chaffee, Roger B (1935–67) US astronaut. He worked for the **Apollo* project on three systems: flight-control communication, instrumentation, and attitude and translation control. He died, along with Gus *Grissom and Edward *White, in a fire aboard *Apollo 1* during a ground test on 27 January 1967. He was selected as an astronaut in 1963.

Challenger A NASA *space shuttle that was used on nine successful flights from 4 April 1983 until 30 October 1985. On its tenth launch on 28 January 1986, a booster rocket failed and the vehicle broke up, killing the seven members of the crew. *Challenger* began as a high-fidelity structural test article (STA-099) that was converted into a shuttle in 1979, the second shuttle constructed and then known as OV-99. On 18 June 1983 it carried the first

(Image © NASA)
Space shuttle *Challenger* viewed from the research module SPAS (shuttle pallet satellite) during STS-7. It was on this mission that Sally Ride became the first US woman to fly in space.

(Image © NASA)

Smoke and clouds of vaporized propellant engulf the *Challenger* orbiter during the disastrous flight of 28 January 1986. Fragments from the break-up leave trails in the air while the orbiter's solid rocket booster has detached and continues to fly out of control.

US woman into space, Sally Ride, and on 30 August 1983 the first African American, Guion Bluford.

Other *Challenger* flights included the first in-orbit satellite repair in April 1984, the first seven-person crew in October 1984 and the first eight-person crew in October 1985. *Challenger* was named after the British naval research vessel HMS *Challenger* that sailed the world in the 1870s.

http://science.ksc.nasa.gov/shuttle/missions/51-l/docs/rogers-commission/table-of-contents.html Report of the independent commission that reviewed the circumstances surrounding the *Challenger* disaster for US president Ronald Reagan.

C

***Challenger* Center (in full: *Challenger* Center for Space Science Education)** A non-profit educational centre that aims to educate people about space. The centre was established in the wake of the *Challenger* space shuttle disaster in January 1986, by the bereaved families of the crew members. Founded on 24 April 1986, the first *Challenger* Center is located in Alexandria, Virginia. More than 50 *Challenger* Centers have since been opened throughout the USA and two in Canada. The first European centre was established at the University of Leicester, England in 1997.

Many centres include simulators of NASA's mission control and a space station, enabling students to experience some of the realities of a space mission. The centres also train teachers, provide curriculum materials for classroom use, and hold summer space camps for students.

chamber pressure The pressure of hot gas within a rocket's combustion chamber. Pressure within a large solid rocket booster is monitored by four main combustion chamber pressure sensors. If an engine loses its chamber pressure, its valves remain at the last commanded position to keep the same mixture ratio. When the space shuttle's two boosters separate, the chamber pressure of each is 3.5 kg per sq cm or less to ensure they are not still burning.

Chandra X-ray Observatory A NASA satellite for X-ray astronomy, the third of NASA's series of four 'Great Observatories', launched by the space shuttle *Columbia* in July 1999. It carries a grazing-incidence X-ray telescope to observe X-rays of 0.1–10.0 keV energy from high-energy regions of the Universe such as the remnants of exploded stars, regions around black holes, ultra-hot gas in clusters of galaxies, active galactic nuclei, and quasars. The X-rays are studied by four instruments: the High Resolution Camera (HRC), the Advanced CCD Imaging Spectrometer (ACIS), High Energy Transmission Grating Spectrometer (HETGS), and the Low Energy Transmission Grating Spectrometer (LETGS).

Chandra operates from a highly elliptical orbit which takes it more than a third of the way to the Moon before returning to its closest approach to the Earth of 16 000 kilometres. It takes 64 hours and 18 minutes to complete an orbit and uninterrupted observations up to 55 hours are possible.

The observatory is named after the Indian-born US astrophysicist Subrahmanyan Chandrasekhar, who won the 1983 Nobel Prize for Physics for his theoretical studies of the physical processes concerning the structure and evolution of stars.

Chandrayaan-1 A lunar orbiter planned for launch by the Indian Space Research Organization in 2007 or 2008. From an altitude of 100 km in polar orbit it will map the chemical composition of the Moon's surface, search for surface or sub-surface ice, particularly at the poles, and prepare a 3-D map of the Moon's topography.

Chang-Diaz, Franklin Ramon (1950–) US astronaut and director of the Advanced Space Propulsion Laboratory at the Johnson Space Center. He has logged 1 601 hours in space and more than 19 hours of space walks during seven shuttle flights. These include the *Atlantis* mission to deploy the **Galileo* spacecraft (October 1989), the *Discovery* flight (June 1998) that made the final docking with the Russian space station *Mir*, and the *Endeavour* flight (June 2002) to the *International Space Station* during which he took three space walks to install the Canadian Mobile Base System. He was selected as an astronaut in 1980.

Chang-Diaz was a visiting scientist at the Massachusetts Institute of Technology from 1983 to 1993, leading a programme to develop plasma propulsion technology for future crewed missions to the planet Mars.

channelization A method of displaying data measurements downlinked from a spacecraft. The value of each measurement, such as the spacecraft pressure or fuel tank temperature, will always be in a specific channel that can be called up. Some spacecraft have thousands of measurements, but each is always associated with a particular channel.

charge-coupled device (CCD) A device for forming images electronically, using a layer of silicon that releases electrons when struck by incoming light. The electrons are stored in pixels and read off into a computer at the end of the exposure. CCDs are used in digital cameras, and have now almost entirely replaced photographic film for applications such as astrophotography, where extreme sensitivity to light is paramount.

Chawla, Kalpana (1961–2003) US astronaut. She was the principal operator of the robot arm on the space shuttle mission launched in November 1997 to carry out microgravity experiments. From 1988, Chawla worked for NASA's *Ames Research Center for research into fluid dynamics before her selection as an astronaut in 1995. She died when the space shuttle *Columbia* broke up on re-entry in February 2003 after a 16 day mission to the *International Space Station*.

checkout The on-board procedures to check that a spacecraft's systems and subsystems are functioning properly. When NASA's space shuttle has been launched and begins to cruise in orbit, the designated checkout periods begin, usually 1 minute and 18 seconds after launch, over the Hawaii tracking station. Among the checks, *telemetry is analysed to see how well the shuttle survived the launch and establish its general health. Questionable components are tested, scientific instruments are turned on and calibrated, and celestial references are monitored to ensure *attitude control.

Cheli, Maurizio (1959–) Italian astronaut with the European Space Agency (ESA). He flew on the space shuttle *Columbia* launched in February 1996 to carry out experiments with the *tethered satellite system (the Italian satellite was lost when the tether broke). His other flight was in October 2000 to the

International Space Station, helping expand it and prepare it for the first resident crew. He logged 36 days in space with four space walks totalling more than 26 hours before leaving the space programme. Maurizio was selected by the ESA in 1992, as one of the second group of European astronauts.

Chelomei, Vladimir (1914–84) A Soviet rocket designer responsible for developing *booster rockets for the lunar programme. In 1945 he created a cruise missile similar to Germany's V-1 weapon of World War II. By 1955 he was developing cruise missiles for submarines. He began working for the Soviet space programme in the 1960s, and his SL-9 rocket, first launched in 1965, evolved into the three-stage *Proton rocket, the workhorse for Soviet space launches.

Cheng Zen rocket (CZ) A Chinese three-stage rocket, evolved from the military's DF-3 intermediate-range ballistic missile. CZ-1, which was 30 m long, launched China's first satellite, **China 1*, in April 1970. The CZ-4B version, which launched the *Zi Yuan 2* remote-sensing satellite in September 2000, was 45.8 m long.

chest pack A *life-support system worn on the chest of a spacesuit. Astronauts taking space walks during NASA's *Gemini* project wore chest packs that received oxygen from the spacecraft via a 15-m umbilical cord and reduced its high pressure. This chest pack also had a fan that circulated the oxygen and a heat-exchange cooling device.

Chiao, Leroy (1960–) US astronaut. A mission specialist, he logged more than 25 hours of extravehicular activity during the second and third of his space flights. His third mission was the *Discovery* flight of October 2000 to expand the *International Space Station*. He was selected as an astronaut in 1990.

Chibis vacuum suit A Soviet *spacesuit introduced in 1978 for cosmonauts about to return from 126 days on the *Salyut 6* space station mission. It was a low-pressure suit that forced the body's blood into the legs, an effect similar to that of Earth's gravity. This caused the cardiovascular system, slowed by weightlessness, to work harder and strengthen the heart in preparation for a return to gravity. Similar units were worn regularly by crews on the **Mir* space station.

Chilton, Kevin Patrick (1954–) US astronaut. He piloted the first flight of the space shuttle *Endeavour* in May 1992, carrying out repairs to the International Telecommunications Satellite. He also flew the *Endeavour*'s mission of April 1994 that carried the Space Radar Laboratory. In March 1996, Chilton commanded the *Atlantis* mission that docked with the Russian space station *Mir*. He logged 704 hours in space before leaving NASA in 1998 to become the US deputy director for political military affairs for Asia, the Pacific, and the Middle East. He was selected as an astronaut in 1987.

***China 1* (or *Mao 1*)** China's first satellite, launched in April 1970 by the *Cheng Zen rocket. It was an experimental satellite that played the patriotic song 'The East Is Red'. It weighed 173 kg.

***Chinastar 1* (or *Zhongwei 1*)** A telecommunications satellite built by the US firm Lockheed Martin for the China Orient Telecomm Satellite Company, part of China's Telecommunications Ministry. Launched in May 1998, it cost US$100 million and has a lifespan of approximately 15 years. The satellite, in a geosynchronous orbit, permits voice and data transmission, and a direct-broadcast television service to China, India, Korea, and other areas of Southeast Asia.

chondrite A type of *meteorite characterized by *chondrules (small spherules), typically about 1 mm in diameter, made up of silicate material.

chondrule A small, round mass of the silicate minerals olivine and orthopyroxene. Chondrites (stony *meteorites) are characterized by the presence of chondrules.

Chondrules are thought to be mineral grains that condensed from hot gas in the early Solar System, most of which were later incorporated into larger bodies from which the *planets formed.

Chrétien, Jean-Loup Jacques Marie (1938–) French astronaut. He flew as a research-astronaut on two **Soyuz* missions: the first linked with *Salyut 7* (June 1982), the second docked with the Russian space station *Mir* for 22 days (November 1988). During this flight, Chrétien became the first Western European to make a *space walk. In September 1997, he docked again with *Mir*, as a mission specialist aboard the space shuttle *Atlantis*. Chrétien was selected as an astronaut in 1980.

Chryse Planitia (or Plains of Gold) A flat region of Mars where NASA's **Viking* Lander 1 sampled soil after arriving in July 1976. Its remote-controlled arm collected soil and tested it, finding it more sterile than the Earth's and containing no sign of life forms.

CID Acronym for *cold-ion detector.

circularization NASA term for the moment a space shuttle enters its circular orbit around the Earth. This occurs 45 minutes after launch, and is controlled by the thrust from the *orbital manoeuvring system.

cislunar space The space between the Earth and Moon. *Apollo* astronauts in the 1960s noted that cislunar space gave the lunar day an impression of 'day-night', with a sunlit lunar surface contrasted against a black sky.

Clark, Laurel Blair Salton (1961–2003) US astronaut and physician. A mission specialist, she was selected as an astronaut in 1996. She previously worked in the Payloads/Habitability Branch of the Astronaut Office. Clark died when the space shuttle *Columbia* broke up on re-entry in February 2003 after a 16-day mission to the *International Space Station*.

Clarke, Arthur C(harles) (1917–) English science-fiction and non-fiction writer. He originated a plan for a system of communications satellites in *geostationary orbit in 1945. His works include 'The Sentinel' (1951), filmed in 1968 by US director Stanley Kubrick as *2001: A Space Odyssey*, and the non-fiction *Interplanetary Flight* (1950) and *The Exploration of Space* (1951). He became chair of the British Interplanetary Society in 1946.

Clarke orbit An alternative name for *geostationary orbit, an orbit 35 900 km high, in which satellites circle at the same speed as the Earth turns. This orbit was first suggested by English writer Arthur C Clarke in 1945.

clean room A sanitized interior area used by NASA to prepare a spacecraft for launch. This is where the engineering components and instruments of the spacecraft are put together and tested by computer programs. The **Cassini* spacecraft, for instance, was taken to a 21 × 21-m/70 × 70-ft 'high bay' that was certified as a class 100 000 clean room. In such a clean room, the air is filtered so that there is less than one particle of dust bigger than 1 micron per 100 000 cu ft (2832 cu m) of air. Astromaterials such as Moon rocks and meteorites are handled and studied in clean-room laboratories to prevent terrestrial contamination. The Lunar Sample Laboratory, for example, is a class 10 000 clean room and the recently constructed Genesis Laboratory is a class 10 clean room.

Cleave, Mary Louise (1947–) US astronaut. She has flown on two *Atlantis* space shuttle missions, the first to deploy three communications satellites (November 1985), and the second to deploy the **Magellan* space probe (May 1989). She was selected as an astronaut in 1980.

Clementine A US lunar probe launched on 25 January 1994. It spent two months surveying the Moon and discovered an enormous crater on the Moon's far side before being sent on to intercept the asteroid Geographos. The mission was abandoned in June 1994 after a faulty thruster failed to shut off during orbital manoeuvres.

Clervoy, Jean-François (1958–) French astronaut with the European Space Agency (ESA). His three NASA space shuttle flights include the May 1997 docking of *Atlantis* with the Russian space station *Mir*, during which Clervoy coordinated more than 20 experiments, and the December 1999 *Discovery* flight to upgrade the *Hubble Space Telescope*, during which he made three eight-hour space walks. He went on to be the senior adviser astronaut for ESA's *Automated Transfer Vehicle (ATV). He was selected as one of the second group of French astronauts in 1985, becoming an ESA astronaut in 1992.

clevis On a solid rocket booster, a deep groove in the rim of the bottom rocket segment used to join segments together. The segment above has a protrusion ('tang') that fits into it, and the joint is sealed by two *O-rings.

Clifford, (Michael) Richard Uram (1952–) US astronaut. His three space shuttle fights include the March 1996 docking of *Atlantis* with the Russian space station *Mir*. His six-hour *space walk during this mission was the first to be undertaken during the docking of a shuttle with a space station. He logged 665 hours in space, leaving NASA in 1997 to become the space station flight operations manager for the Boeing Defense and Space Group. Clifford was assigned to the Johnson Space Center in 1987 and selected as an astronaut in 1990.

CloudSat A NASA satellite that will use millimetre-wavelength radar to measure the vertical structure of clouds and cloud properties from space, providing information that will increase the accuracy of storm and flood warnings as well as an improved understanding of climatic change. *CloudSat*, due for launch in 2005, will fly in orbital formation with other Earth-observing spacecraft, including NASA's *Aqua* and *Aura*, launched 2002 and 2004 respectively, and the forthcoming NASA–CNES (France) *CALIPSO* and the CNES *PARASOL*.

CLT Abbreviation for *command-loss timer.

Cluster A European Space Agency two-year project to study the interaction of the *solar wind, a stream of atomic particles from the Sun's corona, with the Earth's *magnetosphere, from an array of four identical satellites that can provide a three-dimensional view. The first pair of the *Cluster II* space weather satellites was launched in July 2000, and the second pair the following month.

The Ariane 5 rocket carrying the first *Cluster* probes exploded due to a software fault 37 seconds after lift-off on 4 June 1996. New probes had to be built for launch in 2000.

Cluster works in conjunction with the **Solar and Heliospheric Observatory*.

http://sci.esa.int/cluster/ All aspects of the Cluster mission, its objectives, science payload, mission profile, and launch. The pages are frequently updated with images and news of Cluster's latest results. There is also good background information about the solar effects on near-Earth space.

CM Abbreviation for *command module.

CMD Abbreviation for *command system.

CNES Abbreviation for *Centre National d'Etudes Spatiales.

CNO cycle The carbon–nitrogen–oxygen cycle, alternative name for *carbon cycle.

COAS Acronym for *crewman optical alignment sight.

coastal zone colour scanner (CZCS) A multi-channel scanning radiometer carried on NASA's **Nimbus 7* meteorological satellite to map and measure concentrations and temperatures of the Earth's coastal waters and ocean currents. Launched in October 1978, the instrument measured chlorophyll concentrations, sediment, salinity, and temperatures. It began to experience problems in 1984 and was remotely disabled in December 1986.

coast period A time during which a spacecraft or satellite moves without power. During a space shuttle's launch of a satellite, the satellite's coast period after release is at least five minutes, before its thrusters ignite at the *apogee of its orbit. Normally, a satellite is deployed into its orbit 3 hours and 28 minutes after lift-off.

Coats, Michael Lloyd (1946–) US astronaut. His three space shuttle missions include piloting the first flight of *Discovery* in August 1984 and commanding the shuttle's March 1989 mission to deploy a *Tracking and Data-Relay Satellite. Coats was selected as an astronaut in 1978, and was the acting chief of NASA's Astronaut Office May 1989–March 1990. He left NASA in 1991.

C

COBE Contraction of *Cosmic Background Explorer*, US astronomical satellite launched in 1989 to study *cosmic background radiation.

cockpit The area containing the flight controls. In the space shuttle, it is the forward area of the *flight deck, where the pilot and commander sit (the commander on the right). Most instruments are duplicated so both can fly the shuttle. Besides several indicators for attitude, airspeed, and direction, both the pilot and commander have a pistol-grip control column to activate thrusters and change the spacecraft's attitude. While in orbit, however, the manoeuvring controls in the *crew station are often used.

The launch of the space shuttle *Atlantis* in May 2000 marked the replacement of cockpit dials and switches by computer displays, a system known informally as the 'glass cockpit'. The system will eventually be fitted to all orbiters. A further upgrade to cockpit controls is planned for 2005, with the creation of a 'smart cockpit' intended to reduce the pilot's workload.

Cockrell, Kenneth Dale (1950–) US astronaut. He logged over 1 560 hours in space during five space shuttle flights, including the *Atlantis* mission in February 2001 to assemble and supply the *International Space Station (ISS)*, and the *Endeavour* flight in June 2002 to the *ISS* to exchange resident crews and bring the Canadian mobile base for the robot arm. He went on to become a manager in the Flight Crew Operations Directorate at the Johnson Space Center. He joined the Johnson Space Center as an instructor pilot for the *T-38 jet trainer in 1987, and was selected as an astronaut in 1990.

cold-ion detector (CID) A scientific instrument that detects ions. Electronic sparks that can damage a spacecraft's electronic systems are generated by the Sun's hot plasma as the spacecraft passes through the Earth's radiation. To research the phenomenon, the *Mullard Space Science Laboratory built a cold-ion detector for the European Space Agency's Space Technology Research Vehicle STRV-1a, launched in 1994 into geostationary orbit.

Coleman, Cady (born Catherine Grace Coleman) (1960–) US astronaut. She has flown twice on the *Columbia* space shuttle: on the second US **Microgravity Science Laboratory* mission in October 1995 and on the deployment of the **Chandra X-Ray Observatory* in July 1999. She has logged more than 500 hours in space. Coleman was selected as an astronaut in 1992.

Collins, Eileen Marie (1956–) US astronaut and the first female pilot and space flight commander. Collins piloted two space shuttle missions, *Discovery* in February 1995 and *Atlantis* in May 1997, that involved rendezvous with the Russian space station *Mir*. She then led the *Columbia* mission launched in July 1999 to deploy the **Chandra X-Ray Observatory*, and in July–August 2005 commanded *Discovery* on the first space shuttle flight after the *Columbia* disaster of 2003. Collins was selected as an astronaut in 1990.

Collins, Michael (1930–) US astronaut. He performed two space walks on the *Gemini 10* mission in 1966 and piloted the *Apollo 11* command module, which circled the Moon while the first crewed spacecraft landed there in 1969.

Collins was a test pilot for the US Air Force before being selected as an astronaut in 1963. He retired from NASA in 1970. He was the first director of the Smithsonian Institution's National Air and Space Museum in Washington, DC, 1971–8. He was vice-president of LTV Aerospace and Defense Co. 1980–5, and founded his own company in 1985. He wrote *Carrying the Fire* (1974), an account of his experiences in the space programme.

Columbia A NASA *space shuttle that was used on 27 successful flights from 12 April 1981 to 1 March 2002. Known as OV-102 (for Orbiter Vehicle-102), it

(Image © NASA)

Space shuttle *Columbia* takes off from Kennedy Space Center in 1993. Mission STS-58 was designed to study the performance of the human body in space. *Columbia*, the longest-serving of NASA's five space shuttles, is named after the sloop used by Robert Gray to explore the coast of British Columbia in 1792.

was NASA's oldest shuttle, being the first to fly into Earth orbit. Its flights included carrying the first science payload in November 1981, the first four-person crew in November 1982, the first six-person crew in November 1983, and the first to be commanded by a woman, Eileen Collins, in July 1999. On 1 February 2003, the vehicle broke up and burned on re-entry and the crew of seven died. The apparent cause was a hole in the heat shield of a wing, caused just after lift-off by foam insulation that had broken off.

Columbia was named after an 18th-century sloop from Boston, Massachusetts, that made the first American circumnavigation of the globe.

Columbia The command module of the *Apollo 11* spacecraft that took three US astronauts to the Moon in 1969. *Columbia* was piloted by astronaut Michael *Collins, who remained in lunar orbit while Neil *Armstrong and Buzz *Aldrin descended to the surface in the lunar module, known as *Eagle*.

Columbus A research laboratory built by the European Space Agency which will form a permanent part of the *International Space Station* (*ISS*). Columbus is a pressurized cylinder 6.9 m long and 4.5 m wide with room for three astronauts to conduct investigations in biology, medicine, and materials science in conditions of near-weightlessness. Additional equipment can be mounted on the outside of the laboratory for further observations and experiments in the vacuum of space. Columbus is Europe's largest single contribution to the *ISS* and will be launched in the cargo bay of the space shuttle.

combustion The burning of rocket fuel. This occurs in the *combustion chamber when the fuel and oxidizer mix under extreme high pressure and temperature.

combustion chamber A chamber on a liquid-fuel rocket in which the fuel is ignited. The *fuel and oxidizer are pumped into the chamber to be burned at a high temperature and pressure, producing hot gases that are propelled through an outlet to give thrust to the rocket.

combustion instability An unsteady burning of rocket fuel that can cause structural damage to the thrust chamber. Combustion instability was a problem experienced by many of the first launch vehicles, and also with the prototype ascent engine of the *Apollo* lunar module. These were usually caused by poor fuel injection and corrected by redesigns. To test for combustion instability, NASA engineers developed tiny 'bombs' to interupt the fuel flow and see if the engine could recover.

comet A small, icy body orbiting the Sun, usually on a highly elliptical path. A comet consists of a central nucleus a few kilometres across and is made mostly of ice mixed with dust. As a comet approaches the Sun its nucleus (made of ice and gas) heats up, releasing gas and dust, which form a coma up to 100 000 km wide. The famous periodic Halley's Comet was studied during its last appearance in 1986 by the European spacecraft *Giotto* and the Soviet twin-probe *Vega*. In 1997 NASA launched rockets to study the bright comet, Hale–Bopp, as it flew within 196 million km of the Earth. The **Rosetta* spacecraft was launched in 2004 to land on a comet in 2014. The **Deep Impact* probe collided with the nucleus of comet Tempel 1 on 4 July 2005 to study its structure.

http://impact.arc.nasa.gov/index.html Overview of, and the latest news on, asteroid and comet impact hazards from NASA's Ames Space Science Division, with the last Spaceguard Survey Report and a list of future near-Earth objects (NEOs).

http://encke.jpl.nasa.gov/ Links to information on comet observation, including definitions and explanations of technical terms, recent news and observations, photographs, and links to the NASA and Jet Propulsion Laboratory home pages.

http://www.jpl.nasa.gov/sl9/ Description of the comet's collision with Jupiter in 1994, the first collision of two Solar System bodies ever to be observed. The site includes background information, the latest theories about the effects of the collision, and even some animations of Jupiter and the impact.

http://www.pbs.org/wgbh/nova/spacewatch/ Companion to the US Public Broadcasting Service television programme *Nova*, this page provides information on comets: what they are made of, where they come from, and why they are important to us. It includes images and information about comets Hale–Bopp and Hyakutake, including a number of images from the *Hubble Space Telescope*. You can also view images of Venus. There is a recipe for making your own comet, courtesy of the Jet Propulsion Laboratory, and a list of resources for further research.

***Comet Nucleus Tour* (*Contour*)** A NASA space probe in the Discovery programme launched on 3 July 2002, intended to fly past comets Encke in 2003 and Schwassmann–Wachmann-3 in 2006. However, contact was lost on 15 August when the probe fired its solid rocket engine to enter solar orbit, apparently because the probe broke up during the firing.

command and service module (CSM) A connected unit comprising the *command module and *service module in flight, normally identified with the *Apollo* missions. When the *lunar module disengaged to land two astronauts on the Moon, the third remained in the CSM orbiting the Moon. Somewhat confusingly, the CSM was normally referred to as the command module, especially when disengaged from the lunar module.

commander (CDR) An astronaut in charge of the crew of a spacecraft. In the space shuttle, the commander sits to the right of the pilot in the cockpit, and also flies the vehicle.

command-loss timer (CLT) A spacecraft timer that protects against faults in uplink commands (those transmitted from ground to space). It is part of the **command and data-handling subsystem**. Each time its spacecraft receives a command from the ground, the CLT is set to a selected period of time, such as a week. If it ever runs down to zero, this indicates a failure in the spacecraft's receiver or other command components. The CLT then issues commands for actions, such as switching to a back-up system, to restore the broken link.

C

command module (CM) A module of a spacecraft occupied by the crew, who operate the main controls located there. It normally refers to the conical command modules of the *Apollo* missions. They were each 3.9 m in diameter at the widest point and 3.5 m high, a spacious compartment compared to the earlier *Mercury* capsules. The command module was launched sitting above the *service module, and the two remained connected in space to comprise the *command and service modules (CSM).

command sequence A program sequence loaded into a spacecraft's computers to take it through normal flight operations. For the space shuttle, a command sequence is loaded prior to launch to program the vehicle for the launch, and additional sequences are then uplinked by the *command system during the shuttle's flight, to replace the launch commands.

command system (CMD) A *deep-space network system for sending digital data to spacecraft. Flight project teams use the command system to load a spacecraft's computers with software for the flight *command sequence, or for other on-board operations and activities.

Committee on Earth-Observation Satellites (CEOS) An international committee that harmonizes the recording, cataloguing, and networking of Earth-observation satellite data. Established in 1984, CEOS also helps coordinate international mission plans and the exchange of policy and technical information to encourage compatibility between Earth-observation systems.

Its members, who represent all nations active in space, include civil agencies responsible for Earth-observation satellite programmes, and user organizations such as the World Meteorological Organization and the Global Climate-Observing System. Associate members include government organizations involved in Earth-observation programmes.

Committee on Space Research (COSPAR) An interdisciplinary scientific organization promoting worldwide research to develop and utilize space vehicles, rockets, and balloons. It was established in 1958 by the International Council of Scientific Unions (ICSU) to continue the cooperative programmes in satellite and rocket research that had begun during the 'International Geophysical Year' of 1957–8. The ICSU resolution stated that COSPAR would provide a means by which the international scientific community could 'exploit the possibilities of satellites and space probes of all kinds for scientific purposes, and exchange the resulting data on a cooperative basis'.

Committee on the Peaceful Uses of Outer Space (COPUOS) A United Nations (UN) committee set up by the UN General Assembly in 1959 to review and support international cooperation over the peaceful use of outer space. To achieve this aim the committee encourages the international exchange of information on outer space matters, helps coordinate space research programmes within nations, and studies any legal problems arising from the exploration of outer space.

COPUOS originally consisted of 18 member states and by 2001 had a total of 61, making it one of the largest UN committees. A number of intergovernmental and non-governmental organizations have observer status. The committee meets annually with its two standing subcommittees, the Scientific and Technical Subcommittee and the Legal Subcommittee. Its secretariat is the *Office of Outer Space Affairs (OOSA).

communications carrier assembly (or Snoopy helmet) A soft cap worn by NASA astronauts beneath a bubble *space helmet, containing a headset and microphone. During an *extravehicular activity, the cap allows an astronaut to talk with crew members in the spacecraft or with other astronauts on a space walk.

communications satellite A relay station in space for sending telephone, television, telex, and other messages around the world. Messages are sent to and from the satellites via ground stations. Most communications satellites are in *geostationary orbit, appearing to hang fixed over one point on the Earth's surface.

The first satellite to carry TV signals across the Atlantic Ocean was *Telstar*, which was launched into low Earth orbit by the USA on 10 July 1962. The world is now linked by the Intelsat system of communications satellites. Other satellites are used by individual countries for internal communications, or for business or military use. A new generation of satellites, called **direct-broadcast satellites**, are powerful enough to transmit directly to small domestic aerials. The power for such satellites is produced by solar cells. The total energy requirement of a satellite is small; a typical communications satellite needs about 2 kW of power, the same as an electric heater.

Communications Satellite Corporation (Comsat) A private organization that coordinates international communications that use artificial satellites. It was established in 1962 by the US Congress to create an international satellite communications system for television, telephone, and telefax transmissions. Comsat's first satellite, **Early Bird*, was launched in 1965. The corporation is a member of *Intelsat (International Telecommunications Satellite Organization).

Complex 14 A launch site at the *Kennedy Space Center for the *Mercury* spacecraft (*see* MERCURY PROJECT), including the first US orbital flight by John *Glenn. Located 5 km south of *Complex 39, it is now out of use.

Complex 39 A launch site at the *Kennedy Space Center for the space shuttles. The complex's Pad A and Pad B were originally built for the **Apollo* project, but were modified for shuttle launching.

***Compton Gamma-Ray Observatory* (GRO)** A NASA satellite launched in April 1991 by the space shuttle *Atlantis* to observe the high-energy Universe. Named after the American physicist Arthur H. Compton, it was the second of NASA's series of four 'Great Observatories'. It had four instruments larger and more sensitive than any previous orbiting gamma-ray telescopes. They were the Imaging Compton Telescope (COMPTEL), the Energetic Gamma-ray Experiment Telescope (EGRET), the Oriented Scintillation Spectrometer Experiment (OSSE), and the Burst and Transient Source Experiment (BATSE). The instruments investigated solar flares, gamma-ray bursts, pulsars, quasar emissions, nova and supernova explosions, black holes, and interactions of cosmic radiation with interstellar matter. The observatory was commanded to burn up in the atmosphere over the Pacific Ocean on 4 June 2000.

http://cossc.gsfc.nasa.gov/ Dedicated to the nine-year mission of the *Compton Gamma-Ray Observatory*, outlining its objectives and providing information on the spacecraft's scientific instruments and results. Among the many resources are images and links to other gamma-ray astronomy sites.

Conestoga rocket A rocket developed for NASA's Commercial Experiment Transporter (COMET) programme that supported the microgravity industry. The Conestoga had a main engine with additional ones strapped around it. The first and only launch of the COMET spacecraft was in October 1995 but the Conestoga broke into pieces 16 seconds after lift-off. The rocket and programme were cancelled.

Configuration Control Board The NASA body responsible for the final approval of changes in the design of a spacecraft.

Conrad, Pete (born Charles Conrad Jr) (1930–99) Former US astronaut. He flew two *Gemini* missions in 1965–6, became the third man on the Moon on *Apollo 12*, which he commanded in 1969, and commanded the first **Skylab* space station mission in May 1973. During this mission he made a heroic space walk to save the project by freeing a jammed solar panel.

Conrad's *Gemini 5* flight established a space endurance record, and he commanded *Gemini 12* to a world altitude record. He was selected as an astronaut in 1962, left NASA in 1973, and later became a vice-president of the McDonnell Douglas aircraft company.

Constellation-X A proposed NASA X-ray astronomy mission that will consist of four X-ray satellites in formation, each containing a 1.6-m diameter telescope for measuring the spectra of cosmic X-ray sources. The four satellites would work together to produce the observing power of one giant telescope, in similar fashion to optical telescopes such as the Very Large Telescope. One particular target of interest will be the super-massive black holes at the centres of active galaxies. Constellation-X could be launched in 2011.

consumable On a spacecraft, anything that is consumed and can be used up, such as oxygen, fuel, and electricity.

contaminant control cartridge A replaceable cartridge on a *life-support system or extravehicular mobility unit (*see* EXTRAVEHICULAR ACTIVITY) that removes carbon dioxide from the astronaut's air supply.

controlled crash A programmed hard impact of a space probe with the surface of a planet or other body. This is often because fuel has run out, as when NASA's **Near-Earth Asteroid Rendezvous* spacecraft was directed to a controlled crash with the asteroid *Eros in February 2001.

control moment gyroscope A gyroscope used for the *attitude control of a spacecraft, part of the **attitude and orbit control subsystem**. The

International Space Station has four control moment gyroscopes, but only two are required for the station's attitude control.

control-stick steering (CSS) A semi-automatic, or *fly-by-wire, spacecraft steering system, regulated by a computer. Electronic control-stick steering is one of three pilot options on the space shuttle. The other two options are *auto and *direct control (DIR).

convolutional coding A coding system used by most interplanetary spacecraft to ensure error-free data transmission. The system uses symbols instead of data bits to improve the capacity of a radio channel, and the technique is called forward error correction (FEC). Convolutional coding is used with the Viterbi decoding system, a decoding algorithm developed by US computer entrepreneur Andrew Viterbi.

Cooper, (Leroy) Gordon, Jr (1927–2004) US astronaut. He circled the Earth 22 times in *Faith 7* in 1963. When the automatic controls failed, he piloted the craft manually back to Earth, splashing down within 8 km of the recovery ship. He was also the command pilot of *Gemini 5* on an eight-day endurance mission in August 1965.

Cooper was a test pilot with the US Air Force before being selected in 1959 as one of the *'Original Seven' *Mercury* astronauts. He retired from NASA in 1970. In 1989 he became chief executive officer of Galaxy Group Inc., an aircraft engineering design group.

Coordinated Universal Time An alternative name for *Universal Time Coordinated, Greenwich Mean Time (GMT) as maintained to scientific standards by atomic clocks.

***Copernicus* (or Orbiting Astronomical Observatory 3 (OAO-3))** A probe monitoring pulsars and other bright X-ray stars. A collaborative effort between NASA and the UK's Science and Engineering Research Council, it was launched in August 1972 and operated until February 1981. The main body measured 3 × 2 m, and the observatory consisted of an ultraviolet telescope from Princeton University, New Jersey, and four X-ray detectors that primarily took background measurements for an X-ray astronomy experiment developed by the *Mullard Space Science Laboratory of University College London, England. *Copernicus* discovered several pulsars.

COPUOS Acronym for *Committee on the Peaceful Uses of Outer Space.

core segment A pressurized module area holding the control instruments for *Spacelab*. It was used in two forms in the payload bay of the space shuttle, as a **short module** (on its own) or as a **long module** (when connected with the *experiment segment). The core segment's dimensions were 4 × 2.7 m.

Cornerstone A series of missions in the European Space Agency's *Horizon 2000 programme. This includes the *Rosetta* mission launched on 2 March 2004 to land on a comet in 2014.

corona The faint halo of hot (about 2 000 000 °C) and tenuous gas around the Sun, which boils from the surface. It is visible at solar *eclipses or through a

coronagraph, an instrument that blocks light from the Sun's brilliant disc. Gas flows away from the corona to form the *solar wind. NASA's **Reaven Ramaty High-Energy Solar Spectroscopic Imager* mission was launched in 2001 to study the evolution of energy in the corona.

C

Coronas (or Complex Orbital Near-Earth Observations of the Solar Activity) Joint Russian-Ukrainian project involving two orbiting solar observatories to investigate the Sun's internal structure and solar flares.

Coronas-I was launched in March 1994 but failed after only a few months. *Coronas-F* was launched in July 2001 to study the dynamic processes of solar activity in the Sun's active regions, flares, and mass ejections. Its instruments include X-ray and gamma-ray scintillation spectrometers.

COROT A satellite of the French National Space Agency, CNES, designed to search for planets of other stars and to study the structure of stars through their vibrations (known as stellar seismology or asteroseismology). The craft's name is a contraction of COnvection, ROtation & planetary Transits. *COROT* will carry a 0.3-m telescope to monitor the changes in a star's brightness that comes from a planet crossing in front of it. *COROT* will also be able to detect 'starquakes' that send ripples across a star's surface, altering its brightness. The exact nature of the ripples allows astronomers to calculate the star's mass, age and chemical composition. *COROT* is due for launch in 2006.

Cos-B A satellite launched by the European Space Agency in 1975 to study gamma rays in space. It functioned successfully until 1982.

***Cosmic Background Explorer* (*COBE*)** A US astronomical satellite launched in 1989 to study *cosmic background radiation. It confirmed the Big Bang theory of the universe's creation when its Far Infrared Absolute Spectrophotometer (FIRAS) measured the afterglow of the Big Bang. In 1992 it confirmed that radiation has a black body spectrum of temperature 2.73 K (−270.4 °C) and revealed ripples in the background radiation believed to mark the first stage in galaxy formation. Its final instrument was turned off on 23 December 1993.

http://space.gsfc.nasa.gov/astro/cobe/cobe_home.html Describes the goals and workings of the *Cosmic Background Explorer* satellite, with a detailed account of the various instruments the spacecraft carries. Information on cosmology in general is available on the *COBE* Educational Resource page, which has links to tutorials, documents, and images from *COBE*.

cosmic background radiation (or 3° radiation) The electromagnetic radiation left over from the original formation of the universe in the *Big Bang between 10 and 20 billion years ago. It corresponds to an overall background temperature of 2.73 K (−270.4 °C), or 3°C above absolute zero. In 1992 the US **Cosmic Background Explorer* (*COBE*) satellite detected slight 'ripples' in the strength of cosmic background radiation that are believed to mark the first stage in the formation of galaxies.

Cosmic background radiation was first detected in 1965 by US physicists Arno Penzias and Robert Wilson, who in 1978 shared the Nobel Prize for Physics for their discovery. In 2001 NASA launched the **Wilkinson Microwave*

Anisotropy Probe to continue the work of *COBE*. Further studies will be made by the *Planck mission.

cosmic-dust analyser (CDA) An instrument group on the **Cassini* spacecraft designed to measure dust particles in Saturn's rings. The three CDA instruments are: the **precision dust velocity sensor** to measure the speed, direction, and charge of a particle; the **impact plasma sensor** to measure the mass and speed of particles; and the **chemical analyser** to measure the atomic composition.

http://www.jpl.nasa.gov/jupiterflyby/science/cda.html Documents the Jupiter fly-by of 2000. The instruments carried by *Cassini*, including the cosmic-dust analyser, are described. There are pictures of the instrument and links to further information on cosmic dust and particle sensors.

Cosmic Microwave Background Polarization (CMBPOL) A possible NASA mission for launch after 2007 to test the *inflation model of cosmology. It is hoped the spacecraft instruments will detect cosmological background gravitational waves that were produced when the universe was less than one second old.

cosmic radiation The streams of high-energy particles and elctromagnetic radiation from outer space, consisting of electrons, protons, alpha particles, light nuclei, and gamma rays, which collide with atomic nuclei in the Earth's atmosphere and produce secondary nuclear particles (chiefly mesons, such as pions and muons) that shower the Earth. Space shuttles carry dosimeter instruments to measure the levels of cosmic radiation.

Cosmic radiation of lower energy seems to be galactic in origin, while that of high energy is of extragalactic origin. The galactic particles may come from *supernova explosions or *pulsars. At higher energies, other sources must exist, possibly the giant jets of gas that are emitted from some galaxies.

cosmic-ray subsystem (CRS) The particle telescopes on the two **Voyager* probes that measure galactic and solar *cosmic radiation. CRS data correctly predicted that before the end of 2003 the two probes would cross the 'terminal shock', where the solar wind suddenly slows down from supersonic to subsonic speed.

cosmodrome A Russian term for spacecraft launching sites. Russia's main launch facilities are the *Baikonur Cosmodrome and the *Plesetsk Cosmodrome.

cosmogony (Greek: *cosmos* 'universe' and *gonia* 'creation') The study of the origin and evolution of cosmic objects, especially the Solar System.

cosmological principle A hypothesis that the expansion of the *universe is perceived to be the same by any observer at any point within it; that is, that the universe is not expanding from any centre but all galaxies are moving away from one another.

cosmology A branch of astronomy that deals with the structure and evolution of the universe as an ordered whole. Cosmologists construct 'model universes' mathematically and compare their large-scale properties with those of the observed universe.

Modern cosmology began in the 1920s with the discovery that the universe is expanding, which suggested that it began with an explosion, the *Big Bang. An alternative—now discarded—view, the *steady-state theory, claimed that the universe has no origin, but is expanding because new matter is being continually created.

There are a number of differences in the conclusions that can be drawn from the steady-state and Big Bang theories. For example, the number of galaxies per unit volume should not change with distance if the steady-state theory is correct, but should increase with distance if an evolutionary theory is correct, for in looking over a distance we are also looking back in time with the universe gradually getting more compact. The latest counts of faint radio sources do seem to indicate an increase in the number per unit volume with distance; they support an evolutionary model. Another piece of evidence for the Big Bang theory is the cosmic background radiation, which was first detected in 1965 and can be interpreted as the radiation predicted as a necessary consequence of the Big Bang.

cosmonaut A Western term used for any astronaut from the former USSR (now Russia and other former Soviet states). Cosmonauts are trained at the *Gagarin Cosmonaut Training Centre.

cosmonautics A term used for *astronautics in countries of the former USSR.

Cosmos Soviet (later Russian) artificial satellite. The first was launched on 16 March 1962, and over 2 400 had been launched by 2004.

COSPAR Abbreviation for *Committee on Space Research.

Council for the Central Laboratory of the Research Councils (CCLRC) A UK research council established by royal charter in April 1995. Its space facilities include the Chilbolton Observatory, the Earth-Observation Data Centre of the National Environment Research Council, and the Rutherford Appleton Laboratory's Space Test Chamber and satellite ground station.

countdown The time allotted for the sequence of events that prepares a spacecraft for launching. The time is counted backwards, so T-150 is 150 minutes to the launch, which is T-0. Countdowns can cover two days and have several rest periods known as *built-in holds. The last ten counts to blast-off are usually audible.

count rate The number of particles emitted per unit time by a radioactive source. It is measured by a counter, such as a Geiger counter, or ratemeter.

course correction A correction of a spacecraft's course during its *coast period, using its thrusters to make the adjustment. A correction made during a space probe's flight in order to maintain its proper trajectory or speed is known as a **mid-course correction**. The corrections required are generally small, since trajectories are carefully calculated to conserve fuel.

Covey, Richard Oswalt (1946–) US astronaut. His three space shuttle missions include piloting the *Discovery* flight of August 1985 to deploy three communications satellites and to capture and repair another. He was selected as an astronaut in 1978 and left NASA in 1994.

crater A bowl-shaped depression in the ground, usually round and with steep sides. Craters are formed by explosive events such as the eruption of a volcano or the impact of a meteorite.

The Moon has more than 300 000 craters over 1 km in diameter, mostly formed by asteroid and meteorite bombardment; similar craters on Earth have mostly been worn away by erosion. Craters are found on all of the other rocky bodies in the Solar System.

Craters produced by impact or by volcanic activity have distinctive shapes, enabling geologists to distinguish likely methods of crater formation on planets in the Solar System. Unlike volcanic craters, impact craters have raised rims and central peaks and are circular, unless the meteorite has an extremely low angle of incidence or the crater has been affected by some later process.

crawler-transporter A huge, square NASA vehicle that transports the space shuttle from the *Vehicle Assembly Building to the launch pad at the *Kennedy Space Center. Two crawler-transporters were originally used to move the *Apollo*–Saturn Moon rockets.

The crawler-transporter, which moves at 1.6 kph when loaded, is the largest tracked vehicle known, having eight tracks. It is 40 m wide, 35 m long, and weighs 2 721 tonnes. To transport the shuttle, it moves a mobile launcher platform into the Vehicle Assembly Building and then travels to the launch pad with the shuttle on the platform.

Creighton, John Oliver (1943–) US astronaut. His three space shuttle flights included the September 1991 *Discovery* mission to deploy the **Upper-Atmosphere Research Satellite*. He was selected as an astronaut in 1978 and left NASA in 1992.

crew altitude protection system (CAPS) NASA's general name for the protective clothing worn by a space shuttle crew during launch and re-entry. This includes a *g-suit and a helmet containing a microphone and headphones for two-way communications. Oxygen is supplied to the system from the shuttle's *environmental control system.

crewman optical alignment sight (COAS) An optical device used on a space shuttle to align the navigational *inertia measurement unit (IMU) to within 1.4°, enabling the *star tracker to align the IMU more correctly. COAS has two crossing lines like a '+'. The device is normally mounted in the aft flight station, and the shuttle is manoeuvred manually by a flight crew member until a selected star is in the centre of the crossing lines. Two known stars are sighted and the shuttle's computer software determines the coordinates that can realign the IMU to within the required 1.4°.

COAS is also used to visually track targets, for instance during close or rendezvous operations. During the shuttle's ascent and deorbit thrusting periods, the device is mounted at the commander's station to display the shuttle's proper attitude orientation (*see* ATTITUDE CONTROL).

crew patch A NASA patch, or badge, for a particular space mission, worn on a crew member's flight suit above the heart. The *Apollo 11* patch, for example, showed an eagle landing on the Moon with a half Earth in the black background.

crew return vehicle (CRV; or X-38) A proposed NASA 'lifeboat' spacecraft for the **International Space Station*, to return crew members to Earth in an emergency. It was intended to be attached permanently to the station and could also be used for short flights in the station's vicinity. A prototype of the CRV was tested at NASA's *Dryden Flight Research Center, but the project was cancelled in 2001 due to cuts in NASA's budget.

crew station A crescent-shaped area in the space shuttle containing controls for the entire craft. Located behind the cockpit in the aft *flight deck, it contains manoeuvring controls, payload controls, systems monitors, radar displays, closed-circuit television screens, and two windows to view the cargo bay.

Crippen, Robert Laurel (1937–) US astronaut and director of the Kennedy Space Center (KSC). His four space shuttle missions include the first orbital flight of *Columbia* in April 1981 and the second flight of *Challenger* in June 1983. He was selected in 1966 for the navy's *Manned Orbital Laboratory* programme, becoming an astronaut in 1969. He became director of the KSC in 1992.

critical density In cosmology, the minimum average density that the universe must have in order for it to stop expanding at some point in the future.

The precise value depends on *Hubble's constant and so is not fixed, but it is approximately between 10^{-29} and 2×10^{-29} g/cm^3, equivalent to a few hydrogen atoms per cubic metre. The density parameter (symbol Ω) is the ratio of the actual density to the critical density. If $\Omega < 1$, the universe is open and will expand forever. If $\Omega > 1$, the universe is closed and the expansion will eventually halt, to be followed by a contraction (*Big Crunch). Current estimates from visible matter in the universe indicate that Ω is about 0.01, well below critical density, but unseen dark matter may be sufficient to raise Ω to somewhere between 0.1 and 2.

crossrange The permitted distance NASA's space shuttle can safely range outside its *re-entry path. It is up to 2 000 km/1 240 m on either side.

CRS Abbreviation for *cosmic-ray subsystem.

cruise phase A period during a spacecraft's flight in which it undergoes routine on-board operations, such as receiving new command sequences, establishing *attitude control, tracking the Earth with its *high-gain antenna, and deploying *booms and other appendages. The spacecraft also begins collecting data for its mission during this phase.

The cruise phase comes between the *launch phase and the *encounter phase. It may last only for a few months or for years. At NASA, it is normally managed from the *Space Flight Operations Facility at the Jet Propulsion Laboratory.

Crustal Dynamics Project (CDP) An international programme to measure motion within the Earth's crust. Established by NASA in 1979, the CDP has used satellites to collect data about tectonic motions of the Earth's plates in North America, Europe, Australia, and elsewhere. Of special interest to the CDP and its international partners has been the effect of earthquakes and other geophysical phenomena on crustal dynamics.

CRV Abbreviation for *crew return vehicle.

cryogenic Having a very low temperature (approaching absolute zero: 0 K/−273.15 °C). Some rocket propellants, such the fuel used by the Titan 4 rocket for its *Centaur upper stage, are cryogenic.

CSA Abbreviation for *Canadian Space Agency.

CSG Abbreviation for *Centre Spatial Guyanais.

CSM Abbreviation for *command and service module.

CSS Abbreviation for *control-stick steering.

C-type rocket (Russian name: Cosmos 1 rocket) A Western name for Russia's two-stage rocket used to launch satellites for military and scientific missions. It was first launched in 1964.

Culbertson, Frank Lee, Jr (1949–) US astronaut. His space shuttle missions include the *Discovery* mission in September 1993, which he commanded, to deploy the *Advanced Communications Technology Satellite. He was also programme manager of the series of shuttle dockings with the Russian space station *Mir* 1995–2000. In 2001 he became commander of the third resident crew of the *International Space Station (ISS)*. He spent one year as the deputy programme manager of the *ISS* programme before leaving NASA. Culbertson logged 146 days in space. He was selected as an astronaut in 1984.

Cunningham, R(onnie) Walter (1932–) US astronaut. He was the pilot in 1968 of *Apollo 7*, the first crewed flight of the *Apollo* project, in which the service module docked with a Saturn rocket.

Curbeam, Robert Lee, Jr (1962–) US astronaut. His shuttle flights include preparations for the *International Space Station (ISS)*. On the August 1997 flight of *Discovery* he helped test technology for the *ISS*. In February 2001, he flew with the *Atlantis* mission to deliver and attach the *Destiny* science laboratory module, the centrepiece of the *ISS*. In the spring of 2002 he served as deputy associate administrator of safety and mission assurance at NASA headquarters in Washington, DC. He was assigned to a 2004 shuttle flight to the *ISS* which was cancelled after the *Columbia* disaster in Febuary 2003. Curbeam was selected as an astronaut in 1994.

Currie, Nancy Jane (1958–) US astronaut. A flight engineer, she is safety and mission assurance manager of the space shuttle programme at the Johnson Space Center. Before assuming this position in September 2003, she was chief of the Robotics Branch of NASA's Astronaut Office. Her four shuttle missions include the *Endeavour* flight to assemble the *International Space Station* in December 1998, when she operated the robot arm to connect the Unity and Zarya modules, and the *Columbia* flight in March 2002 to service the *Hubble Space Telescope*. She was assigned to the Johnson Space Center in 1987 and selected as an astronaut in 1990.

cut-off A time when a rocket engine's propellant is turned off.

CXO Abbreviation for **Chandra X-Ray Observatory*.

CZ Abbreviation for *Cheng Zen rocket.

CZCS Abbreviation for *coastal zone colour scanner.

Daedalus A futuristic project proposed by the British Interplanetary Society to send a robot probe to nearby stars. The probe, 20 times the size of the Saturn V Moon rocket, would be propelled by thermonuclear fusion (in effect, a series of small hydrogen-bomb explosions). Interstellar cruise speed would be about 40 000 kps.

dark matter The theoretical matter that, according to certain modern theories of *cosmology, is thought to make up over 90% of the mass of the universe but so far remains undetected. Measurements of the mass of galaxies using modern theories showed large discrepancies in the expected values, which led scientists to the conclusion that a theoretical substance that cannot be seen had to account for a significant proportion of the universe. Dark matter, if shown to exist, would account for many currently unexplained gravitational effects in the movement of galaxies.

Theories of the composition of dark matter include unknown atomic particles (cold dark matter) or fast-moving neutrinos (hot dark matter) or a combination of both. Other theories postulate that massively dense objects such as *black holes form the majority of the 'missing' mass in the universe.

In 1993, astronomers identified part of the dark matter in the form of stray planets and *brown dwarfs, and, possibly, stars that have failed to ignite. These objects are known as massive astrophysical compact halo objects (MACHOs) and may make up approximately half of the dark matter in the Milky Way's halo.

In July 2003, the first detailed map of the dark matter distribution in a galaxy cluster was reported. An international team of astronomers studied galaxy cluster CL0024+1654, one of the largest structures in the known universe, using the **Hubble Space Telescope*. The cluster is mainly composed of dark matter that is only detectable by analysing the gravitational interaction between objects. The researchers targeted 39 regions of the cluster and were able to show gravitational warping, known as weak gravitational lensing, of images of galaxies beyond the cluster, caused by the presence of dark matter. The astronomers used their accumulated data to create the first dark matter map of a galaxy cluster.

DARPA Acronym for *Defense Advanced Research Projects Agency.

Darwin A proposed European Space Agency (ESA) mission that will search for Earth-like planets around other stars and analyse their atmospheres for the chemical indicators of life. Darwin will consist of eight spacecraft flying in formation at the L2 *Lagrangian point of the Earth's orbit, 1.5 million km from Earth in the opposite direction to the Sun. Six of these craft will contain telescopes of about 1.5 m aperture and will be arranged in a ring, feeding infrared light to a central hub. The central hub will combine the signals, cancelling out light from the central star while allowing light from any accompanying planets to be seen. The eighth spacecraft is the communications link with Earth, which sits beneath the formation of other satellites to monitor their positions and issue corrections. Darwin is similar to NASA's proposed *Terrestrial Planet Finder, and the two missions may be combined.

Data-Relay and Technology Mission (DRTM) A satellite programme of the European Space Agency, the aim of which is to offer new communications services to Europe and to demonstrate and test in orbit new satellite technologies, including an optical intersatellite data-relay link and the use of *ion engine propulsion to correct a satellite's position whilst in geostationary orbit. The first satellite in this mission, **Artemis*, was launched in 2001.

Data-Relay Satellite (DRS) Either of two satellites scheduled for launch by the European Space Agency (ESA) to provide control, communication, and monitoring for crewed and uncrewed spacecraft. The satellites will be in *geostationary orbit for a projected ten-year lifespan to provide wide coverage of satellites in *low Earth orbit. The ESA's **Artemis*, launched on 12 July 2001, will be employed with the DRS satellites on the Data-Relay and Technology Mission (DRTM), which will provide the equivalent services of NASA's *Tracking and Data-Relay Satellite (TDRS).

data systems operations team (DSOT) A NASA multi-mission team responsible for managing spacecraft data, keeping it flowing through a string of computers and communications links. The DSOT is part of the deep-space mission system (DSMS).

Davis, (Nancy) Jan (1953–) A US astronaut and director of safety and mission assurance at the Marshall Space Flight Center. Her first two space shuttle flights, on *Endeavour* in September 1992 and on *Discovery* in February 1994, involved operating the **Space Habitation Module*. During her third flight, on *Discovery* in August 1997, she was the payload commander who operated the Japanese Manipulator Flight Demonstration robotic arm. Davis joined the Marshall Space Flight Center as an aerospace engineer in 1979 and became an astronaut in 1987.

Dawn A NASA spacecraft that will visit the asteroids Ceres and Vesta to map their surfaces and measure their size, shape, and composition. Being relics from the early days of the Solar System, these asteroids should contain evidence of the conditions present during the birth of the planets. *Dawn* is due for launch in 2006, reaching Vesta in 2010 and orbiting it for 11 months. *Dawn* will then move on to Ceres, reaching it in 2014 for another 11 months of study from orbit.

DBS Abbreviation for *direct-broadcast satellite.

DC-XA reusable launch vehicle An experimental aircraft that was a predecessor to the *space shuttle. It was produced by NASA and the Douglas Aircraft Company (later McDonnell Douglas). Without a pilot, the rocket-propelled DC-XA took off vertically, flew under remote control, and landed as a normal aircraft. The 13-m DC-XA made three successful test flights in 1966 at the White Sands Missile Range, New Mexico, but on the fourth flight a landing gear failed to deploy and it crashed on the runway. NASA then moved on to other prototypes in its *X-series.

decay An unwanted descent in altitude of a satellite's orbit. This is usually due to a decrease in speed caused by *aerodynamic drag. To counteract this effect, thrusters on a satellite must be fired occasionally to maintain its correct **orbital station**. A satellite that descends low enough into the atmosphere to burn up is called a decayed satellite.

deck An area of a space vehicle considered to be a floor, although 'up' and 'down' are uncertain orientations in space. To help astronauts determine 'down', space vehicles have their decks painted in a darker colour than their ceilings. A space shuttle is flown by its pilot and commander from the **flight deck**, which also has seats for two more crew members. Below the flight deck is the **mid-deck**, providing room for three additional crew members and used for sleeping, food preparation, and storage. The **lower deck** is an equipment and stowage area.

'Deck' was taken from the naval term for a ship's floor.

decontamination After the return of a crewed planetary space mission, the removal of harmful substances from an astronaut, spacecraft, and any samples collected. When the *Apollo 11* crew returned from the first landing on the Moon, many scientists believed lunar micro-organisms might infect the Earth. In preparation, the US government formed an Interagency Committee on Back Contamination, and NASA built its *Lunar Receiving Laboratory to quarantine the crew and lunar rocks until the 21st day after the Moon landing.

Deep Impact The NASA probe was launched 12 January 2005 and arrived at comet P/Tempel 1 on 4 July 2005. This was the first spacecraft to impact the surface of a comet and the first mission to investigate beneath the surface of a comet's nucleus. Data is still being analysed.

The impact was recorded by cameras on the deployed impactor and fly-by spacecraft, and broadcast on television. The spacecraft's copper mass impactor, weighing 370 kg, probably created a crater 25 m deep and 100 m in diameter on the nucleus, which is 6 km in diameter.

The mission's science objectives are to observe how the crater forms; measure its depth and diameter, the composition of its interior, and its ejected material; and determine the changes produced by the impact on the comet's natural outflow of gases.

http://deepimpact.jpl.nasa.gov/ Well-designed site about the *Deep Impact* mission to the comet P/Tempel 1, profiling the mission's scientific objectives, mission plan, and hardware. There are views of the spacecraft, and animations depicting the comet encounter and impactor deployment.

deep space The far regions of space in the outer Solar System and beyond. Space probes, such as NASA's *Deep Space 1*, are used to investigate these areas, and they are tracked by the *deep-space network.

Deep Space The name of two NASA space probes, *Deep Space 1* launched in October 1998, and *Deep Space 2* launched aboard the **Mars Polar Lander* in January 1999. They were the first two missions under NASA's New Millennium Program to test new space technologies for future missions.

The 12 experimental instruments of *Deep Space 1* included a xenon ion propulsion system, a high-efficiency solar array, and an autonomous navigation system. The spacecraft successfully concluded the tests, and in September 1999 its mission was extended. On September 2001, *Deep Space 1* flew within 2 000 km of the comet Borrelly to return images of the comet's nucleus. This was only the second time that a comet's nucleus had been studied in close-up. The spacecraft's ion engines were turned off on 18 December 2001 to end the mission. Contact with *Deep Space 2*, which had two miniature probes, was lost on 3 December 1999, the date the probes were to penetrate below Mars's surface. Communication with the *Mars Polar Lander* spacecraft also ended abruptly at that point.

(Image © NASA)

The 70-m antenna of the Canberra deep-space communications complex. The Australian facility is one of three sites that make up NASA's deep-space network; the others are in Goldstone, California, and Robledo near Madrid, Spain.

http://nmp.jpl.nasa.gov/ds1/ Information on the scientific goals of *Deep Space 1* and the advanced technologies tested on it. There are articles which provide a general overview of the mission, technical reports for downloading, and a detailed mission log.

deep-space communications complex (DSCC) Any of three centres used by NASA's *deep-space network (DSN) to communicate with spacecraft. A DSCC uses *downlinks to receive scientific data from a spacecraft and *uplinks to navigate the craft, control its operating modes, and load and reprogram its computers.

The three complexes are located approximately 120° apart in longitude: at Goldstone near Barstow, in California's Mojave desert; Tidbinbilla near Canberra, Australia; and Robledo near Madrid, Spain. This allows constant observation of a spacecraft as the Earth rotates, since each DSCC has an 8- to 14-hour period of contact with a spacecraft, overlapping with the next DSCC. The DSCCs are located away from populated areas to avoid any electrical interference from radio and television stations that could obstruct a spacecraft's signals.

deep-space mission system (DSMS; formerly ground data system (GDS)) A *deep-space network system of computers, software, communication networks, and procedures, which receives and processes data from spacecraft instruments. The front-end processing of data involves receiving, enhancing, and interpreting it, while the back-end processing provides scientists with access to the data.

deep-space network (DSN) NASA's international network of ground tracking and communication stations for space probes. It is the largest scientific telecommunications system in the world. The network consists of three *deep-space communications complexes (DSCCs) at roughly equidistant locations around the Earth for continuous tracking.

Each DSCC maintains the equipment of its *deep-space station. DSN is managed and operated by the Telecommunications and Mission Operations Directorate (TMOD) of the *Jet Propulsion Laboratory.

http://deepspace.jpl.nasa.gov/dsn/ Clearly explaining the workings of the deep-space network (DSN), this site has articles, fact sheets, video clips, and a pictorial history of the DSN.

deep-space station (DSS) The antenna at each of the three deep-space communications complexes (DSCCs) that form the *deep-space network, where data from a spacecraft is received and processed.

Defense Advanced Research Projects Agency (DARPA) A military organization established in 1958 as the first US response to the Soviet launch of **Sputnik*. By 2001, DARPA had a budget of about US$2 billion and employed some 240 personnel, 140 of whom were technical staff. The latter are rotated every three to five years to provide fresh thinking and new perspectives. DARPA has maintained its founding principles of assuring a US lead in state-of-the-art technology for military capabilities and of avoiding technological surprises from adversaries.

The organization has changed its name three times since 1958. It was originally known as the Advanced Research Projects Agency (ARPA), adopting its present name in 1972. It reverted to the original ARPA in 1993, changing back again to DARPA in 1996.

Defense, Department of (DOD) A US government department presided over by the secretary of defense, with headquarters in the Pentagon. It approved early military launches the *Atlas rockets which were later adapted for use by NASA. Early astronauts were normally military pilots, and the DOD loaned NASA hundreds of officers for the **Apollo* project. Some space shuttle flights have been made for the DOD, with the details kept secret.

Deimos One of the two moons of the planet Mars. It is irregularly shaped, 15 × 12 × 11 km, orbits at a height of 24 000 km every 1.26 days, and is not as heavily cratered as Mars's other moon, Phobos. Deimos was discovered in 1877 by US astronomer Asaph Hall, and is thought to be an asteroid captured by Mars's gravity.

NASA's **Viking 2* probe visited Deimos in 1977, flying within 29 km of the Moon's surface and taking detailed photographs of its pitted and boulder-strewn landscape.

Delta rocket A US rocket used to launch many scientific and communications satellites since 1960. Its design is based on the Thor ballistic missile. Several increasingly powerful versions have been produced as satellites became larger and heavier. Solid-fuel boosters were attached to the first stage of the rocket in order to increase lifting power.

NASA's twin rovers were launched to Mars in June and July 2003 atop a Delta 2 rocket.

http://www.boeing.com/defense-space/space/delta/deltahome.htm
Illustration of the design and capabilities of the modern Delta rocket family. The site includes information on the history of the Delta rocket and news of forthcoming launches. Via the Delta communications pages, visitors can access Boeing's impressive image gallery.

delta-V A velocity change by a spacecraft. A delta-V is used for such adjustments as an *orbit trim manoeuvre, an interplanetary *trajectory correction manoeuvre, *three-axis stabilization, and *spin stabilization.

demonstration test facility (DTF) A laboratory or building used to test space components. For example, NASA demonstrates and tests the compatibility of its *deep-space network (DSN) with various spacecraft communications in the demonstration test facility known as DTF-21, at its Jet Propulsion Laboratory.

deorbit To remove a spacecraft from orbit, using its *retrorockets for re-entry into the atmosphere. On the space shuttle, the pilot and commander check computers 20 minutes before firing the retrorockets, to verify that the deorbit and landing software programs are working. The firing of the retrorockets is known as the **deorbit burn**.

Mission control alerts the pilot and commander that the vehicle is 'go for deorbit burn.' They rotate the shuttle to fly backwards in its deorbit-burn attitude. The commander then activates the proper command in the computers and, about an hour before touchdown, the engines fire for three minutes to achieve deorbit. They then point the vehicle forward for landing.

Department of Trade and Industry (DTI) A UK government department involved in supporting space exploration, especially through the *British National Space Centre. The DTI spends approximately £90 million annually on space research and development to support programmes in the UK and those of the European Space Agency.

deployment The release of an artificial satellite or other object into space. Early satellites were deployed directly from rockets. The space shuttle releases satellites, often several on one flight, from its cargo bay after manoeuvring into a **deployment attitude**. This is based on the attitude of the shuttle and the destined orbit and orientation of the satellite. The shuttle releases a satellite destined for a higher *geosynchronous orbit by spinning it on a revolving table before coiled springs project it into space. Other satellites may be deployed simply by an astronaut's gentle push.

desat Abbreviation for *momentum desaturation.

despun A description of a non-spinning part on a spacecraft that is designed to spin for stabilization. A despun antenna, for example, must remain steadily pointed to Earth in order to communicate.

Destiny The US laboratory module that forms the heart of the *International Space Station*, launched by the space shuttle *Atlantis* in February 2001. Destiny

(Image © NASA)

The space shuttle's remote manipulator system is used to move the US Destiny laboratory from the cargo bay of *Atlantis*. It was photographed by astronaut Thomas Jones on extravehicular activity during the manoeuvre.

is 8.5 metres long and 4.3 metres in diameter and consists of three cylindrical sections and two endcones with hatches for connecting to other station components. It holds a series of equipment racks for experiments in human life science, materials research, Earth observations and commercial applications. A window in the centre section allows astronauts to take high-quality photos and video of floods, avalanches, fires and ocean events. Destiny also contains the control centre for the station's robotic arm operations.

DEV Abbreviation for *mechanical devices subsystem.

Dezhurov, Vladimir Nikolaevich (1962–) Russian cosmonaut. He spent 115 days on the Russian space station *Mir*, as crew commander, after being launched on a *Soyuz* mission craft from the Baikonur Cosmodrome in March 1995. He returned to Earth aboard NASA's space shuttle *Atlantis*, landing at the US Kennedy Space Center in July 1995. In 2001 he was assigned to join the third resident crew of the *International Space Station*. Dezhurov was selected as a cosmonaut in 1987.

DHF-2 An alternative name for **STW/China*.

Diamant rocket (English: diamond) A French three-stage launch vehicle developed in the 1960s. Begun as a military project, the Diamant launched France's first satellite (and the first to be launched by a country other than the USA or the USSR), the 41-kg *Asterix*, on 26 November 1965 from Algeria; this was the rocket's first flight.

differenced Doppler A system to measure a spacecraft's changing three-dimensional position as it orbits another planet. Doppler measurements (of the increase or decrease of frequency caused by the vehicle's movement) are taken at ground stations widely separated on Earth. The differences in the two Doppler-shift readings will reveal the spacecraft's velocity and position as they change in space.

DIR Abbreviation for *direct control.

direct ascent mode An early NASA idea for landing astronauts on the Moon. It involved a giant rocket, tentatively named Nova, that could be launched directly to the Moon where it would land softly, then return by propelling itself from the lunar surface back to Earth. This option was discarded in favour of the *lunar-orbit rendezvous mode.

direct-broadcast satellite (DBS) A high-power communications satellite that transmits television broadcasts directly to small parabolic dish aerials. About 25 million analogue DBS receivers were installed in homes throughout Europe and Japan between 1984 and 1998. The first digital DBS service became available in 1994.

direct control (DIR) A pilot control system in the space shuttle that overrides the computers. It is more difficult to use, especially during re-entry, than the two other systems of *auto and *control-stick steering.

direct-sensing instrument A scientific instrument on a spacecraft that can detect nearby phenomena and record their numbers or characteristics. An example is the *heavy-ion counter carried on NASA's **Galileo* probe.

Discoverer The cover name for the US military reconnaissance satellite project Corona. It was jointly funded by the US Air Force and the CIA (Central Intelligence Agency). The first satellite was launched in 1959 and the last, *Discoverer 38*, in 1962.

Discovery The third of the US *space shuttles. It made its first flight on 30 August 1984 with a crew of six. *Discovery* took the second resident crew to the *International Space Station* in 2001. On 26 July 2005 it made the first flight following the 2003 *Columbia* space shuttle disaster, taking cargo and equipment to the *International Space Station*. Prior to this, *Discovery* had made 30 flights.

In 1994 Sergei *Krikalev of Russia became the first cosmonaut in a space shuttle on board *Discovery*, and John *Glenn flew in it in 1998, becoming the oldest man in space.

Discovery program The NASA programme to conduct frequent, low-cost missions to explore the Solar System. Each Discovery mission must be developed, built and launched in less than three years for a total cost of under $299 million. One of the programme's aims is to encourage the use of new technology. The first Discovery mission was the *NEAR Shoemaker asteroid rendezvous in February 1996, and launches have averaged one a year since then.

display and control module A small unit on the chest of an astronaut's *life-support system or extravehicular mobility unit (*see* EXTRAVEHICULAR ACTIVITY). It has displays and controls for water and electrical supplies.

DM Abbreviation for *docking module.

docking Attaching two spacecraft together in orbit, as when a space shuttle joins to the *International Space Station* to deliver supplies or replace the crew. The shuttle has also docked with satellites, recovering them in orbit to repair them. Docking is in contrast to *berthing.

docking module (DM) A module used for the *docking of two or more spacecraft.

When the *Apollo* and *Soyuz* vehicles docked during the **Apollo–Soyuz* Test Project in 1975, a special docking module was constructed to accommodate their different systems. It was 3.15 m long and 1.4 m in diameter. Serving as a passageway between the two spacecraft, it also contained a communication system and electric furnace for experiments.

A small docking module is known as a **docking adapter**. The docking adapter on the **Skylab* space station was joined to the *Apollo* *command module, and had a second docking port in case a rescue vehicle had to be attached. The docking adapter also contained cameras and other instruments for observations of the Earth and Sun.

DOD Abbreviation for Department of *Defense.

Doi, Takao (1954–) A Japanese astronaut who was the first Japanese person to perform an extravehicular activity (EVA). As a mission specialist on a *Columbia* flight in November 1997, he logged more than 12 hours during two

EVAs to test assembly procedures for the *International Space Station* (*ISS*). He is currently in the Astronaut Office on a technical assignment concerning the *ISS*. Doi joined the National Space Development Agency in 1985 and trained with the Johnson Space Center from 1995.

Donatello The third *Multi-Purpose Logistics Module (MPLM) built by the Italian Space Agency for carrying supplies to and from the International Space Station in the cargo bay of the space shuttle. The other two MPLMs are named **Leonardo* and **Raffaello*. *Donatello*'s first mission has not yet been scheduled.

Doppler effect A change in the observed frequency (or wavelength) of waves due to relative motion between the wave source and the observer. The Doppler effect is responsible for the perceived change in pitch of a siren as it approaches and then recedes, and for the *redshift of light from distant galaxies. It is named after the Austrian physicist Christian Doppler.

http://archive.ncsa.uiuc.edu/ Cyberia/Bima/doppler.html Discussion of the Doppler effect, including an explanation of how astronomers use it to calculate the speed with which celestial bodies are moving in relation to one another.

Doppler orbitography and radiopositioning integrated by satellite (DORIS) An instrument used on satellites to indicate their precise trajectory. It employs the Doppler effect, in which a frequency is higher between a transmitter and receiver moving towards one another and lower if they are moving apart. DORIS is a receiver that measures the Doppler shift of radio signals continuously transmitted by 50 ground beacons. Designed and developed by four French groups, among them the *Centre National d'Etudes Spatiales, DORIS has flown on several spacecraft, including the European Space Agency's **Envisat*.

Doppler radar A radar that records changes in apparent frequency due to the relative motion between the instrument and source. *Scatterometers on spacecraft are Doppler radars that investigate phenomena contributing to the Earth's changing climate.

DORIS Acronym for *Doppler orbitography and radiopositioning integrated by satellite.

dosimeter An instrument used to measure radiation. Space shuttle crew members are required to carry dosimeters at all times during a flight, and other units are stored in a mid-deck modular locker. The dosimeters have photographic film that will develop spots if struck by radiation. Before a flight, NASA calculates the mission's radiation dosage and the radiation exposure history of each astronaut. Radiation during a shuttle flight is well below the minimum acceptable levels.

Double Star A joint mission of China and the European Space Agency to study the Earth's magnetosphere. Double Star involves two satellites, *Tan Ce 1* and *Tan Ce 2*, launched on 29 December 2003 and 25 July 2004 respectively. 'Tan Ce' means 'Explorer' in Chinese. *Tan Ce 1* is in a highly elliptical near-equatorial orbit from which it investigates the Earth's magnetic tail, while *Tan*

Ce 2 orbits over the Earth's poles from where it can monitor the development of aurorae. The two satellites work in conjunction with ESA's **Cluster* spacecraft.

double star A pair of stars that appear close together. Many stars that appear single to the naked eye appear double when viewed through a telescope. Some double stars attract each other due to gravity, and orbit each other, forming a genuine *binary star, but other double stars are at different distances from Earth, and lie in the same line of sight only by chance. Through a telescope both types look the same.

Double stars of the second kind (which are of little astronomical interest) are referred to as 'optical pairs', those of the first kind as 'physical pairs' or, more usually, 'visual binaries'. They are the principal source from which is derived our knowledge of stellar masses.

Douglas Aircraft Company The former name (1928–67) of the US company *McDonnell Douglas.

downlink A radio-signal transmission from a spacecraft to the Earth receiving station. An *uplink is the opposite of a downlink transmission.

downrange The direction of a spacecraft's flight path away from its launch site.

drag The resistance to motion experienced by a body passing through a fluid, especially air. Atmospheric drag slows down artificial satellites, space shuttles, and space stations, which all require steering thrusters to keep them at the proper orbiting altitude. Drag also creates the intense heat enveloping a spacecraft during *re-entry.

drogue parachute A small parachute used to slow down a space vehicle, deployed prior to the main parachute. Astronauts in NASA's *Mercury*, *Gemini*, and *Apollo* modules experienced a hard jolt after re-entry as the drogue parachute opened before splashdown (and a harder jolt when the main parachute was deployed). The space shuttle escape suit worn by astronauts on launch and re-entry also includes a drogue and a main parachute.

drop tower A facility used to put objects into free fall for *microgravity experiments on Earth. NASA's 2.2 Second Drop Tower at the *Glenn Research Center began operating in the 1960s. The tower has a free-fall drop of 24.1 m and is used to conduct exploratory microgravity testing for space shuttles and the *International Space Station*. Approximately 90% of the experiments have involved studies of the effects of microgravity on combustion.

Other drop towers include the Drop Tower Bremen at the *Zentrum für Angewandte Raumfahrttechnologie und Mikrogravitation (ZARM; Centre of Applied Space Technology and Microgravity) at the University of Bremen, Germany. The *National Space Development Agency (NASDA) uses a drop tower in Toki, Japan, that has a 100-m free-fall segment, providing a microgravity environment for 4.5 seconds.

DRS Abbreviation for *Data-Relay Satellite.

d

Dryden Flight Research Center NASA's research centre at *Edwards Air Force Base, Rosamund, California. Experimental aircraft have been flown there since the late 1940s. NASA's first flight of the rocket-powered X-15 aircraft took place in March 1960. The prototype for the space shuttle was first tested in 1977, and the first space shuttle, *Columbia*, landed at Edwards in April 1981. Dryden also developed NASA's prototype *crew return vehicle, the X-38, which was cancelled in 2001. The X-43A hypersonic research aircraft, a joint project of Dryden, NASA's Langley Research Center, and industry partners, broke the aeronautical speed record on 27 March 2004, flying at over Mach 7, or about 8 000 kph.

The centre was dedicated in 1976 to Hugh Dryden, a former deputy director of NASA who established the first facility at Edwards.

http://www.dfrc.nasa.gov/ News releases and project information from the Dryden Flight Research Center. The Center's education page has downloadable documents explaining how flight tests are carried out and describing flight testing manoeuvres. The gallery contains videos, graphics, and photographs of experimental aircraft and links to other NASA image galleries and aircraft image archives.

dry weight (or empty weight) The total weight of a rocket without its fuel, or of a spacecraft without its fuel and other *consumables.

DSCC Abbreviation for *deep-space communications complex.

DSMS Abbreviation for *deep-space mission system.

DSN Abbreviation for *deep-space network.

DSOT Abbreviation for *data systems operations team.

DSS Abbreviation for *deep-space station.

DTF Abbreviation for *demonstration test facility.

DTI Abbreviation for *Department of Trade and Industry.

D-type rocket A Western designation for the largest Russian rockets with two or more stages that are used to launch space probes and space stations or their modules. An example is the *Proton rocket, which launched the series of **Salyut* space stations and modules for the *International Space Station*.

Duffy, Brian (1953–) US astronaut. His four space shuttle flights included commanding the *Discovery* mission in October 2000 to take equipment and supplies to the *International Space Station*. He also served as acting deputy director of the Johnson Space Center. Duffy retired from NASA and the US Air Force in 2001 and became vice-president and associate programme manager for the Lockheed Martin Corporation. He was selected as an astronaut in 1985.

Duke, Charles Moss, Jr (1935–) US astronaut. He was the *lunar module pilot of the *Apollo 16* flight in April 1972. During the mission, he and John *Young stayed for more than 71 hours on the Moon's surface, collecting 118 kg of lunar rock and soil. Duke was selected as an astronaut in 1966 and left NASA in 1975.

Dunbar, Bonnie Jeanne (1949–) US astronaut and, from 1995, assistant director of the Johnson Space Center (JSC). Her space shuttle flights include the first mission to dock with the Russian space station *Mir* (on *Columbia* in June 1992), and the eighth shuttle-*Mir* docking (on *Endeavour* in January 1998). Dunbar was a payload officer and flight controller with the JSC when she was selected as an astronaut in 1981.

Duque, Pedro (1963–) Spanish astronaut with the European Space Agency (ESA). He was aboard the *Discovery* space shuttle flight of October 1998 that deployed the *Spartan 201 spacecraft. He was a member of the Expedition Eight crew on the *International Space Station*, launched in October 2003 on *Soyuz* TMA-3 and doing scientific, technical, and educational experiments during a stay of more than a week. He was selected as an ESA astronaut in 1992, supporting two ESA-Russian missions before reporting to the Johnson Space Center from 1996.

dust detector (or dust counter) An instrument carried on a spacecraft to detect and measure the properties of dust particles. It can determine the composition, mass, speed, and direction of the particles, but does not form an image of the source. Dust detectors have been carried on such space probes as **Galileo* and **Cassini*.

dynamic pressure The Earth's atmospheric pressure on a rocket during its acceleration after lift-off.

***Dyna-Soar* (or *X-20*)** NASA's prototype recoverable spacecraft with a delta-winged design, which evolved into the space shuttle. Envisioned in 1958 to follow the experimental X-15 rocket plane, it was designed and developed by NASA, the US Air Force, and the *Boeing Company. The *Dyna-Soar* would be launched into orbit by a Titan III rocket and then glide back to Earth in a 'dynamic ascent and soaring'. It was never flown, and the project ended in 1969. However, in tests, it proved suitable for flying a large cargo bay and its delta wings offered better control than normal aircraft designs.

EADS Abbreviation for *European Aeronautic Defence and Space Company.

Eagle A name given to the lunar module of the *Apollo 11* spacecraft that landed two US astronauts on the Moon on 20 July 1969. *Eagle* was piloted by astronaut Buzz *Aldrin. Mission commander Neil *Armstrong became the first person to walk on the Moon. The descent module of *Eagle* remained on the Sea of Tranquillity while the ascent module took the astronauts back to the orbiting command module, *Columbia*, piloted by fellow astronaut Michael *Collins.

early *Apollo* scientific experiments package (EASEP) The first experiments set up on the Moon. The *Apollo 11* astronauts Buzz *Aldrin and Neil *Armstrong deployed the EASEP from the lunar module equipment bay on 20 July 1969. It consisted of a passive seismometer experiment (PSE) and a lunar ranging retroreflector (LRRR, known as 'LR cubed'). The latter provided a means for astronomers on Earth to measure the distance between their telescopes and the LRRR.

***Early Bird* (or *Intelsat 1*)** The first US commercial communications satellite, launched over the Atlantic Ocean in 1965 by *Intelsat. On 2 May of that year, the satellite transmitted the first television pictures across the Atlantic, linking nine nations. It became a fully commercial server in June 1965 and was retired in March 1969. *Early Bird* had 240 circuits for telephone calls or television programmes.

A remote-sensing satellite also named *Early Bird* was launched in 1997 by a Russian rocket for the international conservation organization Earthwatch Inc., but failed after five days.

Earth The third planet from the Sun. It is almost spherical, flattened slightly at the poles, and is composed of five concentric layers: inner core, outer core, mantle, crust, and atmosphere. About 70% of the surface (including the north and south polar ice caps) is covered with water. The Earth is surrounded by a life-supporting atmosphere and is the only planet on which life is known to exist.

mean distance from the Sun 149 600 000 km

equatorial diameter 12 756 km

circumference 40 070 km

rotation period 23 hours 56 minutes 4.1 seconds

year (complete orbit, or sidereal period) 365 days 5 hours 48 minutes 46 seconds. The Earth's average speed around the Sun is 30 kps. The plane of its orbit is inclined to its equatorial plane at an angle of 23.5°; this is the reason for the changing seasons

atmosphere nitrogen 78.09%; oxygen 20.95%; argon 0.93%; carbon dioxide 0.03%; and less than 0.0001% neon, helium, krypton, hydrogen, xenon, ozone, and radon

surface land surface 150 000 000 sq km (greatest height above sea level 8 872 m Mount Everest); water surface 361 000 000 sq km (greatest depth 11 034 m Mariana Trench in the Pacific). The interior is thought to be an inner core about 2 600 km in diameter, of solid iron and nickel; an outer core about

(Image © NASA)
The Earth rising above the Moon's horizon, as seen from the *Apollo 11* spacecraft in July 1969.

2 250 km thick, of molten iron and nickel; and a mantle of mostly solid rock about 2 900 km thick. The crust and the uppermost layer of the mantle form about twelve major moving plates, some of which carry the continents. The plates are in constant, slow motion, called tectonic drift

satellite the *Moon

age 4.6 billion years. The Earth was formed with the rest of the *Solar System by consolidation of interstellar dust. Life began 3.5–4 billion years ago

Earth observation (EO) The observation of the Earth's surface and atmosphere by Earth-observation satellites such as the **European Remote-Sensing Satellite* and NASA's **Landsat*. *Remote-sensing instruments are employed to provide detailed geological and meteorological data used for weather forecasting and environmental research and by such industries as agriculture and fishing.

Earth-observing system (EOS) NASA's Earth-observation programme, part of its Earth Science Enterprise programme and the first observing system to provide integrated measurements of the Earth's processes. The satellites have been **Terra*, launched on 18 December 1999 to observe the land, oceans, and the Earth's atmosphere; **Aqua*, launched on 4 May 2002 to study the interaction of oceans with land; and **Aura*, launched on 15 July 2004 to study the ozone, air quality, and climate.

Data is received from a coordinated series of polar-orbiting and low-inclination satellites that provide long-term observations of the land, clouds, oceans, polar ice sheets, atmosphere, and biosphere. Scientific interpretation of the EOS data provides improved weather forecasts, better predictions of climate change, practical information for farmers and fishermen, and other enhanced knowledge about the global environment.

http://eospso.gsfc.nasa.gov/ Huge resource with information about the many projects which make up NASA's Earth Science Enterprise. There are superb image galleries organized by theme and by source, as well as a wide selection of documents, from tutorials to technical reports.

Earth-orbit rendezvous mode (EOR) An early NASA plan for sending astronauts to the Moon. The plan was to use two Saturn rockets, one launching a crew into Earth orbit and another carrying up the propellant for the trip to the Moon. The two would dock in orbit and the complete spacecraft and crew would then travel to the Moon.

The idea was proposed by German rocket engineer Wernher *von Braun in opposition to the *direct ascent mode, which would have required a more powerful rocket. Both plans were eventually scrapped in favour of the *lunar-orbit rendezvous mode.

Earth-received time (ERT) A time that a spacecraft event or transmission is received by a *deep-space station. It is recorded in *Universal Time Coordinated.

EASEP Acronym for *early *Apollo* scientific experiments package.

Eastern Test Range (ETR) A region running southeast from *Cape Canaveral used to launch and test US rockets and to track them over the Atlantic Ocean and Caribbean Sea. The US Air Force is in charge of the range.

eccentricity The extent to which an elliptical orbit departs from a circular one. It is usually expressed as a decimal fraction, regarding a circle as having an eccentricity of 0.

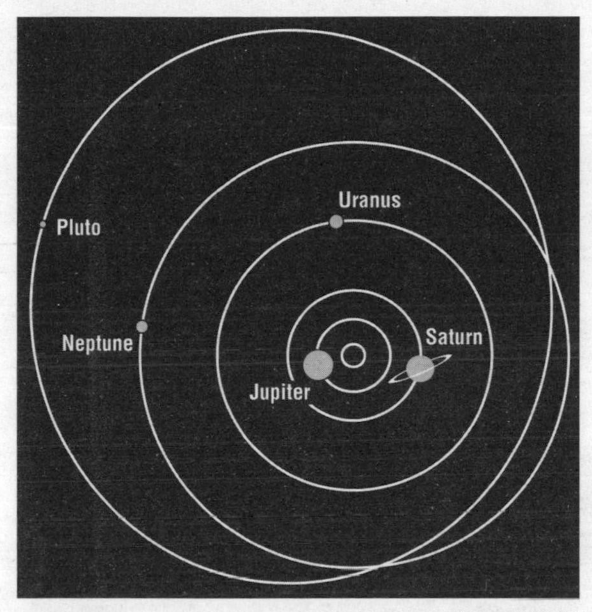

(Image © NASA)
The orbits of the outer planets of the Solar System. The eccentricity of a circular orbit is zero. Pluto is the only planet with a large eccentricity.

Echo A prototype communications satellite, launched by NASA in August 1960. It was an aluminium-coated plastic balloon that proved pictures and sound could be transmitted by reflecting them off an orbiting satellite. *Echo 2* was launched in 1964.

eclipse The passage of one astronomical body through the shadow of another. The term is usually used for solar and lunar eclipses. A **solar eclipse** occurs when the Moon is between the Earth and the Sun (which can happen only at new Moon), the Moon blocking the Sun's rays and casting a shadow on the Earth's surface. A **lunar eclipse** occurs when the Earth is between the Moon and the Sun (which can happen only at full Moon), the Earth blocking the Sun's rays and casting a shadow on the Moon's surface.

During a total solar eclipse the Moon appears to cover the Sun's disc completely and day turns into night. This is known as the umbra. A total solar eclipse can last up to 7.5 minutes, and the Sun's *corona can be seen. Between two and five solar eclipses occur each year but each is visible only from a specific area. During a partial solar eclipse sunlight reaches the Earth from around the edge of the Moon. This is known as the pre-umbra. Lunar eclipses can also be partial or total. Total lunar eclipses last for up to 100 minutes; the maximum number each year is three.

An eclipse can refer to other bodies, for example to an eclipse of one of Jupiter's satellites by Jupiter itself. An eclipse of a star by a body in the Solar System is also called an occultation.

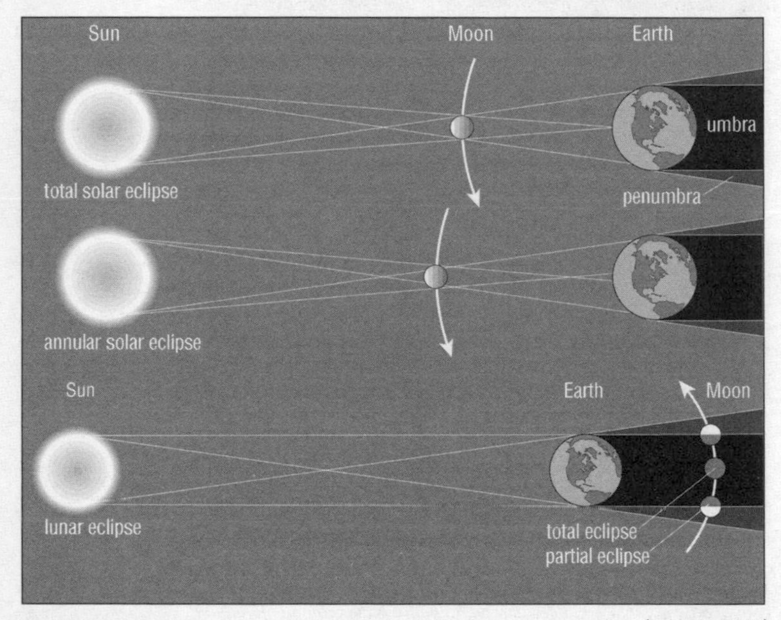

(Image © NASA)

The two types of eclipse: lunar and solar. A lunar eclipse occurs when the Moon passes through the shadow of the Earth. A solar eclipse occurs when the Moon passes between the Sun and the Earth, blocking out the Sun's light. During a total solar eclipse, when the Moon completely covers the Sun, the Moon's shadow sweeps across the Earth's surface from west to east at a speed of 3 200 kph.

http://aa.usno.navy.mil/data/docs/LunarEclipse.html Part of a larger site on astronomical data maintained by the US Naval Observatory, this site provides data on recent and upcoming lunar eclipses for any location around the world. The data available includes the local time of each eclipse 'event'; the altitude and azimuth of the Moon at each of the events; plus the time of moonrise immediately preceding, and the time of moonset immediately following, the eclipse. Sections of Frequently Asked Questions and research information are also included.

http://www.eclipse.org.uk/default_hi.htm Pages dedicated to the eclipse that was visible over the UK on 11 August 1999. This site contains detailed information on the science of eclipses, and includes images of the event.

eclipsing binary A binary (double) star in which the two stars periodically pass in front of each other as seen from Earth.

When one star crosses in front of the other the total light received on Earth from the two stars declines. The first eclipsing binary to be noticed was Algol.

ecliptic The path, against the background of stars, that the Sun appears to follow each year as it is orbited by the Earth. It can be thought of as the plane of the Earth's orbit projected onto the *celestial sphere.

The ecliptic is tilted at about 23.5° with respect to the celestial equator, a result of the tilt of the Earth's axis relative to the plane of its orbit around the Sun.

ECMWF Abbreviation for *European Centre for Medium-Range Weather Forecasts.

ECS Abbreviation for either **European Communications Satellite* or *environmental control system.

EDR Abbreviation for *experiment data record.

educator resource centre (ERC) Any of many NASA centres that provide teachers with information and demonstrations on space-related subjects, usually without cost. The pre-service and in-service training involves science, technology, mathematics, and geography, and the demonstrations include NASA television. ERCs are located near NASA centres, colleges, universities, museums, and other non-profit organizations.

Edwards, Joe Frank, Jr (1958–) US astronaut. He was aboard the space shuttle *Endeavour* that made the eighth docking with the Russian space station *Mir* in January 1998. He retired from NASA and the US Navy on 30 April 2000 and went on to be involved in technology, aviation, and business development with companies around the world. Edwards was selected as an astronaut in 1994.

Edwards Air Force Base A US Air Force base in California, situated on *Rogers Dry Lake, often used as a landing site by the *space shuttle. It is home to the NASA's *Dryden Flight Research Center.

Both Edwards and the *Cape Canaveral site in Florida have three-mile runways suitable for shuttle landings. However, Edwards has the advantage of being without coastal fog and the dry lake bed provides extra safety in case of brake failure or a shift in the wind. Shuttles landing at Edwards must be transported back to Florida for the next launch.

egress flight The outward flight of a spacecraft. This includes the space shuttle's trajectory into orbit and the *Apollo* flights towards the Moon. It is the opposite of an *ingress flight.

EI Abbreviation for *entry interface.

Einstein, Albert (1879–1955) German-born US physicist whose theories of *relativity revolutionized our understanding of matter, space, and time. Einstein established that light may have a particle nature. He was awarded the

Nobel Prize for Physics in 1921 for his work on theoretical physics, especially the photoelectric law. He also investigated Brownian motion, confirming the existence of atoms. His last conception of the basic laws governing the universe was outlined in his unified field theory, made public in 1953.

Einstein Observatory US astronomical satellite launched in 1978. It operated until 1981, returning the first high-resolution images of X-ray sources. Originally known as *HEAO-2*, it was renamed in honour of physicist Albert Einstein to mark the centenary of his birth in 1879.

Eisele, Donn Fulton (1930–87) US astronaut. He was aboard the first crewed *Apollo* flight, *Apollo 7*, in October 1968. The mission provided the first live television pictures of a spacecraft crew. He was selected as an astronaut in 1963 and left NASA in 1972.

ejecta Any material thrown out of a *crater due to volcanic eruption or the impact of a *meteorite or other object. Ejecta from impact craters on the *Moon often form long bright streaks known as rays, which in some cases can be traced for thousands of kilometres across the lunar surface.

ejection seat A device for propelling a pilot out of an aircraft or spacecraft to parachute to safety in an emergency. The **Mercury* project used escape rockets but the **Gemini* project introduced two ejection seats. Crews on the first four space shuttle flights were strapped into ejection seats but these were abandoned for launches as being too complicated for a full crew. Following the **Challenger* disaster, NASA has worked on developing other escape systems.

E-layer (formerly Kennelly–Heaviside layer) The lower regions (90–120 km) of the ionosphere, which reflect radio waves, allowing their reception around the surface of the Earth. The E-layer approaches the Earth by day and recedes from it at night.

ELDO Acronym for *European Launcher Development Organization.

electromagnetic field A region in which a particle with an electric charge experiences a force. If it does so only when moving, it is in a pure **magnetic field**; if it does so when stationary, it is in an **electric field**. Both can be present simultaneously. For example, a light wave consists of an electric field and a magnetic field travelling simultaneously at right angles to each other.

electromagnetic radiation The transfer of energy in the form of *electromagnetic waves.

electromagnetic spectrum A complete range, over all wavelengths and frequencies, of *electromagnetic waves. These include (in order of decreasing wavelength) radio and television waves, microwaves, infrared radiation, visible light, ultraviolet light, X-rays, and gamma radiation.

The colour of sunlight is made up of a whole range of colours. A glass prism can be used to split white light into separate colours that are sensitive to the human eye, ranging from red (longer wavelength) to violet (shorter wavelength). The human eye cannot detect electromagnetic radiation outside this range. Some animals, such as bees, are able to detect ultraviolet light.

electromagnetic waves Oscillating electric and magnetic fields travelling together through space at a speed of nearly 300 000 kps. Visible light is composed of electromagnetic waves. The **electromagnetic spectrum** is a family of waves that includes radio waves, infrared radiation, visible light, ultraviolet radiation, X-rays, and gamma rays. All electromagnetic waves are transverse waves. They can be reflected, refracted, diffracted, and polarized.

Radio and television waves lie at the **long wavelength–low frequency** end of the spectrum, with wavelengths longer than 10^{-4} m. Infrared radiation has wavelengths between 10^{-4} m and 7×10^{-7} m. Visible light has yet shorter wavelengths from 7×10^{-7} m to 4×10^{-7} m. Ultraviolet radiation is near the **short wavelength–high frequency** end of the spectrum, with wavelengths between 4×10^{-7} m and 10^{-8} m. X-rays have wavelengths from 10^{-8} m to 10^{-12} m. Gamma radiation has the shortest wavelengths (less than 10^{-10} m). The different wavelengths and frequencies lend specific properties to electromagnetic waves. While visible light is diffracted by a diffraction grating, X-rays can only be diffracted by crystals. Radio waves are refracted by the atmosphere; visible light is refracted by glass or water.

http://www.colorado.edu/physics/2000/waves_particles/index.html Great explanation of the electromagnetic spectrum. The pages are very well presented in a conversational style with large, clear text. The site makes good use of Java applets to illustrate the principles behind electromagnetism. Subsequent pages develop the subject further, examining topics such as electric force and electric force fields.

elevon (contraction of elevator and aileron) The movable flaps on the back edges of the space shuttle. Each delta wing has two elevons that are used to control the vehicle's pitch and roll as it settles into denser air on re-entry.

elliptical galaxy One of the main classes of galaxy in the Hubble classification, characterized by a featureless elliptical profile. Unlike spiral galaxies, elliptical galaxies have very little gas or dust and no stars have recently formed within them. They range greatly in size from giant ellipticals, which are often found at the centres of clusters of galaxies and may be strong radio sources, to tiny dwarf ellipticals, containing about a million stars, which are the most common galaxies of any type. More than 60% of known galaxies are elliptical.

ELV Abbreviation for *expendable launch vehicle.

encounter The first observation by a spacecraft of a target of its mission. Data collection begins the mission's *encounter phase.

encounter phase The period during a spacecraft's flight in which it begins collecting data, for which the mission was intended. This may last from less than a week up to, for a satellite in orbit, several years. The collecting period includes a *far encounter phase and *near encounter phase.

At NASA, an encounter phase is normally managed from the *Space Flight Operations Facility at the Jet Propulsion Laboratory.

Endeavour The fifth of the US *space shuttles. It made its first flight on 7 May 1992. *Endeavour* was built as a replacement for **Challenger*, destroyed in an explosion shortly after lift-off in 1986. It was used for the mission to repair the **Hubble Space Telescope* in 1993 and for the first assembly of the *International Space Station* (*ISS*) in 1998 that connected the *Unity* module to the *Zarya* module, as well as five more trips to the *ISS* before the *Columbia* disaster in 2003. It made the last shuttle flight before that tragedy, being launched on 23 November 2002 to change the resident crews and install truss hardware. This was *Endeavour*'s 19th flight.

e

end effector NASA jargon for the 'hand' of the remote-control arm of *Canadarm. The end effector device, which has three wires, twists around and then locks onto the object it is capturing, such as a satellite to be lifted and deployed.

Energetic X-Ray Imaging Survey Telescope (EXIST) A mission proposed by NASA for launch after 2007 to survey the hard X-ray band comprising high-energy X-rays of 5–600 keV that lies between the X-ray and gamma-ray wavelengths of the universe. EXIST will use a wide-field telescope array to conduct the first high-sensitivity imaging of the whole sky to survey those wavelengths, and will continue the monitoring of gamma-ray bursts.

Energiya rocket (or Energia rocket) The most powerful Soviet *booster rocket, first launched on 15 May 1987. Used to launch the Soviet space shuttle **Buran* in 1968, the SL-17 Energiya booster was capable, with the aid of strap-on boosters, of launching payloads of up to 190 tonnes into Earth orbit. It stood 60 m tall and was the first Soviet rocket to use liquid-hydrogen fuel.

http://www.astronautix.com/lvfam/energia.htm Illustrated guide to the Energiya family of boosters and launch vehicles. There is detailed historical background as well as technical information.

Energiya Rocket and Space Complex A prime contractor for *Rosavia Kosmos. It developed the powerful *Energiya rocket that launched the shuttle **Buran*. Energiya also owns and operates the mission control centre in *Kaliningrad and operated the **Mir* space station.

engineering data The data concerning a spacecraft's environment (such as the interior pressure and temperature) and the functional health of its computers and other systems. This is repeatedly transmitted to Earth, ensuring that a temporary loss of telemetry is not critical. Although engineering data is used to obtain *science data, it is normally given less priority than the latter.

engineering model A full-scale, non-flight model of a spacecraft replicating the *flight model in almost every way, to enable full-scale tests to be completed on all operating systems and instruments before the final assembly and pre-flight preparation of the flight model is completed.

Engineering Test Satellite (ETS) Any of a series of Japanese satellites used for research into technologies to improve communications and broadcasting satellites and other future space developments. Japan's *National Space

Development Agency launched *ETS-1* (also called *Kiku 1*) in 1975 to acquire information about its *N rocket, orbit injection, and tracking and control. *ETS-7* (*Kiku 7*) was launched in 1997 for rendezvous docking experiments, having a chase satellite (*Hikoboshi*) and target satellite (*Orinhime*). *ETS-8* is scheduled for launch in 2005 to verify the world's largest geostationary satellite bus (*see* SPACECRAFT BUS) technology.

England, Anthony Wayne (1942–) US astronaut. He flew with the *Spacelab 3* (*see* SPACELAB) mission launched on the *Challenger* space shuttle in July 1985. He was involved in operating the *Spacelab* systems and the *Canadarm. He worked for the United States Geological Survey before being selected as a scientist-astronaut in 1967. England left NASA in 1988.

Engle, Joe Henry (1932–) US astronaut. He flew on two space shuttle missions: the second orbital test flight of *Columbia* in November 1981 and the *Discovery* flight of August 1985 to deploy three communications satellites. Engle was selected as an astronaut in 1966.

Enterprise The first space shuttle to be built and flown. It was originally designed as a prototype to test the shuttle's performance while airborne and during landing.

It had a dummy engine and plastic plates instead of thermal protection tiles, and had to be attached piggyback to a Boeing 747 aircraft in order to become airborne. After four uncrewed and three crewed captive flights, the *Enterprise* was released on 12 August 1977 for the first of five independent piloted flights. In 2003, it was placed in the Smithsonian Institution's National Air and Space Museum.

The craft was named after the spaceship of the *Star Trek* television series.

entry interface (EI) The moment when a spacecraft enters the Earth's denser atmosphere, which NASA sets at an altitude of 121 km. An *ablation shield is needed at this point to protect the vehicle from heat that would otherwise destroy it.

environmental control system (ECS) A system that controls the interior environment of a crewed spacecraft. The system removes excess heat, absorbs exhaled carbon dioxide, generates and circulates re-oxygenated air, removes contamination, controls humidity, and maintains the correct pressure level. An ECS protects astronauts from such problems as the *bends, *hypoxia, and toxicity.

Environmental Science Services Administration (ESSA) A US government agency that funded the development of *Television Infrared Observation Satellites during the 1960s. Under ESSA the satellites were to be named *ESSA-1*, *ESSA-2*, and so on, and operated by the agency. The series was renamed NOAA when the *National Oceanic and Atmospheric Administration (NOAA) took over the project.

Environmental Systems Commercial Space Technology Center (ESCSTC) A centre in Gainsville, Florida, that develops and tests technologies for use in long-duration human space flight. Located in the Department of

Environmental Engineering Sciences at the University of Florida, it is a cooperative undertaking between NASA and the university, as well as additional academic and industrial partners. The projects concentrate on the needs of three technical areas defined by NASA: air revitalization, solid-waste recovery, and water recovery. The research concentrates on reducing the mass and volume of hardware, devising low power consumption, and providing for minimum supervision by the crew. ESCSTC also helps find commercial applications for the technologies that it develops.

environmental test A test conducted on a spacecraft before its launch at the *Kennedy Space Center. After being assembled in the *clean room, a spacecraft is brought to the environmental test laboratory. The rigorous testing includes the craft being placed on a shaker table, where it is subjected to vibrations resembling those of a launch. The craft is also enclosed in a thermal-vacuum chamber to test thermal resistance.

The testing laboratory is the final stage before a spacecraft is transferred to the launching pad.

Envisat A European Space Agency (ESA) environment satellite, an Earth-observation satellite that measures the atmosphere, ocean, land, and ice. It was launched on 28 February 2002 into a polar orbit on a five-year mission. The data, added to previous measurements by ESA's *European Remote-Sensing Satellite (ERS) series, will be used to monitor environmental and climatic changes.

Envisat's instruments include an *advanced along-track scanning radiometer, an *advanced synthetic aperture radar, a *microwave radiometer, *global ozone measurement by occultation of stars, and a *scanning image absorption spectrometer for atmospheric cartography.

EO Abbreviation for *Earth observation.

EOR Abbreviation for *Earth-orbit rendezvous mode.

EOS Acronym for *Earth-observing system.

equipment bay An area of a spacecraft used to house permanent equipment or to carry temporary scientific equipment. A space shuttle has three avionics equipment bays (two forward, one aft) on its mid-deck and an equipment bay below the mid-deck floor that houses the main components of the waste management and air-revitalization systems.

Other examples are the *International Ultraviolet Explorer* (*IUE*) satellite with a main equipment bay at its base for most of the high-powered electronic equipment, and the *Ulysses* space probe's upper equipment bay that holds its reaction control system. The *Apollo 11* lunar module used its equipment bay to store the early *Apollo* scientific experiments package (EASEP).

ERC Abbreviation for *educator resource centre.

ergometer NASA name for a stationary exercise bicycle used on an orbiting spacecraft. Physical exercise is crucial for astronauts spending lengthy periods in *microgravity, which causes loss of muscle strength and may lead to reduced bone density. The ergometer also gives a workout to the cardiovascular system.

An ergometer was installed on the *Skylab* space station, in the aft compartment.

Eros An asteroid discovered in 1898 by German astronomer Gustav Witt. It was the first asteroid to be discovered that has an orbit coming within that of Mars. It passes within 22 million km of the Earth. It is elongated, measures 33 × 13 × 13 km, rotates around its shortest axis every 5.3 hours, and orbits the Sun every 1.8 years.

NASA's **Near-Earth Asteroid Rendezvous* spacecraft went into orbit around Eros on 14 February 2000. Pictures showed the asteroid to be heavily cratered, indicating that it is probably older than was previously thought. The surface was discovered to contain aluminium, iron, magnesium, and silicon.

ERS Abbreviation for **European Remote-Sensing Satellite*.

ERT Abbreviation for *Earth-received time.

ESA Abbreviation for *European Space Agency.

escape harness A vestlike outer garment, worn by NASA space shuttle astronauts during a launch countdown. Donned in the *white room, the harness can be used during a launch-pad fire, explosion, or other emergency to assist rescuers in pulling the crew out of the vehicle.

escape suit A space garment designed to allow astronauts to evacuate their spacecraft in the event of an emergency. High-altitude operation suits were worn by the crews of NASA's first four space shuttle launches (April 1981–June 1982). This protection was discontinued from the 5th to the 25th flight, but the *Challenger* accident in 1986 prompted the adoption of a new design of *launch/entry suits, which can be used for high-altitude bail-outs.

The bright orange one-piece suit has an oxygen supply, parachute, life raft, life jacket, and additional survival equipment. It can be inflated with air to create thermal protection, and has special pressure at the legs to provide anti-g protection against the effects of high acceleration.

escape tower rocket A solid rocket on top of the *Mercury* spacecraft, used in an emergency escape. The rocket had a 25 200 kg thrust and was supported by a 3-m tall steel frame. An astronaut would rotate an abort handle to fire it and propel the capsule away from the launch vehicle.

On the launch pad, the rocket could blast the *Mercury* capsule upward for approximately 0.8 km for parachute deployment. The use of escape tower rockets was discontinued for space shuttle launches when it was decided that proper use of the rocket was too complex, and that weight taken up by the rocket could be better used for payloads. Russia used an escape tower system for *Soyuz* spacecraft launches.

escape velocity The minimum velocity required for a spacecraft or other object to escape from the gravitational pull of a planetary body. In the case of the Earth, the escape velocity is 11.2 kps; the Moon, 2.4 kps; Mars, 5 kps; and Jupiter, 59.6 kps.

ESCSTC Abbreviation for *Environmental Systems Commercial Space Technology Center.

ESL Abbreviation for *Expert Support Laboratory.

ESOC Acronym for the *European Space Operations Centre, the European Space Agency (ESA) mission control centre at Darmstadt, Germany.

ESRO Acronym for *European Space Research Organization.

ESSA Acronym for *Environmental Science Services Administration.

ESTEC Acronym for the *European Space Technology Centre, the European Space Agency (ESA) research centre at Noordwijk, the Netherlands.

ESTRACK A network of eight tracking stations worldwide, operated by the European Space Agency. The stations provide links between satellites in orbit and the ground configuration control room (GCCR) in the operations control centre (OCC) at Darmstadt, Germany. One of the stations is mobile and can be transported anywhere in the world. The seven permanent ground stations are located at Kiruna, Sweden; Redu, Belgium; Maspalomas, Spain; Villafranca, Spain; Malindi, Kenya; Perth, Australia; and Kourou, French Guiana.

ETR Abbreviation for *Eastern Test Range.

EUMETSAT Contraction of *European Organization for the Exploitation of Meteorological Satellites.

EURECA Contraction of **European Retrievable Carrier*.

EuroMir A European–Russian cooperative programme agreed in 1992, allowing European Space Agency (ESA) astronauts an opportunity to fly long-duration science missions aboard Russia's **Mir* space station. Four astronauts were selected in 1993 to train for two planned missions. The ESA astronaut for EuroMir 94 was the German Ulf *Merbold, who was launched on *Soyuz* TM-20 on 4 October 1994 for a mission lasting 31 days. The EuroMir 95 astronaut was another German, Thomas Reiter, who was launched with two cosmonauts aboard *Soyuz* TM-22 on 3 September 1995, for a 179-day mission. A EuroMir 97 mission was discussed but not agreed by ESA member states. Several other European astronauts flew missions to *Mir* under agreements signed by Russia with individual countries.

Europa The fourth-largest moon of the planet Jupiter, diameter 3 140 km, orbiting 671 000 km from the planet every 3.55 days. It is covered by ice and was originally thought to be criss-crossed by thousands of thin cracks some 50 000 km long. These are now known to be low ridges.

NASA's *Galileo* spacecraft began circling Europa in February 1997. One of the first discoveries from the data it sent back was that the 'cracks' covering the surface of the moon are in fact low ridges. These features lend credence to the idea that Europa possesses a hidden subsurface ocean. NASA plans to

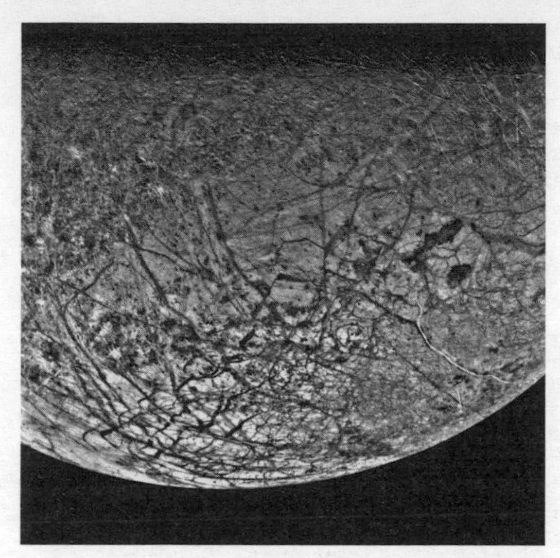

(Image © NASA)
The surface of Europa, here imaged by the *Voyager 2* probe in 1979, is criss-crossed by a complex network of low ridges.

launch the Europa Orbiter in 2008 to arrive in 2010 and search for water beneath Europa's icy surface, using a radar sounder and altimeter.

http://www.solarviews.com/solar/eng/europa.htm Comprehensive latest information on Jupiter's moon. There is a full discussion of evidence that Europa may have liquid water under its frozen surface. There are also a number of colour images of Europa.

Europa rocket A launch vehicle developed by the *European Launcher Development Organization (ELDO) during the 1960s. Based on the UK's Blue Streak rocket, the Europa rocket had some failures, which contributed to ELDO's demise.

European Aeronautic Defence and Space Company (EADS) A company that developed and produces the *Ariane rockets. Its diverse space work has included contributions to the **Rosetta* and **Mars Express* probes. Astrium, an EADS joint venture with the UK's BAE Systems, is the prime industrial contractor for the *International Space Station* (*ISS*), responsible for the *Columbus* laboratory, Europe's main contribution. Astrium helped develop the *X-38* crew rescue vehicle, and the integrated cargo carrier (ICC). A projected nine uncrewed supply vehicles, known as the Automated Transfer Vehicles (ATVs), are scheduled to be built by Astrium for the *ISS* from 2003 to 2013.

European Centre for Medium-Range Weather Forecasts (ECMWF) An international organization using meteorological satellites to supply medium-range weather forecasts to its European member states. Located in

Reading, England, the ECMWF has working arrangements with the *European Organization for the Exploitation of Meteorological Satellites (EUMETSAT), the World Meteorological Organization (WMO), and the African Centre for Meteorological Applications for Development (ACMAD).

The ECMWF was established in 1973 and was originally a European Cooperation in Science and Technology (COST) project. Its member states are Austria, Belgium, Denmark, Finland, France, Germany, Greece, Ireland, Italy, Luxembourg, the Netherlands, Norway, Portugal, Spain, Sweden, Switzerland, Turkey, and the UK. It also has cooperative agreements with Croatia, the Czech Republic, Hungary, Iceland, Romania, Slovenia, and Serbia and Montenegro.

European Communications Satellite (ECS) Any of a series of geostationary satellites that have provided services to Europe in such areas as television and digital telephony. The system is owned by the *European Telecommunications Satellite Organization and operated by the European Space Agency.

ECS-1 was launched on 16 June 1983 and operated for more than 13 years, while *ECS-2*, launched in 1984, lasted 9 years. *ECS-3* was destroyed by a launch failure. *ECS-4*, launched in 1987, was still operating in 2001, after more than 12 years, although *ECS-5*, launched in 1988, was retired in 2001. A new European communications satellite, **Artemis*, was launched on 12 July 2001. It also relays communications to the Earth from other satellites.

European Launcher Development Organization (ELDO) A European organization formed in February 1964 to develop rocket-launching capabilities. The member nations were Belgium, France, Germany, Italy, the Netherlands, and the UK. After the mixed success of its *Europa rocket, ELDO disbanded and its activities were taken over in 1975 by the new *European Space Agency.

European Organization for the Exploitation of Meteorological Satellites (EUMETSAT) An intergovernmental organization that establishes, maintains, and uses meteorological satellites for 18 European states. Established in 1986, its headquarters are in Darmstadt, Germany, and its primary ground station is in Fucino, Italy, with a back-up in Weilheim, Germany. Since 1995, EUMETSAT has owned and operated the geostationary **Meteosat* weather satellites, which were developed by the European Space Agency and first launched in the late 1970s. The first of the second generation *Meteosat* satellites was launched in August 2002. The first *Metop* satellite, which is the first European polar-orbiting satellite dedicated to operational meteorology and climate monitoring, is to be launched in 2005 with two more satellites scheduled in the same series.

EUMETSAT's member states are Austria, Belgium, Denmark, Finland, France, Germany, Greece, Ireland, Italy, Luxembourg, the Netherlands, Norway, Portugal, Spain, Sweden, Switzerland, Turkey, and the UK. It also has seven cooperating states, Hungary, Slovakia, Poland, Croatia, Serbia and Montenegro, Slovenia, and Romania.

European Remote-Sensing Satellite (ERS) A series of European Earth-observation satellites with a *remote-sensing facility. The European Space Agency (ESA) launched *ERS-1* in 1991 and *ERS-2* in 1995 to observe land surfaces, oceans, and polar caps. The extensive data collected by the satellites was used by over 2 000 scientists to forecast weather, monitor crops, and to assess climate trends, marine pollution, and the destruction of tropical rainforests. The spacecraft also monitored natural disasters, such as earthquakes and floods.

The satellites, which worked in combination for about a year, were succeeded by ESA's **Envisat*. *ERS-1* ended its life in 2000 after running out of fuel, having lasted three times longer than predicted. In 2004, *ERS-2* remained in a polar orbit at a height of 800 km, orbiting every 100 minutes. Both satellites carried an *active microwave instrument (AMI), a *radar altimeter, and an *along-track scanning radiometer. A *global ozone measurement experiment was added to ERS-2.

European Retrievable Carrier (EURECA) A European Space Agency (ESA) satellite that carried out 71 experiments on microgravity, solar observations, and material technology from July 1992 to July 1993. It also carried the Wide-Angle Telescope for Cosmic Hard X-Rays (WATCH) developed by the Danish Space Research Institute, detecting 19 cosmic gamma-ray bursts.

EURECA, weighing 4 500 kg, was the largest spacecraft ever built and flown by the ESA, and the first that returned with samples. It acquired the name 'Retrievable Carrier' because it was deployed by NASA's space shuttle *Atlantis* and retrieved by *Endeavour*.

European Space Agency (ESA) An organization of 16 European countries that engages in space research and technology. It was founded in 1975, with headquarters in Paris, France. The participating countries are Austria, Belgium, Denmark, Finland, France, Germany, Greece, Ireland, Italy, Luxembourg, the Netherlands, Norway, Portugal, Spain, Sweden, Switzerland, and the UK.

ESA astronauts have participated in 43 space missions and stayed on the **Mir* space station, and will help crew the **International Space Station* (*ISS*). The ESA has developed the *Ariane rockets (used to launch most ESA satellites), the **Spacelab* scientific workshop, the **Giotto* space probe, and various scientific and communications satellites. It also created components for the *ISS*, including a robotic arm. The ESA's *European Remote-Sensing Satellite *ERS 2* was launched in 1995 to work in tandem with *ERS-1* (launched in 1991 and retired in 2000), improving measurements of global ozone. The ESA's **Envisat* satellite was launched on 28 February 2002. Its **Mars Express*, launched in June 2003, has discovered water ice and methane on that planet, and the *Venus Express* is scheduled for a November 2005 launch.

http://www.esa.int/ Introduction to the European Space Agency, using the format of an informative and colourful slide show detailing the history of the ESA and its current projects and future developments. Included are a large gallery of still images showing past space explorations, plus a video archive.

European Space Operations Centre (ESOC) A spacecraft operating facility controlled by the *European Space Agency (ESA), located in Darmstadt, Germany. The centre is responsible for satellite control and monitoring and data retrieval. ESOC is often involved in a mission from its conception, taking over operations after a launch. It runs the ESA Operations Control Centre

(OCC) with its main control room (MCR) of 14 working positions and large wall displays for special operations such as deep-space missions.

The OCC also has a flight dynamics room (FDR) for orbit and attitude calculations, a dedicated control room (DCR) for routine spacecraft monitoring and control, and a ground configuration control room (GCCR) for the *ESTRACK network. *Space debris research is carried out by a Mission Analysis Section. A dedicated mission support system (DMSS) of 32 workstations with hardware and software is used for mission control and data processing.

European Space Research Organization (ESRO) One of the predecessors of the *European Space Agency (ESA). Established in 1962, its member states included ten European nations and Australia. NASA launched seven ESRO satellites from 1968 to 1972. In 1975 the ESA replaced ESRO and the *European Launcher Development Organization (ELDO).

European Space Technology Centre (ESTEC) A facility of the *European Space Agency that tests a spacecraft and its payload and instruments for extreme stresses before the launch. ESTEC has *clean rooms with a Large European Acoustic Facility (LEAF) for testing launcher noise levels and a large space simulator (LSS) to test space temperature conditions. There are also electromagnetic and electrostatic facilities and a compact payload test range. A hydra multi-axis shaker and electrodynamic shakers are used to simulate vibrations of the launcher.

ESTEC also has several laboratories, including microgravity and materials, mechanical, thermal, software, and robotic facilities. It is based at Noordwijk, the Netherlands.

http://www.estec.esa.nl/ News of European Space Technology Centre (ESTEC) space science and technology research programmes. Downloadable Adobe documents provide in-depth information about the spacecraft testing facilities used to ensure that spacecraft are built to withstand launches and the space environment.

European Telecommunications Satellite Organization (EUTELSAT) A European agency that provides satellite communications for 150 countries, with its 1 400 television channels reaching 110 million homes by 2004. The agency now has 24 satellites in service. One of the new satellites, *EUROBIRD*, was launched in March 2001 to cover western and central Europe for a period of 12 years, providing broadcasting and telecommunications services on 24 channels. EUTELSAT also provides coverage for Africa, many parts of Asia, and (by interconnectivity) the Americas.

In 1991 a private company was created from the original intergovernmental agency to control all activities and assets, and a new intergovernmental organization was set up to monitor the company's practice, such as non-discrimination and fair competition.

EUTELSAT Contraction of for *European Telecommunications Satellite Organization.

EUVE Abbreviation for **Extreme-Ultraviolet Explorer*.

EVA Abbreviation for *extravehicular activity.

Evans, Ronald Ellwin, Jr (1933–90) US astronaut. He was the *command module pilot of the *Apollo 17* mission of December 1972, circling the Moon while astronauts Gene *Cernan and Jack *Schmitt landed to explore the lunar surface. Evans mapped the Moon from his orbit and later carried out a space walk to retrieve film from the cameras. Evans was selected as an astronaut in 1966 and left NASA in 1977.

exhaust The discharged used gas from rocket engines. Exhaust from a launch vehicle provides the power for a spacecraft's blast-off and *thruster exhaust is used for its later orbital adjustment of attitude (*see* ATTITUDE CONTROL) and *deorbit burn.

Exhaust velocity is the velocity of the exhaust from a rocket engine. A greater thrust is achieved by a higher velocity, which can be produced by the correct mixture ratio of *fuel and oxidizer.

EXIST Acronym for *Energetic X-Ray Imaging Survey Telescope.

exobiology The study of life forms that may possibly exist elsewhere in the universe and of the effects of extraterrestrial environments on Earth organisms. Techniques include space probe experiments designed to detect organic molecules, and the monitoring of radio waves from other star systems.

ExoMars A planned European probe to Mars and the first significant mission of the *Aurora programme. It will consist of a Mars orbiter carrying a descent module with a Mars rover. The rover will explore the surface of Mars to study the rocks and search for possible signs of past or present life. *ExoMars* is intended for launch in 2009.

Exosat An *X-ray astronomy satellite launched in May 1983 by the European Space Agency (ESA). The satellite weighed 400 kg and for three years observed X-rays from objects including galaxies, supernova remnants, and binary stars.

expansion ratio For a rocket engine, the ratio of gas expansion in the *nozzle area to expansion in the *throat area. In the space shuttle's orbital manoeuvring system (OMS), the nozzle-throat ratio is 55:1.

Expedition One A designation for the first resident crew of the **International Space Station* (*ISS*). The members of the Expedition One crew were astronaut Bill *Shepherd (commander) and cosmonauts Yuri *Gidzenko (pilot) and Sergei *Krikalev (flight engineer). The crew was launched aboard a *Soyuz* spacecraft from the Baikonur Cosmodrome on 31 October 2000 and arrived at the *ISS* on 2 November. The crew returned aboard the space shuttle *Discovery* on 20 March 2001 after spending 140 days in space.

expendable launch vehicle (ELV) A launch system with rockets that are not recoverable after they burn out and separate from a spacecraft.

experiment data record (EDR) A record of the data collected during a scientific experiment on a spacecraft. This data is recorded in an EDR file, from which it can be extracted and converted to a common data format (CDF) to generate a formatted file of archive data. These data sets, such as the Voyager 2 images of the planet Uranus, are available to researchers on CD-ROM and magnetic tape.

experiment segment A pressurized module area for scientific experiments on **Spacelab*. Measuring 4 × 2.7 m, the segment was able to accommodate four experimenters at the same time, and the first *Spacelab* mission carried 77 experiments. The experiment segment was connected to the *core segment to create a *long module.

Expert Support Laboratory (ESL) A laboratory for scientific input and expert investigations concerning spacecraft experiments. For example, the *European Space Agency provided ESL facilities for its **Envisat* research projects; these scientists investigated methods for determining gas densities for atmospheric measurements.

Explorer A series of US scientific satellites. *Explorer 1*, launched on 31 January 1958, was the first US satellite in orbit and discovered the *Van Allen radiation belts around the Earth. *Explorer 42* in 1970 was the first satellite to study X-rays from space, and *Explorer 48* in 1972 was the first to study gamma rays from space.

'explosive' bolt A bolt that is disengaged electrically in order to separate components in space, such as a spacecraft from its launch vehicle. Contrary to its name, there is no explosion involved in the process; the bolt is removed with only a slight mechanical shock, and this action does not produce loose particles.

Extrasolar Planet A planet around a star other than the Sun, also known as an exoplanet. Two main methods have been used to discover them. One looks for a cyclical Doppler shift in the light from a star that would be caused as it orbits its common centre of gravity with one or more planets. This technique led to the first discoveries of extra-solar planets in the late 1990s. The second method looks for minute changes in the light output of the parent star caused by an orbiting planet passing in front of it. This has also proved successful. Space missions such as **COROT* and **Kepler* will expand the search for extra-solar planets.

http://www.obspm.fr/encycl/encycl.html Semi-technical site containing regular updates on discoveries and links to the home sites of various planetary search teams.

extravehicular activity (EVA; or space walk) The movement and work carried out by an astronaut outside a spacecraft. The term usually refers to activity undertaken from a space shuttle or space station in orbit, but Moon walks—the only true space walks—are also EVAs. An EVA is necessary for operations carried out in space, such as repairs to spacecraft, that cannot be accomplished using a remote-control arm. The Soviet cosmonaut Alexei *Leonov completed the first EVA (10 minutes) on 18 March 1965. The US astronaut Edward *White undertook the first EVA for NASA (36 minutes) on 3 June 1965.

During an EVA the astronaut wears an **extravehicular mobility unit** (EMU), a combined spacesuit and *life-support system. The **extravehicular life-support system** (ELSS) was an early *chest pack worn by astronauts aboard *Gemini* spacecraft. It was attached to a 7.5-m umbilical cord and supplied communications, spacesuit pressure, and emergency oxygen in the event of failure of the umbilical supply. The *Apollo* spacesuits had a **portable life-support system**, while space shuttle suits have a **primary life-support system**.

The **extravehicular visor assembly** (EVVA) is a transparent shell that fits over a *space helmet to reduce light and heat, and to protect against solar radiation and micrometeoroid impacts. A coating of gold on the sun visor reflects the light and heat, while adjustable eyeshades provide additional protection from glare. The assembly also contains a television camera and lighting attachment. **Skylab* astronauts wore a similar *Skylab* extravehicular visor assembly (SEVA).

http://www.spaceflightnow.com/station/stage5a/ fdf/evastats.html Breakdown of space walk duration by space programme (*Gemini*, *Apollo*, and so on) and general extravehicular activity statistics. A downloadable Adobe document has more information about the space walks made during specific missions.

***Extreme-Ultraviolet Explorer* (*EUVE*)** A NASA spacecraft which investigated the relatively unexplored extreme-ultraviolet spectrum of the universe. *EUVE* was launched in June 1992 with a science payload consisting of three scanning telescopes and an extreme ultraviolet spectrometer/deep survey instrument. Besides mapping the ultraviolet sky, the mission made specific spectroscopic observations of targets suggested by scientists, investigated the emissions of stars, and studied the *interstellar matter.

The payload was designed and built at the Space Sciences Laboratory at the University of California at Berkeley. The science operations were completed in January 2001 and the EUVE transmitters turned off in February. The spacecraft re-entered the atmosphere in January 2002.

http://archive.stsci.edu/euve/ Still being updated with the latest scientific results from the Extreme-Ultraviolet Explorer (EUVE), this site outlines the achievements of the nine-year EUVE mission. Its pages show images from the all sky survey and targets of interest, and describe the EUVE spacecraft and its instrumentation.

Eyharts, Léopold (1957–) French astronaut with the European Space Agency (ESA). While aboard the Russian space station *Mir* in February 1998, he carried out scientific experiments on behalf of France. Later that year he began training at NASA's Johnson Space Center to prepare for a space shuttle flight. He went on to handle technical assignments in the Space Station Operations Branch of the Astronaut Office. Eyharts was selected as an astronaut by the Centre National d'Etudes Spatiales in 1990 and was selected by the ESA in 1992.

F-1 engine A rocket engine on the Saturn V launch vehicle used for the *Apollo* missions in the 1960s. The F-1, sponsored by the US Air Force, had a 680-tonne/748-ton thrust that created an 'earthquake' vibration and noise for miles around. It was powered by immense kerosene and liquid-oxygen tanks.

The Saturn V first stage, built by the Boeing Company, had five F-1 engines amounting to 3 400 tonnes of thrust (180 million horsepower). During launch, the F-1 turbopump delivered 57 000 L of fuel and 95 000 L of oxidizer each minute. It carried the *Apollo* vehicle to an altitude of 60 km in 160 seconds at a speed of 13 360 kph.

Three early F-1 engines failed in 1962 during static testing due to combustion instability, but a special injector was designed to solve the problem.

Fabian, John McCreary (1939–) US astronaut. He flew on the second mission of the space shuttle *Challenger* in June 1983 and was part of the *Discovery* mission in June 1985 that deployed three communications satellites. Fabian was selected as an astronaut in 1978 and left NASA in 1986.

Faget, Max (1921–2004) A US aeronautical engineer who contributed to the designs for almost all of NASA's crewed spacecraft, covering the **Mercury* project (1961–3) and the **Apollo* project (1961–72) as well as the space shuttle. He also devised the *Little Joe rocket. He joined the Langley Aeronautical Laboratory (later the *Langley Research Center), Virginia, in 1946, where he conducted research on the *Mercury* heat shield. Faget joined NASA in 1958 and became director for engineering and development at the Johnson Space Center. He left NASA in 1981 and founded Space Industries Inc., in 1982.

failure mode NASA terminology for any failure of equipment or spacecraft. Anything from a rip in a spacesuit to a computer problem is termed a failure mode.

fairing A strong protective covering placed over an artificial satellite or other *payload on top of a rocket during its launch. The fairing is jettisoned when the rocket ascends above the atmosphere.

Faith 7 The *Mercury* spacecraft on the final flight of NASA's **Mercury* project. US astronaut Gordon *Cooper piloted the mission, launched on an *Atlas rocket on 15 May 1963 and returning to splashdown the following day. *Faith 7* made 22 orbits in 34 hours and 19 minutes. This was NASA's first evaluation of the effects on an astronaut of spending a day in space.

far encounter phase (FE) The initial phase of a spacecraft's encounter with a target, especially a planet. This begins the high-priority, data-gathering period of operations for which the mission is intended. The far encounter phase begins when a planet's full disc no longer fits within the field of view of the spacecraft's instruments. Images and other observations are then made of parts of the planet rather than its whole disc. During this encounter, new features may be observed and the spacecraft's instruments can then be reprogrammed to record them during the *near encounter phase (NE) that follows FE.

***Far-Ultraviolet Spectroscopic Explorer* (*FUSE*)** NASA spacecraft launched in June 1999 to explore the universe using high-resolution spectroscopy in the far-ultraviolet spectral region. *FUSE* is part of NASA's *Origins Program. Its instruments include a spectrograph, an electronic camera, and telescope mirrors. In one investigation, the mission will search for deuterium, an isotope of hydrogen, in three regions—the *interstellar matter, gas clouds in the galaxy, and intergalactic clouds—to understand how galaxies evolve and the condition of the universe when it was a few minutes old.

Johns Hopkins University, Baltimore, Maryland, was the leading developer and operator of the mission, the first time a university has played such a role in a major mission. Partners are the University of Colorado at Boulder, the University of California at Berkeley, the Canadian Space Agency (CSA), and France's Centre National d'Etudes Spatiales (CNES).

http://fuse.pha.jhu.edu/ Guide to the *Far-Ultraviolet Spectroscopic Explorer* mission. Clear illustrated pages outline the project's aims and explain how the telescope works. The site is regularly updated with mission news, and there are lots of pictures, a best of *FUSE* gallery, and animations.

FAST Acronym for **Fast Auroral Snapshot Explorer*.

***Fast Auroral Snapshot Explorer* (*FAST*)** NASA satellite that studies the Earth's *aurorae. Launched on 21 August 1996 from a Pegasus rocket, *FAST* is the second mission of NASA's Small Explorer Satellite Program (SMEX). The primary objective of the mission is to study the microphysics of space plasma and the accelerated particles that cause aurore. It measures the electric and magnetic fields, energetic electrons and ions, plasma waves, and thermal plasma density and temperature. Its instruments include 16 electrostatic analysers (ESAs). *FAST* crosses the doughnut-shaped auroral zones on the poles four times in each orbit, collecting 'snapshots' of high-resolution data.

fault-protection program Computer software that allows a spacecraft autonomous monitoring and control. A fault-protection program, which runs in one or more subsystems, detects problems, and reduces their impact or requests *safing to shut down specific components. It can also re-establish the spacecraft's failed communications with the ground.

FCS Abbreviation for *flight control system.

FDF Abbreviation for *flight data file.

FE Abbreviation for *far encounter phase.

ferret satellite A spy satellite that intercepts foreign military and governmental radio transmissions.

firing room Any of four rooms at NASA's *launch control centre, used for all aspects of the *checkout and launch phase for space vehicles. The rooms are located on the third floor of the centre, each having a checkout, control, and monitor subsystem (CCMS). Firing Rooms 1 and 3 are designed for the full control of launch and orbit operations. Firing Room 2 is normally used for the development and testing of computer software. Engineering analysis and support of the checkout and launch operations is carried out in Firing Room 4.

Fisher, Anna Lee Tingle (1949–) US astronaut and physician. She was a mission specialist aboard the *Discovery* space shuttle flight of November 1984 that deployed three communications satellites. Fisher became chief of the Space Station Branch of NASA's Astronaut Office and went on to the Shuttle Branch, awaiting a future flight. She was selected as an astronaut in 1978.

Fisher, William Frederick (1946–) US astronaut and physician. He flew as a mission specialist on the *Discovery* space shuttle mission of August 1985 that deployed three communications satellites. Fisher held several NASA administrative positions before being selected as an astronaut in 1980. He left NASA in 1991.

fish stringer A term used by astronauts for a tool tether used during *extravehicular activity. It has seven hooks spaced at intervals along its length so that tools, which have loops attached, can be held securely in place when not in use. Space shuttle missions are usually equipped with four or five fish stringers.

flame bucket A deep open pit built under a launch pad to receive and channel the hot gases from a rocket during lift-off. The flame bucket has one thick, metal side that is bent towards one of three open sides through which it deflects the exhaust gases.

flame deflector An angular construction beneath a launch pad that diverts the rocket's exhaust to the side. At NASA's Kennedy Space Center, flame deflectors are made of steel and covered with a high-temperature concrete surface. Those for a space shuttle are 12 m high.

An ascending rocket below an altitude of 92 m will have its exhaust channelled by the flame deflectors into a 149-m concrete-and-brick flame trench that transects the launch pad. Above the flame deflectors are 16 nozzles that release water to suppress the noise, the effects of which could damage the shuttle and its payloads.

flare The action of lifting the nose of a space shuttle to prepare for landing. This occurs 519 m above the ground, about 0.8 km from the runway and some 15 seconds before touchdown, as the shuttle slows to 483 kph.

flight control The control of a spacecraft or rocket using built-in computers on the craft or *uplinks from ground stations. This includes control of the *flight path and of the spacecraft's operational equipment. Many flight control commands are initiated by the space mission's **flight director** and carried out by a team of **flight controllers**, who have responsibility for one or more aspects of a space flight. Orbiting space shuttle and space station astronauts have regular conferences with flight controllers in Houston, Texas, and Moscow, Russia.

Known informally as 'Flight', a flight director is responsible for all major decisions during a flight, such as shortening a mission because of problems. Decisions made by the Flight are relayed to astronauts by the *capsule communicator.

NASA's first flight director (1961–83) was Christopher Columbus Kraft Jr.

flight control system (FCS) A system that flies a spacecraft. The space shuttle can be flown automatically or by using *control-stick steering (CSS) to send electrical signals to the computers that activate the *thrusters, *elevons, and *rudder.

flight data file (FDF) A massive file of procedures for a space shuttle or space station mission. It contains checklists for normal, back-up, and emergency procedures, as well as activity plans for the crew, maps of the Earth, star charts, and other information. The crew reviews the file during their period of isolation in the final days before the launch, and it remains immediately available during the mission, being stowed in containers on the flight deck next to the seats of the pilot and commander.

flight deck The area on a space shuttle containing the flight-deck control and display instruments, as well as seats for the commander and the pilot. Sitting behind the commander and pilot during launch and landing are two other crew members: a mission specialist and a payload specialist. At the rear of the flight deck, where two *crew stations face the cargo bay, are controls for deploying satellites and the robotic arm *Canadarm.

flight model A flight-ready version of a spacecraft. It is preceded by the *engineering model and *structural model of the spacecraft, built during the development of the flight project.

flight path The actual or planned course of a spacecraft. The specific direction is the imaginary line followed by the centre of the spacecraft's mass.

flight plan The written instructions for the general control of the spacecraft and for planned events. A space shuttle flight plan is hundreds of pages long. Crew members study a flight plan, including its recent revisions and additions, during their isolation in the final days before launch, and the mission commander, in particular, reviews it during the hours before launch.

flight simulator A computer-controlled device to train pilots and astronauts by simulating the experience of flying. A space shuttle crew will undergo hundreds of 'launches' in NASA's flight simulator at the *Johnson Space Center before the real lift-off.

flight suit A suit worn by crew members within a spacecraft during flight. Space shuttle crew members on flights 1–5 wore one-piece blue flight suits, with many pockets for personal items and equipment. These were worn during launch and re-entry, and provided no protection in the event of depressurization. They were also worn in orbit.

From shuttle flight 26 onwards, astronauts wore orange pressure suits during launch and landing. However, crew members now wear casual clothes while in orbit, such as shorts, T-shirts, trousers, and sweaters. These are usually custom-made in colours and patterns specific to the mission and each includes the flight badge. *International Space Station* crews also dress casually in space.

flight system Any spacecraft-operating system. It might contain several dozen *subsystems. A single flight system may have subsystems on board the spacecraft as well as on Earth, for example a telecommunications system with transmitter and receiver subsystems in orbit and at ground stations. The two terms 'flight system' and 'flight subsystem' are used inconsistently, however, since some flight subsystems also contain systems.

flotation collar An inflated circular plastic device that prevents a NASA spacecraft from sinking after *splashdown. Several bright orange flotation collars are installed around the spacecraft by US Navy divers. The spacecraft also automatically deploys three **flotation balloons** to keep it upright.

fluid shift A shift in the distribution of human body fluids due to the effects of *microgravity. In the early stages of a space flight, an astronaut experiences a puffy face, stuffy nose, and headaches due to the shift of fluids from the lower body and the dilation of cranial veins. Body mechanisms are then activated that lead to a significant loss of water and body mass. During longer periods in space, however, the cardiovascular system adapts to the microgravity environment. When an astronaut returns to Earth, the fluid shifts back to the lower body, and this can cause *orthostatic hypotension and loss of consciousness.

flyaround A spacecraft flight that circles a planet or other astral body without going into orbit. In February 2001, the space shuttle *Atlantis* performed half of a flyaround of the **International Space Station* in order to obtain photographic and video documentation of the newly expanded facility.

flyback booster A space shuttle booster rocket in development that will use liquid fuel instead of solid fuel. In 2004, the system was being developed by NASA and the Boeing Company, in preparation for flight demonstrations. The flyback boosters will separate from the shuttle at an altitude of approximately 49.3 km, descend to an altitude of approximately 7 620 m, and fire their jet engines, to be guided by computers back to the Kennedy Space Center to land on a runway. The shuttle's present solid-fuel boosters expend their fuel about two minutes after launch. The boosters then separate from the shuttle and fall into the Atlantic Ocean from which they are recovered and refurbished for a future mission.

fly-by A spacecraft flight that explores a planet or other astral body by passing it without being captured into its orbit and without having its flight path altered by the body's gravitational pull. The craft therefore has only a limited amount of time to gather data. Fly-by spacecraft, which follow either an escape trajectory or a continuous solar orbit, must be capable of surviving long interplanetary periods. Deep-space fly-bys have included encounters by the **Voyager* probes, **Mariner* spacecraft, and **Pioneer* probe series.

fly-by-wire A description of a spacecraft's control system that uses electronic assistance to regulate its flight. The space shuttle's semi-automatic *control-stick steering is a fly-by-wire option. When the astronaut moves the *hand controller, it sends electrical impulses to move the *elevons and *rudder, unlike the older mechanical method, which consisted of a direct link from hand to aircraft flaps.

Astronauts aboard *Mercury* spacecraft also used the fly-by-wire system to reverse their capsule if they wanted to avoid orbiting backwards.

Foale, (Colin) Michael (1957–) English-born US astronaut. He has had five space missions. In 1995 he flew on the space shuttle *Discovery* mission that made the first rendezvous with the Russian space station *Mir*, and in 1997 he spent 134 days on *Mir*, helping to carry out repairs and maintenance to the mechanical systems. In December 1999 he served on the *Discovery* mission to repair and upgrade the **Hubble Space Telescope*. He was a member of the fifth resident crew on the *International Space Station*, being launched on *Soyuz* TMA-3 on 18 October 2003 for a six-month stay. Foale joined NASA in 1983 as a payload officer at the Johnson Space Center and was selected as an astronaut in 1987.

focus Either of two points lying on the major axis of an elliptical *orbit on either side of the centre. One focus marks the centre of mass of the system and the other is empty. In a circular orbit the two foci coincide at the centre of the circle and in a parabolic orbit the second focus lies at infinity. *See* *KEPLER'S LAWS.

footprint Of a satellite or spacecraft, the area of the Earth or other body that is being imaged by its instruments, or over which its signals can be received.

foot restraint Any device that secures an astronaut's foot for stability in the microgravity atmosphere. In the space shuttle, Velcro foot-loop restraints are used by placing one or both feet in the loop. They are located on the 'floor' of the aft flight deck work stations, galley, waste collection system, and mid-deck lockers. Foot restraints on the shuttle's *gunwales are also used by astronauts on extravehicular activities.

Forrester, Patrick Graham (1957–) US astronaut. A *mission specialist, he was assistant to the director of NASA's Flight Crew Operations Branch before flying on the space shuttle *Discovery* in August 2001 on the mission that delivered the third resident crew to the *International Space Station* (*ISS*). His next flight to the *ISS* was postponed in 2004 following the 2003 *Columbia* space shuttle disaster. Forrester joined NASA's Johnson Space Center as an aerospace engineer in 1993 and was selected as an astronaut in 1996. He has logged 285 hours in space, with two space walks totalling 11 hours 45 minutes.

Freedom 7 A *Mercury* spacecraft named by the US astronaut Alan *Shepard, who flew it in May 1961 on its sub-orbital flight and became the first US human in space. The flight was officially designated as MR-3, with MR-2 carrying the chimpanzee *Ham into space in January 1961.

Freedom 7 began a *Mercury* naming tradition of including the number of the *'Original Seven' astronauts. *Freedom* was also an early name for the **International Space Station*.

free fall A state in which a body is falling freely under the influence of gravity. In orbit, astronauts and spacecraft are still held by gravity and are in fact falling freely toward the Earth. However, because of their speed (orbital velocity), the amount they fall just equals the amount the Earth's surface curves away, and the effect is apparent weightlessness.

free ride A spacecraft's increased acceleration to obtain orbital speed due to the Earth's rotation. An artificial satellite going into a polar orbit has no free ride and must provide all of its own power.

frequency The number of periodic oscillations, vibrations, or waves occurring per unit of time. The SI unit of frequency is the hertz (Hz), one hertz being equivalent to one cycle per second. Frequency is related to wavelength and velocity by the equation: $f = v/\lambda$ where f is frequency, v is velocity, and λ is wavelength. Frequency is the reciprocal of the period T: $f = 1/T$.

At one end of the electromagnetic spectrum are long radio waves with a frequency in the range $10^4 - 10^5$ Hz and at the other extreme are gamma rays with a frequency in the range $10^{19} - 10^{22}$ Hz.

Frequency modulation (FM) is a method of transmitting radio signals in which the frequency of the **carrier wave** is changed and then decoded.

One kilohertz (kHz) equals 1 000 hertz; one megahertz (MHz) equals 1 000 000 hertz.

frequency and timing system (FTS) The master clock used by NASA's *deep-space network (DSN). It is essential to the DSN's other six systems, such as its *tracking system, *telemetry system, and *monitor system. Synchronization of the FTS is done in microseconds by comparing the FTS data with pulses from satellites of the *global positioning system.

Friendship 7 A *Mercury* space capsule that placed astronaut John *Glenn as the first US astronaut in orbit in February 1962. Glenn chose the name 'Friendship', adding '7' as a reference to the *'Original Seven' astronauts. *Friendship 7* followed the 1961 sub-orbital flights on *Freedom 7* (astronaut Alan *Shepard) and *Libertybell 7* (astronaut Gus *Grissom). Later capsules taking astronauts into orbit were *Aurora 7*, *Sigma 7*, and **Faith 7*.

front-end processor A small computer used to coordinate and control the communications between a large mainframe computer and its input and output devices.

FTS Abbreviation for *frequency and timing system.

fuel A substance burned to produce heat and power, particularly for rocket propulsion. The most common rocket fuel is liquid hydrogen. An oxidizer, usually *liquid oxygen, is also required in order for the rocket fuel to burn in the oxygen-free environment of space. Rocket fuel may be stored as either *liquid fuel or *solid fuel, and is ignited in the *combustion chamber.

fuel cell A cell converting chemical energy directly to electrical energy. It works on the same principle as a battery but is continually fed with fuel, usually hydrogen. Fuel cells are used on spacecraft because batteries are too heavy. Crews can also drink the water that is produced as a by-product. Fuel cells on a space shuttle can produce 12 kW of power, six times as much as those flown during the *Apollo* project.

Hydrogen is passed over an electrode (usually nickel or platinum) containing a catalyst, which splits the hydrogen into electrons and protons. The electrons pass through an external circuit while the protons pass through a polymer electrolyte membrane to another electrode, over which oxygen is passed. Water is formed at this electrode (as a by-product) in a chemical reaction involving electrons, protons, and oxygen atoms. A current is generated between the electrodes.

fuel tank A container for the propellant of a rocket or spacecraft. The space shuttle has a large main external fuel tank and two strap-on solid rocket boosters, which are reusable.

The main space shuttle tank weighs 751 tonnes and supplies the three main engines with liquid oxygen and liquid hydrogen. It is divided into two sections connected by a propellant feed, with the liquid-oxygen tank on top and the liquid-hydrogen tank below. It was originally painted white but is now dark and unpainted to reduce the weight by 270 kg.

A ruptured main fuel tank caused the *Challenger* disaster of 28 January 1986 that killed the seven crew members.

Fuglesang, Christer (1957–) Swedish astronaut with the European Space Agency (ESA). He was assigned to the Russian mission control centre from September 1995 to February 1996 as a coordinator for ESA's EuroMir 95 mission, a 180-day mission in which an ESA astronaut joined the Russian space station *Mir*. He joined NASA's Johnson Space Center in 1996 to train as a mission specialist for space shuttle flights. His first scheduled flight in 2004 was postponed because of the fatal *Columbia* flight the previous year. Fuglesang worked at the European Organization for Nuclear Research (CERN), near Geneva, Switzerland, before being selected as an ESA astronaut in 1992.

Fullerton, (Charles) Gordon (1936–) US astronaut. He was the pilot of the third flight of the space shuttle *Columbia* in March 1982, and commanded the *Challenger* flight that carried the second *Spacelab* mission in June 1985. Fullerton became a crew member of the **Manned Orbital Laboratory* programme in 1966 until its termination in 1969. He was subsequently transferred to NASA as an astronaut. After logging 380 days in space, he ended his astronaut service in 1986 and in 1988 became a research pilot at NASA's *Dryden Flight Research Center.

FUSE Acronym for **Far-Ultraviolet Spectroscopic Explorer*.

Gagarin, Yuri Alexeyevich (1934–68) A Soviet cosmonaut who on 12 April 1961 became the first human in space, aboard the spacecraft *Vostok 1* (*see* VOSTOK). He completed one orbit of the Earth, taking 108 minutes from launch to landing. He died in a plane crash while training for the *Soyuz 3* mission.

Born near Gzhatsk in the Smolensk region in the former USSR, Gagarin became a Soviet Air Force pilot and parachutist in 1957. He was named a 'Hero of the Soviet Union' and became an international spokesperson for his nation's space programme. The Russian space station *Salyut* (Russian 'Salute') was named to honour his memory.

http://www.allstar.fiu.edu/aerojava/gagarin.htm Biography of the peasant's son who became 'the Columbus of the cosmos'. It traces Gagarin's education and training as a pilot prior to describing the courage with which he piloted *Vostok 1*.

Gagarin Cosmonaut Training Centre (in full: Yuri Gagarin Cosmonaut Training Centre) A Russian centre used to train cosmonauts, located in *Star City near Moscow, Russia. It includes centrifuges, a neutral buoyancy tank, mock-ups of *International Space Station* (*ISS*) modules, and simulators representing various spacecraft. NASA astronauts routinely train at the centre for joint projects, which have included the *Apollo–Soyuz* Test Project, space shuttle flights, and the *ISS*. This includes practice with Russian equipment and experiments, as well as Russian language courses. Astronauts from many other nations, from the UK to China, have also been trained at the centre.

http://howe.iki.rssi.ru/GCTC/gctc_e.htm Tour of the Cosmonauts Training Centre, its facilities, and its museum. The site includes historical background and a discussion of the centre's role today, with lots of interesting photographs of cosmonaut training in progress.

Gaia A European Space Agency observatory to be launched in 2011 to measure some 1 billion stars and provide detailed three-dimensional pictures of their distribution and motions. The measurements are to be precise to a few millionths of a second of arc. The data will, among other results, provide information about the evolution and formation of stars in our galaxy, the distribution of dark matter, and the categorization of rapid evolutionary stellar phases. Many thousands of extra-solar planets are expected to be discovered and their orbits and masses determined. *Gaia* will carry two astrometric telescopes and a 'Spectro' telescope comprising a radial velocity spectrograph and a medium-band photometer.

galactic halo An outer, sparsely populated region of a galaxy, roughly spheroid in shape and extending far beyond the bulk of the visible stars. In our own Galaxy, the halo contains the globular clusters, and may harbour large quantities of *dark matter.

galactic plane The plane passing through the *Sun and the centre of our Galaxy defining the mid-plane of the galactic disc. Viewed from the Earth, the galactic plane is a great circle (galactic equator) marking the approximate centre line of the *Milky Way.

galaxy A grouping of millions or billions of stars, held together by gravity. It is believed that there are billions of galaxies in the *universe. There are different types, including spiral, barred spiral, and elliptical galaxies. Our own galaxy, the *Milky Way, is about 100 000 light years across (a light year is the distance light travels in a year, about 9.5 billion km), and contains at least 100 billion stars.

The galaxies are moving away from our own Galaxy in all directions. The universe is thus expanding in all directions. The evidence for this comes from examining light from the galaxies by splitting the light into a spectrum. A feature known as the redshift appears, where the light is shifted towards the red end of the spectrum due to an increase in wavelength, owing to the source's velocity.

Spiral galaxies, such as the Milky Way, are flattened in shape, with a central bulge of old stars surrounded by a disc of younger stars, arranged in spiral arms like a Catherine wheel.

Barred spirals are spiral galaxies that have a straight bar of stars across their centre, from the ends of which the spiral arms emerge. The arms of spiral galaxies contain gas and dust from which new stars are still forming.

Elliptical galaxies contain old stars and very little gas. They include the most massive galaxies known, containing a trillion stars.

The Milky Way is a member of a small cluster, the *Local Group. The Sun lies in one of its spiral arms, about 25 000 light years from the centre.

***Galaxy Evolution Explorer* (*GALEX*)** NASA's Space Ultraviolet Small Explorer launched on 28 April 2003 on a 28-month mission to map the global history of the galaxy and investigating the causes of star formation during 80% of the universe's life, when galaxies evolved and most stars were formed. *GALEX's* science team is led by scientists at the *California Institute of Technology (Caltech).

GALEX Contraction of **Galaxy Evolution Explorer*.

Galilean satellites The four largest moons of Jupiter (*Io, *Europa, *Callisto, and *Ganymede). The moons were seen through a telescope for the first time by the Italian astronomer Galileo Galilei in 1610. The largest Galilean satellite is Ganymede with a diameter (5 262 km) similar in size to that of the planet Mercury.

http://www.jpl.nasa.gov/galileo/ganymede/discovery.html Article documenting Galileo Galilei's discovery of Jupiter's four largest satellites, now known as the Galilean satellites. There are pictures (and translations) of original manuscripts as well as modern Voyager views.

***Galileo* probe** A NASA space probe to the giant planet Jupiter, launched from the space shuttle *Atlantis* on 18 October 1989. En route to Jupiter *Galileo* picked up speed by flying past Venus, in February 1990, and the Earth twice, in

December 1990 and again in December 1992. The probe flew past the asteroids Gaspra, in October 1991, and Ida, in 1993, taking close-up photographs. It also witnessed the crash of Comet Shoemaker–Levy into Jupiter in July 1994.

On arrival at Jupiter in December 1995 a small probe was released which plunged into the atmosphere of Jupiter and radioed back information during its descent for 57 minutes until it was destroyed by the atmospheric pressure. *Galileo* itself went into orbit around Jupiter, surveying its ever-changing cloud formations and moons for nearly eight years until it was commanded to burn up in the planet's atmosphere on 21 September 2003.

http://www2.jpl.nasa.gov/galileo/ Full details of the *Galileo* mission to the Solar System's largest planet, featuring images of Jupiter's turbulent atmosphere and its moons.

galley In a spacecraft, the area used for food stowage, preparation, and dining. In the space shuttle, this is a compact space located in the mid-deck. It includes an oven, injectors for hot and cold water, trays, towels, refrigerated food, drinks, and a pantry for food storage. On some flights, payloads take up the galley space and the food preparation system becomes limited to food warmers and drink dispensers.

'Galley' comes from the nautical term for the corresponding area on a ship.

gamma radiation A very high-frequency, high-energy electromagnetic radiation, similar in nature to X-rays but of shorter wavelength, emitted by the nuclei of radioactive substances during decay or by the interactions of high-energy electrons with matter. Cosmic gamma rays have been identified as coming from pulsars, radio galaxies, and quasars, although they cannot penetrate the Earth's atmosphere.

Gamma rays are stopped only by direct collision with an atom and are therefore very penetrating; they can, however, be stopped by about 4 cm of lead or by a very thick concrete shield. They are less ionizing in their effect than alpha and beta particles, but are dangerous nevertheless because they can penetrate deeply into body tissues such as bone marrow. They are not deflected by magnetic or electric fields.

gamma-ray astronomy The study of celestial objects that emit gamma rays (energetic photons with very short wavelengths). Much of the radiation detected comes from collisions between hydrogen gas and cosmic rays in our Galaxy. Some sources have been identified, including the Crab Nebula and the Vela pulsar (the most powerful gamma-ray source detected).

Gamma rays are difficult to detect and are generally studied by use of balloon-borne detectors and artificial satellites. The first gamma-ray satellites were *SAS II* (1972) and *COS-B* (1975), although gamma-ray detectors were carried on the *Apollo 15* and *Apollo 16* missions, in 1971 and 1972, respectively. *SAS II* failed after only a few months, but *COS-B* continued working until 1982, carrying out a complete survey of the galactic disc.

The **Compton Gamma-Ray Observatory* was launched in April 1991 and the **International Gamma-Ray Astrophysics Laboratory* (*Integral*) in 2002.

gamma-ray Burst An intense flash of gamma rays from space lasting from a few milliseconds to a few minutes. Such bursts were first detected in the late

1960s by US military satellites designed to monitor nuclear explosions, but more recent satellites such as the *Compton Gamma-Ray Observatory* found that they come from all directions at a rate of about one per day. Hundreds of times brighter than a typical supernova, they are the most powerful explosions in the Universe. There are two main types of burst: those of less than two seconds duration, and those that last longer than two seconds. Current theories suggest that the shorter bursts are caused by two neutron stars merging to form a black hole, whereas the longer bursts are caused by stars with masses greater than about 40 Suns exploding in an ultra-energetic event called a hypernova and collapsing into a black hole.

Gamma-Ray Large Area Space Telescope (GLAST) A planned gamma-ray observatory to make observations at high energies extending up to about 300 GeV. Led by NASA, it has participation from France, Germany, Japan, Italy, and Sweden. *GLAST* will carry two instruments, the main one being the Large Area Telescope (LAT) which will make gamma-ray images of astronomical objects. The second instrument is the *GLAST* Burst Monitor (GBM), which will look for gamma-ray bursts. The GBM has a field of view that will enable it to watch over two-thirds of the sky at a time. *GLAST* is scheduled for launch in 2007.

Gamma-Ray Observatory (GRO) *See* Compton Gamma-Ray Observatory.

gantry Another term for *umbilical tower.

Ganymede The largest moon of the planet Jupiter, orbiting every 7.2 days at a distance of 1.1 million km. It is the largest moon in the Solar System, 5 260 km in diameter (larger than the planet Mercury). Its surface is a mixture of extensively cratered and grooved terrain. Molecular oxygen was identified on Ganymede's surface in 1994. It is thought that Ganymede has a water ice crust and possibly a buried water ocean.

The space probe **Galileo* detected a magnetic field around Ganymede in 1996; this suggests it may have a molten core. *Galileo* photographed Ganymede at a distance of 7 448 km. The resulting images were 17 times clearer than those taken by Voyager 2 in 1979.

Gardner, Dale Allan (1948–) US astronaut. He made two space shuttle flights: on board *Challenger*'s third flight in 1983, which had the first night-time launch and landing (30 August–5 September), and on the second flight of *Discovery* in November 1984. Gardner was selected as an astronaut in 1978 and left NASA in 1986.

Gardner, Guy Spence (1948–) US astronaut. One of his two space shuttle flights was the *Columbia* mission in December 1990 that carried three ultraviolet telescopes and an X-ray telescope. Gardner was selected as an astronaut in 1980, worked at NASA's headquarters in Washington, DC, from 1992, and left NASA in 1995.

Garn, Jake (born Edwin Jacob Garn) (1932–) Former US astronaut, the first US senator in space. He flew as a payload specialist on the space shuttle

Discovery in April 1985. Two communications satellites were deployed during the seven-day mission. When Garn returned to the senate floor, he received a standing ovation and an embrace from Senator John *Glenn, the former astronaut.

Garneau, Marc (1949–) Canadian astronaut and executive vice-president of the Canadian Space Agency from 2001. His space shuttle flights include the *Endeavour* mission of November 2000 to assemble the **International Space Station* and deliver supplies. He logged 667 hours in space. Garneau was selected as a Canadian astronaut in 1983, and became deputy director of the Canadian astronaut programme in 1989. He was selected for astronaut training at NASA's *Johnson Space Center in 1992.

Garriott, Owen Kay (1930–) US astronaut. He was the science-pilot of **Skylab* 3, launched in July 1973, and logged more than 13 hours performing three *extravehicular activities. He was a *mission specialist aboard *Spacelab 1* in November 1983, the first European Space Agency flight. Garriott was selected as an astronaut in 1965 and left NASA in 1986.

GAS Abbreviation for *Getaway Special.

gas chromatograph/mass spectrometer (GCMS) An instrument aboard the *Viking* space probe to the planet Mars that looked for signs of organic life forms. After a robot arm scooped out samples, the automated, miniaturized GCMS could separate any organic compounds in its gas chromatograph section and then identify them in the mass spectrometer. The GCMS was sensitive enough to detect a few parts of any organic compound in one billion parts of soil. However, it (and two other related experiments) found no signs of life on the planet.

gas giant Any of the four large outer planets of the Solar System, *Jupiter, *Saturn, *Uranus, and *Neptune, which consist largely of gas and have no solid surface.

Gaspra (or Asteroid 951) The first asteroid observed in close-up by a spacecraft. The **Galileo* probe made a fly-by in October 1991 on its way to the planet Jupiter. It passed Gaspra at 8 km per second, coming within 1 600 km to take many images of the S-type asteroid, made up of a mixture of nickel, iron, and silicates.

Gaspra's irregular dimensions are approximately 19×12× 11 km. It is estimated to be between 300 million and 500 million years old. Its orbit averages 205 million km from the Sun. The asteroid was discovered in 1916 by the Russian astronomer Grigoriy Neujmin, who named it after a Black Sea holiday resort.

GCF Abbreviation for *Ground Communications Facilities.

GCMS Abbreviation for *gas chromatograph/mass spectrometer.

GDS Abbreviation for *Ground Data System.

Gemar, Sam (born Charles Donald Gemar) (1955–) US astronaut. A *mission specialist, his three space shuttle flights included the *Discovery* mission (September 1991) to deploy the **Upper Atmosphere Research Satellite* and the *Columbia* mission (March 1994) on which *microgravity experiments were conducted. Gemar was selected as an astronaut in 1985.

***Gemini* project** A US space programme (1965–6) in which astronauts practised rendezvous and docking of spacecraft, and working outside their spacecraft, in preparation for the **Apollo* project Moon landings.

Gemini spacecraft carried two astronauts and were launched by *Titan rockets. Flights were designated by the double name *Gemini-Titan*.

The meeting of *Gemini 7* with *Gemini 6* in 1965 was the first rendezvous between two spacecraft.

g

General Dynamics Corporation A US corporation whose space accomplishments include the Atlas and Centaur rockets, as well as the space shuttle's mid-fuselage. It has also contributed to the **Hubble Space Telescope*, the **International Space Station*, and the *Mars Exploration Rover. The Atlas was created by General Dynamics Corporation's Convair Division as an intercontinental ballistic missile and later modified for NASA's space programme. The Centaur rocket was developed by the Space Systems Division to launch the space shuttle and satellites.

The General Dynamics Corporation formed its Convair Division in 1954. In 1985 it separated Convair's space programmes to create the Space Systems Division. The latter was sold to the US company Martin-Marietta in 1994 and the Convair Division was disbanded two years later.

***Genesis* mission** A NASA space probe launched on 8 August 2001 to collect samples of the *solar wind. The spacecraft travelled 1.5 million km towards the Sun to the L1 *Lagrangian point of the Earth's orbit where it unfolded collectors to absorb the solar wind ions for 30 months. It then returned to Earth, but the landing parachute failed to deploy and the sample return capsule crashed into the Utah desert on 8 September 2004. However, scientists were able to retrieve some of the samples for study.

http://genesismission.jpl.nasa.gov/ Description of the mission and its results with educational resources and a look at the difficulties of examining such microscopic samples of extraterrestrial material.

GEO Abbreviation for *geosynchronous orbit.

***Geostationary Operational Environmental Satellite* (*GOES*)** Any of a series of US meteorological satellites operated by the *National Oceanic and Atmospheric Administration for the National Weather Service. *GOES-12* was launched in July 2001 and is the first of the series to carry a Solar X-ray Imager (SXI) to observe events on the Sun that could affect the Earth. The series' primary mission is to monitor dynamic weather events, such as hurricanes, and other environmental conditions on the Earth's surface and in near space. The satellites give an accurate location of a phenomenon by using an imager and sounder.

geostationary orbit (or geostationary Earth orbit) The circular path 35 900 km above the Earth's Equator on which a *satellite takes 24 hours, moving from west to east, to complete an orbit, thus appearing to hang stationary over one place on the Earth's surface. Geostationary orbits are used particularly for communications satellites and weather satellites.

geosynchronous orbit (GEO) The circular orbit of an artificial satellite that seems to remain motionless above the Earth at the same longitude on the Equator, although it may move slightly north or south (hence it is not geostationary). Communications satellites and meteorological satellites have this type of orbit and use their *thrusters to maintain their correct positions. A geosynchronous orbit has a low inclination and circles with the Earth every 23 hours, 56 minutes, and 4 seconds.

geosynchronous transfer orbit (GTO) The initial elliptical orbit of an artificial satellite destined for a circular *geosynchronous orbit or *geostationary orbit. The GTO altitude at its *apoapsis (furthest point from the Earth) is about 37 000 km. The satellite then moves to a circular orbit by rotating parallel to the Earth's Equator at apoapsis and firing its rocket engine.
Launch vehicles are often compared by the weights they can lift into GTO.

***Geotail* (*GTL*)** A Japanese satellite launched by NASA in July 1992 to observe the tail of the Earth's *magnetosphere. The instruments included two sets of plasma instruments and two sets of high-energy particle instruments. *Geotail* was the first satellite launched under the International Solar Terrestrial Physics (ISTP) programme involving the USA, Europe, Russia, and Japan. Japan's Institute of Space and Astronautical Science (ISAS) developed the spacecraft and provided about two-thirds of its scientific instruments while NASA contributed about one-third. Geotail is operated from ISAS and its data is received by both ISAS and NASA.

Gernhardt, Michael Landon (1956–) US astronaut. His space shuttle flights include two **Microgravity Science Laboratory* missions in April 1997 and July 1997, both aboard *Columbia*. In July 2001, he flew aboard *Atlantis* to install a new airlock module on the *International Space Station*, completing three space walks to do so. He has logged 43 days in space and more than 23 hours on four space walks. Gernhardt has been assigned to a future flight with an unknown date because of the 2003 *Columbia* space shuttle disaster. He worked independently on the development of new space-station tools until his selection as an astronaut in 1992.

getaway special (GAS) Any of a group of small scientific experiments carried in a canister in the *cargo bay aboard NASA's space shuttles. The experiments do not require attention from the shuttle crew and are inexpensive and self-contained, with their own electrical supplies. A GAS is normally developed by industry, schools, or individuals.

g-force The force experienced by a pilot or astronaut when the craft in which he or she is travelling accelerates or decelerates rapidly. The unit g denotes the acceleration due to gravity, where 1 g is the ordinary pull of gravity.

Early astronauts were subjected to launch and re-entry forces of up to 6 g or more; in the space shuttle, more than 3 g is experienced on lift-off. Pilots and astronauts wear *g-suits that prevent their blood pooling too much under severe g-forces, which can lead to unconsciousness.

Giacconi, Riccardo (1931–) Italian-born US physicist whose work has been fundamental in the development of *X-ray astronomy. In 1962 a rocket sent up by Giacconi and his group to observe secondary spectral emission (*see* SPECTRUM) from the Moon detected strong X-rays from a source evidently located outside the Solar System. X-ray research has since led to the discovery of many types of stellar and interstellar material. Giacconi and his team developed a telescope capable of producing X-ray images. In 1970 they launched a satellite called *Uhuru*, devoted entirely to the detection of stellar and interstellar X-rays. In 1981 he became the first director of the Space Telescope Science Institute and from 1993 to 1999 was director general of the European Southern Observatory. He was awarded the Nobel Prize for Physics in 2002 for his discovery of cosmic X-ray sources.

giant star A member of a class of stars characterized by great size and *luminosity. Giants have exhausted their supply of hydrogen fuel and derive their energy from the fusion of helium and heavier elements. They are roughly 10–300 times bigger than the Sun with 30–1 000 times the luminosity. The cooler giants are known as *red giants.

Gibson, Edward George (1936–) US astronaut. He was the science-pilot on the third flight to the **Skylab* space station, in November 1973. He spent 84 days living on the space station and performed three space walks, totalling more than 15 hours, to change film in the telescope cameras. Gibson was chosen as a scientist-astronaut in 1965 and left NASA in 1982.

Gibson, Robert Lee (1946–) US astronaut. His five space shuttle flights include commanding the *Atlantis* in June 1995 for the first shuttle docking with the Russian space station *Mir*. He served as chief of NASA's Astronaut Office 1992–4 and briefly as deputy director of NASA's Flight Crew Operations Branch in 1996. Gibson was selected as an astronaut in 1978 and left NASA in 1996 to pursue private business interests.

Gidzenko, Yuri Pavlovich (1962–) Russian cosmonaut. He was commander of the 180-day EuroMir 95 mission, on which European Space Agency astronauts came aboard the Russian space station *Mir* from September 1995 to February 1996. In 2000 he became one of the first resident crew members aboard the *International Space Station*, transported in a *Soyuz* spacecraft launched on 31 October 2000, and returning aboard the space shuttle *Discovery* on 18 March 2001. He was selected as a cosmonaut in 1987.

gimbal A device that holds a rocket engine in place and allows it to swivel. When the rocket and its nozzles tilt in this manner to control direction, they are said to be 'gimballed'. Before a space shuttle is launched, its final test is to gimbal the main engines.

***Ginga* (Japanese: 'galaxy')** A satellite owned by Japan's Institute of Space and Astronautical Science that investigated gamma rays and other space emissions. The craft was launched in February 1987 at the *Kagoshima Space Centre. Before terminating its mission in November 1991, *Ginga* made a number of discoveries, including several evolving black holes.

Giotto A space probe built by the European Space Agency to study *Halley's Comet. Launched by an Ariane rocket in July 1985, *Giotto* passed within 600 km of the comet's nucleus on 13 March 1986. On 2 July 1990, it flew within 23 000 km of the Earth, which diverted its path to encounter another comet, Grigg–Skjellerup, on 10 July 1992.

GLAST Acronym for **Gamma-Ray Large Area Space Telescope*.

Glavkosmos The first partly-commercial Soviet (later Russian) space company, established in 1985 as the marketing agent for launch services. It was supported by the government's export agency, Licensintorg, which provided legal and financial support. Glavkosmos was responsible for the first Soviet commercial launch contract, to use a Proton booster to launch an *International Maritime Satellite Organization (Inmarsat) communications satellite, and was also involved in arranging the first commercial flights to the **Mir* space station by guest cosmonauts from other countries.

Glenn, John Herschel, Jr (1921–) US astronaut and politician. On 20 February 1962, he became the first US astronaut to orbit the Earth, doing so three times in the spacecraft **Friendship 7*. The flight lasted 4 hours 55 minutes. On 29 October 1998, Glenn became the oldest person in space when, at the age of 77, he embarked on a nine-day mission aboard the shuttle *Discovery*.

Glenn was elected to the US Senate as a Democrat from Ohio in 1974; and re-elected in 1980 and 1986.

Born in Cambridge, Ohio, Glenn became a US Marine Corps pilot, flying combat missions in World War II and the Korean War (1950–3). He joined NASA in 1959 and retired in 1964, resigning from the Corps as a colonel in 1965.

Glenn unsuccessfully sought the Democratic presidential nomination in 1984. He left the Senate in 1999, the year NASA renamed its Lewis Research Center in Cleveland, Ohio, as the John H Glenn Research Center.

Glenn Research Center (in full: NASA John H Glenn Research Center at Lewis Field; formerly Lewis Research Centre (1958–99)) NASA facility located in Cleveland, Ohio. It is NASA's leading centre for aeropropulsion and turbomachinery, developing new propulsion, power, and communications technologies for space missions, including the **International Space Station*. It conducts *microgravity research into fluid physics, combustion science, and some *material science, managing many experiments aboard space shuttles and space stations. It also conducts research to develop and transfer technology to US industry.

The centre was established in 1941 as the Aircraft Engine Research Laboratory by the *National Advisory Committee for Aeronautics (NACA). It was later renamed the Flight Propulsion Research Laboratory in 1947, the Lewis Flight Propulsion Laboratory in 1948 (after NACA director George W Lewis), and the NASA Lewis Research Center in 1958; it was named in honour of astronaut John *Glenn in 1999. In 2001, it had more than 3 600 employees, over half of them being scientists and engineers, and more than 150 buildings, as well as additional facilities near Sandusky, Ohio.

http://www.lerc.nasa.gov/ Tutorials and technical reports on jet propulsion and space flight engineering: the focus of the site is the research and engineering capabilities of the centre. There is a well-stocked image gallery with pictures of Glenn, its projects, and its personnel. A clickable aerial view of the facility at Lewis Field allows visitors to take a virtual tour and to find out what goes on in many of the buildings.

glideslope The glide path followed by a space shuttle as it descends in preparation for landing. The path has an angle of 22° to the ground. A shuttle at this point has become an aircraft that depends on its wings and rudder to glide to the runway.

global ozone measurement experiment (GOME) An ozone-monitoring experiment conducted aboard both of the **European Remote-Sensing Satellites*, launched in 1991 and 1995. GOME was the first European ozone-monitoring instrument, measuring the level of ozone in the Earth's atmosphere, investigating ozone and other gases involved in ozone photochemistry, and providing data on the causes of ozone depletion. It was succeeded by the *scanning image absorption spectrometer for atmospheric cartography on the **Envisat*.

global ozone monitoring by occultation of stars (GOMOS) An instrument used by the European Space Agency (ESA) to monitor and measure ozone in the Earth's atmosphere. GOMOS will be carried aboard ESA's **Envisat*, and programmed to take more than 600 profile measurements each day. This is the first time that star *occultation will be used to measure ozone. The data collected by GOMOS are used for global ozone mapping and for monitoring trends in ozone depletion.

global positioning system (GPS) A US satellite-based navigation system, a network of 24 satellites in six orbits, each circling the Earth once every 24 hours. Each satellite sends out a continuous time signal, plus an identifying signal. To fix position, a user needs to be within range of four satellites, one to provide a reference signal and three to provide directional bearings. The user's receiver can then calculate the position from the difference in time between receiving the signals from each satellite.

The position of the receiver can be calculated to within 0.5 m, although only the US military can tap the full potential of the system. Other users can obtain a position to within 100 m. This is accurate enough to be of use to boats, walkers, and motorists, and suitable receivers are on the market.

http://www.navcen.uscg.gov/faq/gpsfaq.htm Good description of the navigational aid known as the global positioning system. There is information on the history of the navigation system, how it works, where to learn more about it, and the acronyms used to describe it.

GLOW Acronym for *gross lift-off weight.

g-meter A spacecraft meter that indicates the *g-force. When the space shuttle re-enters the atmosphere, the first indication is the needle of the g-meter rising from zero. It records a maximum g-force of 1.7 g before the shuttle lands.

go NASA term indicating that a spacecraft action is approved. The mission control centre, for example, will inform a space shuttle crew, 'You are go for deorbit.' The first crew to land on the Moon in 1969 received such messages as '*Apollo 11*, this is Houston. You are go for TLI' (*translunar injection). When a computer's alarm light proved to be a false alert during the *Apollo 11* mission, the message was 'We're go on that alarm.'

GOCE Abbreviation for **Gravity-Field and Steady-State Ocean Circulation Explorer*.

Goddard, Robert Hutchings (1882–1945) US rocket pioneer. He launched the first liquid-fuelled rocket at Auburn, Massachusetts, in 1926. By 1932 his rockets had gyroscopic control and could carry cameras to record instrument readings. Two years later a Goddard rocket achieved the world altitude record with an ascent of 3 km.

Goddard developed the principle of combining liquid fuels in a rocket motor, the technique used subsequently in every practical space vehicle. He was the first to prove by actual test that a rocket will work in a vacuum and he was the first to fire a rocket faster than the speed of sound.

Goddard was born in Worcester, Massachusetts. At Clark University in his home town, and at Mount Wilson, California, he carried out experiments with naval signal rockets, and went on to design and build his own rocket motors. On the USA's entry into World War I, he turned his energies to investigating the military application of rockets. In 1929, instruments, and a camera to record them, were carried aloft for the first time.

The military rockets Goddard developed in World War II were more advanced than the German V2, although smaller. A few days before the end of the war he demonstrated a rocket fired from a launching tube.

Goddard Space Flight Center A NASA field centre at Greenbelt, Maryland, responsible for the development and operation of uncrewed scientific satellites, including the **Hubble Space Telescope* and the *Geostationary Operational Environmental Satellite (GOES). Goddard is also the home of the *National Space Science Data Center, a repository of data collected by satellites.

http://www.gsfc.nasa.gov/ Packed with news and information on space missions and science projects at Goddard, as well as a biography of US rocket pioneer Robert Goddard. Press releases dating back to 1995 and fact sheets are available in the Flight Center press room. The site features a good collection of recently acquired images of Earth from space, with captions and links.

Goddard Trophy (in full: Robert H Goddard Memorial Trophy) A US trophy awarded annually by the *National Space Club for significant contributions in the fields of rocketry and astronautics. Established in 1958, the trophy is named after the rocket pioneer Robert *Goddard. It is presented at the annual Goddard memorial dinner in Washington, DC. Recipients include the former US presidents Lyndon B Johnson and Ronald Reagan.

Godwin, Linda Maxine (1952–) US astronaut. Her four space shuttle flights include the *Atlantis* mission in March 1996 to dock with the Russian space station *Mir*. During the mission she performed a six-hour space walk, the first to be undertaken while a shuttle docked with an orbiting space station. She flew on the *Endeavour* in December 2001 to exchange resident crews on the **International Space Station* and to make a space walk to wrap thermal blankets around solar array gimbals. She has spent 38 days in space with 10 hours on two space walks. She went on to join the CapCom Branch of the Astronaut Office at the Johnson Space Center. Godwin joined NASA's Payload Operations Division in 1980 and was selected as an astronaut in 1985.

GOES Acronym for *Geostationary Operational Environmental Satellite.

going uphill NASA slang for leaving the Earth's atmosphere or entering space.

Goldin, Daniel Saul (1940–) A NASA administrator from 1992 until his resignation in 2001, who had the longest tenure in NASA history. His style of leadership was based on the mantra 'faster, better, cheaper'.

Goldin was with TRW Space and Technology Group in Redondo Beach, California, for 25 years (as vice-president and general manager) before joining NASA. In 1962 he worked at NASA's Lewis Research Center, Cleveland, Ohio, on electric propulsion systems for human interplanetary travel.

In 1999, an independent review panel concluded that Goldin had over-curtailed budgets at NASA to the detriment of mission successes. By the time of his resignation NASA was experiencing a budgetary crisis.

Goldstone One of a series of antenna that make up the NASA *deep-space network (DSN), serving as the main telecommunications link with interplanetary spacecraft. Located in the Mojave Desert, California, and measuring 34 m in diameter, Goldstone is one of the three main antennae of the DSN (the others are located in Canberra, Australia, and Madrid, Spain). The antennae are located at regular intervals about the Earth, allowing the tracking of spacecraft 24 hours a day. Goldstone is managed by the *Jet Propulsion Laboratory.

GOME Acronym for *global ozone measurement experiment.

GOMOS Acronym for *global ozone monitoring by occultation of stars.

Goonhilly A satellite-tracking station in Cornwall, England, owned by the telecommunications company BT (formerly British Telecom). Established in 1962, it is the largest satellite station in the world and is equipped with 25 dishes and a communications satellite transmitter–receiver in permanent contact with most parts of the world.

Gordon, Richard F(rancis), Jr (1929–) US astronaut. In September 1966, he piloted *Gemini 11*, which docked with an Agena rocket. Gordon tethered the rocket to the craft in the course of his 44-minute space walk. On the *Apollo 12* mission of November 1969, the second crewed lunar landing, Gordon piloted the command module, remaining in lunar orbit. He was a US Navy pilot before being selected as an astronaut in 1963, and retired from NASA in 1972.

Gordon was NASA's chief of advanced programmes 1971–2, working on the design of the space shuttle.

Born in Seattle, Washington, Gordon was awarded a Bachelor of Science (BSc) degree by the University of Washington in 1951.

Gorie, Dominic Lee Pudwill (1957–) US astronaut and chief of the Astronaut Safety Branch. He piloted the space shuttle *Discovery*'s final docking mission with the Russian space station *Mir* in June 1998. He was pilot of the radar topography mission undertaken by *Endeavour* in February 2000, which mapped more than 75 million sq km of the Earth's surface. He flew *Endeavour* in December 2001 to the **International Space Station* to exchange the resident crews. Gorie was selected as an astronaut in 1994.

GPS Abbreviation for *global positioning system.

Grabe, Ronald John (1945–) US astronaut. His missions included piloting the first flight of the space shuttle *Atlantis* in October 1985, and its flight of May 1989 to deploy the **Magellan* spacecraft. Grabe was selected as an astronaut in 1981 and left NASA in 1994.

GRACE Acronym for *Gravity Recovery and Climate Experiment.

grain A piece of solid propellant in a rocket. The entire solid propellant is also referred to as the grain. It is shaped evenly around the inside of the rocket to control the thrust as the grain burns.

Granat A Russian satellite, developed in collaboration with other European nations, that carried out X-ray and gamma-ray observations. Launched in December 1989, it made specific investigations until September 1994, when it began a general survey. *Granat* produced very deep images of the centre of the galaxy before ending its life in November 1998.

grappling pin A protruding pin on a spacecraft that is normally used to retrieve it in space. The *Canadarm, the remote-controlled arm on the space shuttle and *International Space Station* (*ISS*), is used for this task. The wires of the Canadarm's *end effector at the end of the arm serve as a grappling snare, closing over the grappling pin to capture a payload in space. The Canadarm is also stowed in the *ISS Destiny* module by attaching each end to a grappling pin.

grappling snare The wires of the *end effector of the *Canadarm payload deployment and recovery system. It is used to describe the function of the end effector when it is attached to a *grappling pin.

gravitational field A region around a body in which other bodies experience a force due to its gravitational attraction. The gravitational field of a massive object such as the Earth is very strong and easily recognized as the force of gravity, whereas that of an object of much smaller mass is very weak and difficult to detect. Gravitational fields produce only attractive forces.

gravitational field strength (symbol *g*) The strength of the Earth's gravitational field at a particular point. It is defined as the gravitational force in newtons that acts on a mass of one kilogram. The value of g on the Earth's surface is taken to be 9.806 N kg^{-1}.

The symbol g is also used to represent the acceleration of a freely falling object in the Earth's gravitational field. Near the Earth's surface and in the absence of friction due to the air, all objects fall with an acceleration of 9.8 m s^{-2}.

gravitational force (or gravity) One of the four fundamental *forces of nature, the other three being the electromagnetic force, the weak nuclear force, and the strong nuclear force. The gravitational force is the weakest of the four forces, but acts over great distances. The particle that is postulated as the carrier of the gravitational force is the graviton.

gravitational lensing The bending of light by a gravitational field, predicted by German-born US physicist Albert Einstein's general theory of relativity. The effect was first detected in 1917, when the light from stars was found to bend as it passed the totally eclipsed Sun. More remarkable is the splitting of light from distant quasars into two or more images by intervening galaxies. In 1979 the first double image of a quasar produced by gravitational lensing was discovered and a quadruple image of another quasar was later found.

gravitational potential energy The energy possessed by an object when it is placed in a position from which, if it were free to do so, it would fall under the influence of gravity. If the object is free to fall, then the gravitational potential energy is converted into kinetic (motion) energy. The gravitational potential energy E_p of an object of mass m kilograms placed at a height h metres above the ground is given by the formula:

$$E_p = mgh$$

where g is the gravitational field strength (in newtons per kilogram) of the Earth at that place.

In a hydroelectric power station, gravitational potential energy of water held in a high-level reservoir is used to drive turbines to produce electricity.

gravitational waves The hypothetical waves carrying gravitational energy, predicted by German-born US physicist Albert *Einstein's theory of general relativity to be emitted from an accelerating massive body, but never discovered. Spacecraft can conduct radio science experiments that search for gravitational waves. These might be found by measuring over a length of time the tiny Doppler shifts of a spin-stabilized or reaction-wheel-stabilized spacecraft travelling through interplanetary space. If a gravitational wave were to go through the solar system, the spacecraft's distance would be observed to increase and then decrease by millimetres.

gravity (or gravitational force) A force of attraction that arises between objects by virtue of their masses. The larger the mass of an object the more strongly it attracts other objects. On Earth, gravity causes objects to have weight; it accelerates objects (at 9.8 m s^{-2}) towards the centre of the Earth, the ground preventing them falling further.

The Earth's gravity also attracts the Moon towards the Earth, keeping the Moon in orbit around the Earth. The Moon's gravity is one-sixth that of the Earth, so objects on the Moon weigh less than on Earth. The Sun contains 99.8% of the mass of the Solar System, and the resulting large force of gravity keeps the planets of the Solar System in orbit around the Sun.

Spacecraft launched from Earth must overcome the force of gravity before entering space. This is achieved by using rocket boosters at various stages of the launch. The spacecraft needs an acceleration of three times that of gravity (3 g).

escape velocity

This is the velocity that a projectile or spacecraft needs to reach in order to escape from a gravitational field. The escape velocity from the surface of the Earth is about 40 000 kph; that from the Moon (with one-sixth the gravitational pull of the Earth) is about 8 500 kph.

relativity

German-born US physicist Albert *Einstein's general theory of *relativity treats gravitation not as a force but as the curvature of space-time around a body. Relativity predicts the bending of light and the *redshift of light in a gravitational field; both have been observed. Another prediction of relativity is that **gravitational waves** should be produced when massive bodies are violently disturbed. These waves are so weak that they have not yet been detected with certainty, although observations of a *pulsar (which emits energy at regular intervals) in orbit around another star have shown that the stars are spiralling together at the rate that would be expected if they were losing energy in the form of gravitational waves.

***Gravity-Field and Steady-State Ocean Circulation Explorer* (*GOCE*)** A European Space Agency (ESA) satellite scheduled for launch in 2006. It is expected to record the most accurate measurement ever of the Earth's gravity and its geoid (hypothetical shape of the Earth as an average sea level that goes over and under the land). *GOCE*, which will orbit the Earth for two years, is part of the ESA's Living Planet Programme.

gravity-field survey The spacecraft measurements of a planet's concentrations of mass. Planets are not perfectly round, having variations in mass, as in a mountain range. A spacecraft uses its radio and ground stations to measure the Doppler shifts in its orbit. After removing factors in the shifts due to planetary movement, solar wind, and other phenomena, the remaining *differenced Doppler measurement indicates the variations in mass distribution at and below the planet's surface. NASA's *Jet Propulsion

Laboratory pioneered the technique, which aids the accurate navigation of spacecraft and helps geologists locate the Earth's petroleum and mineral deposits.

Space probes have also carried out gravity-field surveys, including the *_Mars Global Surveyor_ mission and the _Magellan_ spacecraft survey of the planet Venus.

Gravity Probe B A satellite launched by NASA on 20 April 2004 to test German-born US physicist Albert Einstein's general theory of relativity. Developed by physicists and engineers at NASA and Stanford University, California, the probe will measure minute changes in the direction of spin of four gyroscopes carried on board. This will indicate how the Earth warps space and time, and how its rotation pulls space-time around with it. The satellite's polar orbit is at an altitude of 644 km.

http://einstein.stanford.edu/ Development and progress of the _Gravity Probe B_ mission, with information about the spacecraft design, payload, and scientific goals. There is also a good discussion of the theory of relativity.

Gravity Recovery and Climate Experiment (GRACE) A joint US–German project to measure the Earth's gravitational field very precisely using two satellites in a polar orbit and flying approximately 220 kilometres apart. Slight variations in the strength of the Earth's gravitational field are detected by measuring changes as small as one micrometre in distance between the two satellites as they orbit Earth. Such changes are caused not just by variations of density within the Earth but also by movements of water and ice on the surface of the Earth. The twin _GRACE_ satellites were launched on 17 March 2002 for an intended 5-year mission.

grazing-incidence telescope A telescope used on satellites for observing extreme ultraviolet and X-ray wavelengths which are so energetic that they penetrate normal mirrors instead of being reflected. To overcome this, the mirrors of a grazing-incidence telescope are aligned nearly parallel to the path of the incoming photons, which then glance off them at shallow angles.

Great Observatories Four large astronomy satellites, operating at wavelengths from the infrared via visible light to X-rays and gamma rays. In order of launch they are the *_Hubble Space Telescope_, the *_Compton Gamma Ray Observatory_, the *_Chandra X-ray Observatory_, and the *_Spitzer Space Telescope_.

Great Red Spot A prominent oval feature, 14 000 km wide and some 30 000 km long, in the atmosphere of the planet *Jupiter, south of its equator. It was first observed in 1664. Space probes show it to be an anticlockwise vortex of cold clouds, coloured possibly by phosphorus.

Gregory, Frederick Drew (1941–) US astronaut. His three space shuttle missions included piloting _Challenger_ in April 1985 on a *_Spacelab_ flight. Gregory joined NASA's Langley Research Center as a research test pilot in 1975, was selected as an astronaut in 1978, and joined NASA's headquarters in Washington, DC, in 1992 as associate administrator of the Office of Safety and Mission Quality. In 2002 he became NASA's deputy administrator in Washington, DC.

Gregory, William George (1957–) US astronaut. He piloted the *Endeavour* space shuttle mission in March 1995 that established a duration record of more than 16 days in space. Gregory was selected as an astronaut in 1990. He left NASA and retired from the US Air Force in 1999, and went on to be manager of business development for Honeywell.

Griggs, (Stanley) David (1939–89) US astronaut. During the *Discovery* space shuttle flight of April 1985, he conducted the first unscheduled extravehicular activity in preparation for a satellite rescue attempt. Griggs joined the Johnson Space Center in 1970 and was selected as an astronaut in 1978. He was killed in the crash of a vintage aircraft in June 1989.

Grissom, Gus (born Virgil Ivan Grissom) (1926–67) US astronaut. One of the *'Original Seven' astronauts selected in 1959, Grissom flew the second *Mercury* sub-orbital mission in July 1961, and commanded the first crewed *Gemini* mission, *Gemini 3*, in March 1965. He died in the *Apollo 1* spacecraft fire in January 1967.

Enhanced view of Jupiter's Great Red Spot region. Below the Great Red Spot are two more cyclonic storms. The *Voyager* image was among the first to reveal the incredibly turbulent flow within Jupiter's zonal bands, which from afar look stable and placid.

GRO Abbreviation for **Compton Gamma-Ray Observatory*.

gross lift-off weight (GLOW) The overall weight of a spacecraft at *lift-off, including the main rocket, boosters, propellant, and payload. The gross lift-off weight for the space shuttle *Columbia* is 2 025 000 kg.

Ground Communications Facilities (GCF) NASA's overall communications network. It uses voice and data communications to connect the central communications terminal (CCT) of the Network Operations Control Center (NOCC) at the *Jet Propulsion Laboratory with other facilities, including the three *deep-space communications complexes (DSCCs).

The communications links are by satellites, land lines, submarine cables, and terrestrial microwave. The GCF circuits carry information critical to space missions, such as command, tracking, and monitor and control data. They also carry telemetry from shuttle, satellite, and lunar missions, and data concerning radio and radar astronomy observations.

ground control The control of a spacecraft from a station on Earth. This can be done by *uplinking data to on-board computers.

ground crew A team on the ground who give round-the-clock support to a space mission. Each member of a ground crew has a specific area of responsibility and duties. A space shuttle mission, for example, has separate ground crews for each of three phases of the flight. The launch support team of engineers and scientists at the Kennedy Space Center work with the Johnson Space Center's *mission control centre to prepare the shuttle for its launch and to monitor its ascent to orbit. The three orbit support teams, each having about 50 engineers, scientists, and technicians, are in Building 30 at the Johnson Space Center and monitor the many systems and activities on board an orbiting spacecraft. The landing support team of technicians then receives the shuttle and its crew after landing.

ground data system (GDS) The former name for the *deep-space mission system.

ground support The *ground crews and equipment that support a space mission. The ground support equipment (GSE) includes tools and devices used to test, inspect, repair, assemble, disassemble, and transport a space vehicle or rocket. These are used during the research and development phase, the operational phase, and to support the guidance system. GSE does not include buildings or areas.

ground track The path of a spacecraft over an area of the Earth. It is plotted on a map in NASA's *mission control centre, and the Flight Dynamics Office releases spacecraft vector postings to the general public, who may then plot the progress of space shuttles and satellites.

Grumman Aerospace A US company that built the *Apollo* *lunar module (LM) and the wings for the space shuttle.

Grumman's original LM design in 1962 weighed 11 250 kg, which was increased by additional equipment to 14 850 kg for the Moon landing in 1969.

This work was done under the company's original name of the Grumman Aircraft Engineering Corporation.

Grumman Aerospace of Bethpage, Long Island, New York, built instruments for NASA, such as the Energetic Gamma-Ray Experiment Telescope (EGRET). The company later merged with Northrop to become the Northrop Grumman Corporation. Its Logicon Division provides the base operations for the Kennedy Space Center, including engineering, logistics, and project management.

Grunsfeld, John Mace (1958–) US astronaut. His four space shuttle flights include docking with the Russian space station *Mir* aboard *Atlantis* in January 1997, flying on the *Discovery* mission in December 1999 to upgrade the **Hubble Space Telescope*, and returning on March 2002 on *Columbia* to *Hubble*, taking three space walks to install a new camera and upgrade other equipment. He has logged 45 days in space, with five space walks totalling more than 37 hours. He went on to be chief scientist at NASA's headquarters in Washington, DC. Grunsfeld was a senior research fellow at the California Institute of Technology, involved in X-ray and gamma-ray astronomy, before his selection as an astronaut in 1992.

g-suit The clothing worn by a pilot or astronaut that exerts pressure on the abdomen and lower parts of the body to prevent or retard the collection of blood below the chest during high acceleration.

GTL Abbreviation for **Geotail.*

GTO Abbreviation for *geosynchronous transfer orbit.

guest cosmonaut Russian (formerly Soviet) name for a national of another country who joins a cosmonaut mission. The first guest cosmonaut was the Czechoslovakian Vladimir *Remek, who flew with the *Soyuz 28* mission that docked with the *Salyut 6* space station in March 1978.

Guianese Space Centre English name for the *Centre Spatial Guyanais.

guidance system An internal computer system that regulates the flight control of a spacecraft or rocket. It senses the motions of the vehicle on its three axes and calculates the changes needed.

The *Apollo 11* lunar module had both a primary guidance and navigation system and an abort guidance system as a back-up for an emergency return to lunar orbit.

The space shuttle's guidance system, during launch, ascent, re-entry, and landing, has four computers operating at the same time with a fifth as a back-up. Each computer can perform 325 000 operations a second.

Guidoni, Umberto (1954–) Italian astronaut. He was on board NASA's *Columbia* space shuttle flight in February 1996 to test the *tethered satellite system (the Italian satellite was lost when the tether broke). He also flew on the *Endeavour* mission in April 2001 to deliver the robotic arm, *Canadarm 2, to the *International Space Station*. He has logged 681 hours in space.

Guidoni was involved in the development of the tethered satellite system when selected as an astronaut by the Agenzia Spaziale Italiana (Italian Space

Agency) in 1989. He became a European Space Agency astronaut in 1998 and began training at the Johnson Space Center in 1996.

gunwale On the space shuttle, the upper edge of the payload bay, especially when it is opened to space. Astronauts can use foot restraints positioned there when working on extravehicular activites. The term 'gunwale' is derived from the older usage for the upper edge of a ship or boat.

Gutierrez, Sidney McNeill (1951–) US astronaut. He piloted the *Columbia* space shuttle flight in June 1991 that carried the **Spacelab* Life Sciences mission. In April 1994, he commanded the *Endeavour* flight that carried the Space Radar Laboratory, which took images of 20% of the Earth. Gutierrez was selected as an astronaut in 1984 and retired from NASA in 1994.

gyroscope A mechanical instrument, used as a stabilizing device and consisting, in its simplest form, of a heavy wheel mounted on an axis fixed in a ring that can be rotated about another axis, which is also fixed in a ring capable of rotation about a third axis. Applications of the gyroscope principle include the gyrocompass, the gyropilot for automatic steering, and gyro-directed torpedoes.

The components of the gyroscope are arranged so that the three axes of rotation in any position pass through the wheel's centre of gravity. The wheel is thus capable of rotation about three mutually perpendicular axes, and its axis may take up any direction. If the axis of the spinning wheel is displaced, a restoring movement develops, returning it to its initial direction.

gyrostabilized binoculars One of the sighting aids used by astronauts in space. The 14×40 binoculars have a gyrostabilized system that helps a crew member locate and retain the target, as the image remains stable and clear in spite of hand tremor or other vibrations. NASA's gyrostabilized binoculars are electrically powered by six alkaline AA batteries and operate for up to three hours on one battery pack.

Space shuttle astronauts have access to gyrostabilized and two other types of binoculars: the small 10×40 version with high magnification and wide field of view, and 7×14 binoculars with high magnification and a close focal distance.

Hadfield, Chris Austin (1959–) Canadian astronaut. He is the chief astronaut for the Canadian Space Agency. In November 1995, he was the first Canadian to operate the robotic *Canadarm while in orbit, during his first space shuttle flight on *Atlantis*. He also flew with the *Endeavour* mission in April 2001 to deliver Canadarm 2 to the *International Space Station*. During the same mission he performed two space walks to unfold the device. From 2001 to 2003, he was NASA's director of operations at the Gagarin Cosmonaut Training Centre, and then trained and qualified as a cosmonaut to fly as a flight engineer on future *Soyuz* missions. He was selected as a Canadian astronaut in 1992 and began training at NASA's Johnson Space Center the same year.

***Hagoromo* (Japanese: 'robe of heaven')** A satellite put into orbit around the Moon by Japan. The launch marked Japan as the third nation to put a satellite into lunar orbit. The 11.7-kg satellite was launched in March 1990 from the *Hiten satellite, but *Hagoromo*'s transmitter, used by ground stations to track it, failed almost immediately and contact was never recovered.

Haise, Fred Wallace, Jr (1933–) US astronaut. He was aboard the *Apollo 13* flight to the Moon that had to abort its mission when an oxygen tank exploded. In 1977 he commanded the space shuttle *Enterprise* during test flights on the back of a Boeing 747 aircraft. Haise was assigned to NASA as a research pilot in 1962 before being selected as an astronaut in 1966. He left NASA in 1979.

***Hakucho* (Japanese: 'swan')** Japan's first X-ray astronomy satellite. Launched in February 1979, its original name, *Corsa-B*, was changed to *Hakucho* as a symbol for Cygnus X-1, an intense source of X-rays from the direction of the constellation Cygnus (the Swan). The satellite, whose mission ended in April 1985, was specifically designed to monitor and investigate transient X-ray bursts.

Halley's Comet A comet that orbits the Sun roughly every 75 years, named after English astronomer Edmond Halley, who calculated its orbit. It is the brightest and most conspicuous of the periodic comets, and recorded sightings go back over 2 000 years. The comet travels around the Sun in the opposite direction to the planets. Its orbit is inclined at almost 20° to the main plane of the Solar System and ranges between the orbits of Venus and Neptune. It will next reappear in 2061.

The comet was studied by space probes at its last appearance in 1986. The European probe *Giotto* showed that the nucleus of Halley's Comet is a tiny and irregularly shaped chunk of ice, measuring some 15 km long by 8 km wide, coated in a layer of very dark material, thought to be composed of carbon-rich compounds. This surface coating has a very low *albedo, reflecting just 4% of the light it receives from the Sun. Although the comet is one of the darkest objects known, it has a glowing head and tail produced by jets of gas from fissures in the outer dust layer. These vents cover 10% of the total surface area and become active only when exposed to the Sun. The force of these jets affects the speed of the comet's travel in its orbit.

http://www.solarviews.com/eng/halley.htm Attractive site devoted to the comet, with facts and statistics, images, and information about the spacecraft that have monitored it.

Halsell, James Donald, Jr (1956–) US astronaut. He has flown on five space shuttle missions since 1994, piloting two and commanding three, and has logged more than 1 250 hours in space. He was NASA's director of operations at the Gagarin Cosmonaut Training Centre in 1998. In 2000, he became manager of Shuttle Launch Integration at NASA's Kennedy Space Center and in 2003 became leader of the Space Shuttle Return to Flight Planning Team. He was assigned to a future flight postponed due to the 2003 *Columbia* space shuttle disaster. Halsell was selected as an astronaut in 1990.

Ham A chimpanzee launched into space by NASA in January 1961. The sub-orbital flight, launched with a Redstone rocket, tested the *Mercury* spacecraft. Problems on the flight included a leaking valve causing a severe drop in the capsule's pressure, and the launch escape system firing and inadvertently carrying the rocket to a much higher altitude than planned. Ham splashed down 209 km off target and survived in good health after being weightless in space for nearly seven minutes. He performed well, pulling levers for banana-pellet rewards and receiving electric shocks on the soles of his feet if he failed.

http://www.solarviews.com/eng/history.htm A history of space exploration, incorporating lists of historical publications, chronologies, and summaries of past, current, and future missions.

Hammond, (Lloyd) Blaine, Jr (1952–) US astronaut. His two space shuttle flights include the *Discovery* mission in September 1994 that used lasers for the first time in environmental research. He was later chief of the Safety Branch of NASA's Astronaut Office before leaving NASA. Hammond was selected as an astronaut in 1984.

hand controller A manual control on a space shuttle that rotates like an aircraft's control stick. The pilot uses it for attitude manoeuvres. Movements of the hand controller are instantly translated by the on-board computers into commands to the vehicle's thrusters.

handhold A device on a spacecraft to aid mobility in microgravity. Handholds are located throughout the crew compartment of the space shuttle, such as those to help crew members into and out of their seats.

During launch and re-entry, adjustable mirrors are attached to the handholds between two windows by the commander and two by the pilot. This allows them to see controls in areas that are obscured during these critical times. The mirrors can be removed while in orbit.

Harbaugh, Gregory Jordan (1956–) US astronaut. His four space shuttle missions include the *Atlantis* flight in June 1995, the first mission to dock with the Russian space station *Mir*, and the *Discovery* flight in February 1997 to service the **Hubble Space Telescope*. He logged 818 hours in space and more than 18 hours of extravehicular activity. In 1997, he became manager of the Extravehicular Activity Project Office, and in 2001 he left NASA. Harbaugh was selected as an astronaut in 1986.

hard suit An early hard-shell spacesuit using metals. Prototype hard suits intended for extravehicular activities were made of aluminium and rubber and first produced for the US Air Force in 1955 by Litton Systems. Their RX series was joined in 1966 by the Ames Research Center's AX series of hard suits, which included the segmented Michelin Man design.

hard upper torso The upper part of a spacesuit that is hardened to support the helmet, arms, and lower torso. The hard upper torso of a suit used for extravehicular activities is also a mounting interface for the *primary life-support system, the *display and control module, and the electrical harness for communications.

Harris, Bernard Anthony, Jr (1956–) US astronaut and physician. His two space shuttle flights include the *Discovery* mission of February 1995, as payload commander, when he became the first African American to undertake a space walk. He logged more than 438 hours in space before leaving NASA in 1996 to become chief scientist and vice-president of Science and Health Services. He joined the Johnson Space Center in 1987 as a clinical scientist and flight surgeon, and was selected as an astronaut in 1990.

Hart, Terry Jonathan (1946–) US astronaut. He flew aboard the space shuttle *Challenger* on its April 1984 mission to deploy the *Long-Duration Exposure Facility. Hart was selected as an astronaut in 1978 and left NASA in 1984, when he became the director of Engineering and Operations for the satellite network of the AT&T Corporation.

Hartsfield, Henry Warren, Jr (1933–) US astronaut. His three space shuttle flights included *Discovery*'s first mission, launched in August 1984 to deploy three satellites. In 1993 he became manager of the *International Space Station* Independent Assessment and in 1996 director of NASA's Human Exploration and Development of Space Enterprise Independent Assurance. Hartsfield was assigned to the **Manned Orbital Laboratory* in 1966 until its termination in 1969, when he was transferred to NASA as an astronaut.

hatch A door in a spacecraft. A space shuttle, for example, has hatches connecting the mission and payload stations to the living quarters below. Hatches can also open outside, such as the shuttle's *ingress–egress side hatch.

'Hatch' was adopted from the nautical term for doors leading to cabins or lower levels on ships.

Hauck, Frederick Hamilton (1941–) US astronaut. His three space shuttle flights included commanding the *Discovery* mission of September 1988, the first shuttle to be launched after the *Challenger* disaster in 1986. Hauck was selected as an astronaut in 1978 and left NASA in 1989.

Hawley, Steven Alan (1951–) US astronaut. His five space shuttle flights include the *Discovery* mission in April 1990 to deploy the **Hubble Space Telescope*. He also flew with the *Columbia* mission in July 1999 to deploy the **Chandra X-Ray Observatory*. He went on to be associate director for Astromaterials Research and Exploration Science at the Johnson Space Center. He was selected as an astronaut in 1978.

Hayabusa A Japanese space probe to sample an asteroid, launched on 9 May 2003. It will arrive at asteroid Itokawa in September 2005 and travel alongside it for two months, mapping its surface, before moving in to make landings at three sites to collect samples. The craft will then return to Earth and a re-entry capsule will parachute to the ground near Woomera, Australia, in June 2007.

heading alignment circle The final flight curve taken by a space shuttle to align itself with the runway for landing. The shuttle flies over the runway and 11 km beyond it, then arcs round into the heading alignment circle.

h

HEAO Abbreviation for *High-Energy Astronomy Observatory.

heat shield Any heat-protecting coating or system, such as an *ablation shield, which protects spacecraft from the heat of *re-entry, when temperatures can reach 1 500 °C.

heat sink A surface that absorbs frictional heat as a spacecraft re-enters the Earth's atmosphere. NASA's team for the **Mercury* project first considered using a beryllium heat sink, but decided on an *ablation shield (which vaporizes) as the better surface.

heavy-ion counter (HIC) An instrument that monitors and measures energetic heavy ions in space. The *Galileo* space probe has an HIC that is composed of two Low-Energy Telescopes (LETs), which are solid-state detector telescopes. An HIC uses stacks of silicon wafers to measure the heavy ions that hit the spacecraft, since this can cause changes in the vehicle's electronics. The data can be used to design more resistant electronics for future missions. The *Galileo* HIC has also been used to monitor and investigate this type of radiation around the planet Jupiter and to observe heavy ions associated with solar flares and cosmic rays.

heliocentric orbit The orbit of a spacecraft or space body around the Sun. A spacecraft travelling to a planet is in an orbit around the Sun. Its *trajectory can be determined by converting the known motions of the Earth and the spacecraft into a heliocentric orbit measurement.

Helios One of two spacecraft launched in the mid-1970s to study the Sun, its magnetic field, the solar wind, and gamma radiation. *Helios 1*, the first cooperative deep-space venture between the USA and West Germany, was launched in December 1974, and *Helios 2* in January 1976. Their orbits took them within 45 million km of the Sun, the closest approach of any spacecraft. The rockets were able to withstand a temperature of about 370 °C because of their special heat-dispersal systems. The two spacecraft acted in concert to gather data, and *Helios 2* was especially successful in monitoring gamma radiation. Both spacecraft were built in West Germany and launched by NASA.

heliosphere A region of space through which the *solar wind flows outwards from the Sun. The **heliopause** is the boundary of this region, believed to lie about 100 astronomical units from the Sun, where the flow of the solar wind merges with the interstellar gas.

Helms, Susan Jane (1958–) US astronaut. By 2001 she had made five space shuttle flights, logging over 5 000 hours in space, with one space walk of 8 hours and 56 minutes, a world record. Her fifth flight was on *Discovery* in March 2001 as one of the three members of the second resident crew to be transported to the *International Space Station*. She left NASA to return to the US Air Force in July 2002 to become chief of the Space Control Division of the Requirements Directorate of the Air Force Space Command. She was manager of a programme to develop a CF-18 flight control system simulation for the Canadian Forces when selected as an astronaut in 1990.

hemispherical resonator gyroscope (HRG) A small, highly accurate *gyroscope placed on the **Hubble Space Telescope*. It was developed by NASA's Goddard Space Flight Center and Litton Guidance and Control Systems Space Operations.

Henize, Karl Gordon (1926–93) US scientist-astronaut. He was a mission specialist on the second *Spacelab* mission aboard the space shuttle *Challenger* in July–August 1985.

Henize was a senior astronomer at the Smithsonian Astrophysical Observatory 1956–9, responsible for establishing and operating a global network to track artificial satellites. In 1959 he became an associate professor of astronomy at Northwestern University, and chaired two NASA groups concerned with *Spacelab* telescopes 1974–80. After his flight, he became a senior scientist in NASA's Space Sciences Branch.

Henricks, Terence Thomas (1952–) US astronaut. He was the first person to log more than 1 000 hours as a space shuttle pilot and commander. His flights include a 16-day mission aboard *Columbia* in June 1996, achieving a record duration time for a shuttle flight. He left NASA in 1997 to pursue a career in business. He was selected as an astronaut in 1985.

HEO Abbreviation for *high Earth orbit.

Hermes A spaceplane proposed by the *European Space Agency to ferry astronauts to and from the *International Space Station*. *Hermes* would have been launched on an Ariane 5 rocket and flown back to Earth to land on a conventional runway. The project was abandoned in the early 1990s due to financial pressures.

Herrington, John Bennett (1958–) US astronaut. A mission specialist, he was on the *Endeavour* flight in November 2002 to the **International Space Station* to exchange resident crews, and he installed the P1 Truss on three space walks taking nearly 20 hours. He went on to be a member of NASA's astronaut support personnel team that is responsible for space shuttle launch preparations and post-landing operations. Herrington was selected as an astronaut in 1996.

Herschel Space Observatory A European Space Agency satellite scheduled for launch in 2007. It will carry a 3.5-m infrared telescope for a minimum three-year study of the birth and evolution of stars and galaxies. The observatory, 7 m high and 4.3 m wide, will also carry three scientific instruments: a high-resolution spectrometer, a photometer, and a camera. It will be the only spacecraft so far to cover the far infrared to submillimetre range of the spectrum.

The observatory, named after the German-born English astronomer William Herschel, will be launched with the spacecraft **Planck*. The two craft will separate after launch and operate independently at the L1 *Lagrangian point 1.5 million km from the Earth.

h

HESSI *See* REUVEN RAMATY HIGH-ENERGY SOLAR SPECTROSCOPIC IMAGER.

HETE-2 Acronym for **High-Energy Transient Explorer 2*.

HGA Abbreviation for *high-gain antenna.

HIC Abbreviation for *heavy-ion counter.

Hieb, Richard James (1955–) US astronaut. His three space shuttle missions include the first flight of *Endeavour* in May 1992, during which he made three space walks totalling 17 hours to capture and repair an Intelsat communications satellite. One of these was the first space walk to involve three people. Hieb has logged more than 750 hours in space. He joined NASA in 1979 to work in crew procedure development and crew activity planning and was selected as an astronaut in 1985.

high Earth orbit (HEO) A spacecraft orbit that is above an altitude of approximately 32 180 km. This is called a 'supersynchronous' orbit, above the geosynchronous orbit and the *medium Earth orbit. Such an orbit provides a large observing area, as well as long, uninterrupted viewing intervals. It is often highly inclined and highly elliptical to avoid the Earth's *magnetosphere. The **Chandra X-Ray Observatory* is in HEO.

High-Energy Astronomy Observatory (HEAO) Any of three NASA satellites that observed the X-ray and gamma-ray sky. *HEAO-1* was launched in August 1978 and functioned until January 1979 to provide almost constant monitoring of X-ray sources near the ecliptic poles (points on the celestial sphere along an imaginary line running vertically through the Earth at right angles to the *ecliptic). *HEAO-2*, renamed the **Einstein Observatory*, was launched in November 1978 and ended its mission in April 1981. *HEAO-3* was launched in September 1979; the satellite continued on its course until May 1981, although the mission was rendered defunct when the cryogen freezing mixture in the satellite's detectors was depleted.

high-energy particle detector An instrument on board a spacecraft that measures the energy spectra and composition of trapped energetic electrons and of atomic nuclei. It may use several solid-state detector telescopes.

High-energy particle detectors have been flown on space probes including *Galileo* and the two *Voyager* probes.

High-Energy Transient Explorer 2 (HETE-2) A satellite that detects and pinpoints gamma-ray bursts in space. Launched in October 2000 from the Kawjalein facility, a military launch site on the Marshall Islands in the Pacific Ocean, the mission is a collaboration between the USA, Japan, France, and Italy. *HETE-2*, designed for a two-year lifespan, will also survey the X-ray sky. It carries two X-ray detectors: an X-ray camera and a wide-field X-ray monitor.

high-gain antenna (HGA) A spacecraft antenna that provides a high amplification of radio-frequency signals that it transmits or receives. It is usually a parabolic dish, and the larger the collecting area the greater the increase in amplification. An HGA requires a high directional accuracy. The dish may be fixed to the spacecraft, as was the HGA on the *Magellan* probe, whose attitude had to be manoeuvred to point its antenna towards the Earth. The *Mars Global Surveyor*, however, had a moveable HGA on an articulated arm, enabling the dish to point towards the Earth regardless of the position of the spacecraft.

high-inclination orbit An orbit of a spacecraft that has a high tilted angle from the equator. The Russian space station *Mir* and the *International Space Station* (*ISS*), both at an angle of 51.6°, and NASA's *Hawkeye* (*Explorer 52*), at nearly 90°, all have high-inclination orbits. At such an angle, the radiation danger increases for astronauts, such as those making space walks from the *ISS*. Spacecraft launched into a high-inclination orbit receive no *free ride from the Earth's rotation, and NASA conducts such launches from *Vandenberg Air Force Base in California to avoid population centres.

Hilmers, David Carl (1950–) US astronaut. His four space shuttle flights include the *Discovery* launch in September 1988, the first shuttle mission after the *Challenger* tragedy of 1986. In January 1992, he flew with the *Discovery* mission that conducted 55 experiments in the *Microgravity Science Laboratory*. In October 1992 he left NASA to begin his medical studies at Baylor College of Medicine. He was selected as an astronaut in 1980.

Hinotori A Japanese satellite, launched on 21 February 1981, which observed solar flares at X-ray wavelengths during solar maximum. Its instruments obtained the first images of solar flares in hard X-rays (energies up to 40 keV). *Hinotori* entered the Earth's atmosphere and burned up on 11 July 1991.

Hipparcos A European Space Agency satellite launched on 8 August 1989 to measure the positions, distances, motions, brightnesses, and colours of stars to an accuracy far beyond that attainable from the ground. Its name recalls the ancient Greek astronomer Hipparchus, who compiled the first known star catalogue, and is also an acronym for High Precision Parallax Collecting Satellite. *Hipparcos* observed until 1993. The main *Hipparcos* Catalogue of 118 218 stars was published in 1997. An auxiliary catalogue of lesser accuracy, containing over a million stars, was called the Tycho Catalogue. Further data analysis led to the expanded Tycho 2 Catalogue containing over 2.5 million stars, published in 2000, including 99% of all stars down to magnitude 11, almost 100 000 times fainter than the brightest star, Sirius.

Hire, Kathryn Patricia (1959–) US astronaut. She flew with the *Spacelab* mission aboard the space shuttle *Columbia* in April 1998. During the flight, experiments were conducted concerning the effects of microgravity on the brains and nervous systems of the crew members, as well as on more than 2 000 live animals. She went on to join the Astronaut Support Personnel Team for Kennedy Space Center Operations. Hire was assigned to the Kennedy Space Center as an engineer in 1989 and selected as an astronaut in 1994.

Hiten A Japanese satellite launched in January 1990, whose mission was to test and verify technologies for future lunar and planetary missions. *Hiten* was placed into a highly elliptical Earth orbit so that it passed by the Moon ten times during its lifetime. It carried a small satellite named **Hagoromo* that was put into orbit around the Moon. Among the mission's experiments were those for trajectory control using gravity assist double lunar swingbys, and also tests of telemetry, aerobraking, and the on-board computer. *Hiten* also detected and measured the mass and velocity of micrometeorite particles.

The mission was ended in April 1993 when *Hiten* was intentionally crashed into the Moon. It was Japan's first lunar fly-by, lunar orbiter, and lunar surface impact. Originally launched as *Muses-A*, *Hiten* was renamed after a flying Buddhist angel, and *Hagoromo* named after the veil worn by *Hiten*.

HL Abbreviation for *horizontal lander.

Hobaugh, Charles Owen (1961–) US astronaut. He was selected as an astronaut in 1996, and assigned to the Spacecraft Systems/Operations Branch of NASA's Astronaut Office. In July 2001, he flew his first mission aboard *Atlantis*, to install a new airlock module on the *International Space Station*. His flight in 2003 was postponed due to the *Columbia* tragedy that year.

Hoffman, Jeffrey Alan (1944–) US astronaut. He logged more than 1 211 hours in space during five shuttle flights, including the *Endeavour* mission in December 1993 to restore the **Hubble Space Telescope* to its full capacity. Hoffman was selected as an astronaut in 1978. He left the astronaut programme in 1997 to become NASA's European representative in Paris, France. In 2001, he was assigned by NASA to the Massachusetts Institute of Technology (MIT) as a professor of aeronautics and astronautics, also doing research projects there using the **International Space Station*.

In the 1970s, Hoffman played a leading role in developing the medium-energy X-ray experiment on the **Exosat* satellite for the European Space Agency. With the Massachusetts Institute of Technology (MIT) from 1975 to 1978, he worked on the analysis of X-ray data from the *SAS-3* satellite operated by MIT.

Hohmann transfer orbit (or Hohmann trajectory) A specific fuel-saving *trajectory for a spacecraft moving from one orbit to another. It is used, for instance, to travel from the Earth's orbit to that of the planet Mars. To use the least fuel possible, the vehicle leaves the Earth's orbit at the transfer orbit's

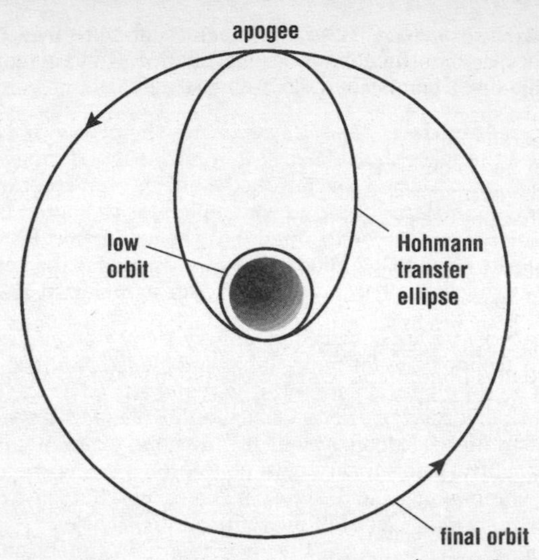

(Image © Research Machines plc)

A spacecraft in low Earth orbit (LEO) can change to a higher orbit in the same plane with minimum fuel expenditure by utilizing the Hohmann transfer orbit. Whilst in LEO the spacecraft fires its engines to increase velocity, causing the trajectory to become elliptic. This ellipse is carefully calculated to attain the desired final altitude. At this point the engines are again fired and the spacecraft transfers to the final orbit.

*perihelion (closest point to the Sun) and joins Mars's orbit at the transfer orbit's *aphelion (point furthest from the Sun). The spacecraft's orbit is named after the German engineer Walter Hohmann, who calculated it in 1916.

hold A pause in a countdown to launch a spacecraft. This delay can be caused by a technical fault, but may be a planned *built-in hold, for example a *ten-minute hold.

hold-down arm A device that restrains a rocket as it begins to fire on the launch pad, automatically releasing it at lift-off. The Saturn V rocket required four hold-down arms on its mobile launcher during assembly, transportation to the launch site, and positioning on the launch pad. Each arm had to sustain a vertical thrust of 725 800 kg.

home institution NASA term for an organization proposing scientific experiments aboard its spacecraft. NASA's *announcement of opportunity is sent to the scientific community, including its own centres, and proposals come from many bodies around the world, including government agencies, universities, and other scientific organizations. Each home institution must be able to support the mission in such areas as personnel and technology if its proposal is accepted.

horizontal lander (HL; or personnel launch system (PLS)) One of two compact, blunt-nosed experimental NASA spacecraft proposed as a small space-taxi vehicle. The horizontal lander is a 'lifting-body' space vehicle, and was designed to use an expendable launch vehicle and to return and land on a runway like a space shuttle. *HL-10* was built by Northrup and first test-flown from a modified B-52 bomber in 1966. By 1970 it had attained an altitude of 27 500 m and a speed of 1 965 kph. *HL-20*, which resembled *HL-10*, was built at Langley Research Center in 1990 as a non-flying full-scale mock-up that was 8.8 m long. It had been planned that *HL-20* would be used to take crews and supplies to the *International Space Station* and to rescue stranded astronauts. However, it has not been developed further.

Horizon 2000 A programme of major science missions initiated in 1984 by the European Space Agency. Each mission is designated 'Cornerstone' and the programme's launches include **Giotto* (1985), **Ulysses* (1990), the **Solar and Heliospheric Observatory* (1995), **XMM-Newton* (1999), and **Rosetta* (2004). A Horizon 2000 Plus programme was formulated in 1994 to continue the Cornerstone missions until 2016, and future proposals include *Gaia* (2011) and BepiColombo (2012).

h

Horowitz, Scott Jay (1957–) US astronaut. His four space shuttle flights include the *Atlantis* mission in May 2000 to construct and supply the *International Space Station*, and a return *Discovery* flight in August 2001 to exchange resident crews and install the Leonardo module. He has logged 1138 hours in space. Horowitz was selected as an astronaut in 1992.

HOTOL (abbreviation for horizontal take-off and landing) A reusable hypersonic spaceplane invented by British engineer Alan Bond in 1983 but never put into production.

HOTOL was to be a single-stage vehicle that could take off and land on a runway. It featured a revolutionary dual-purpose engine that enabled it to carry far less oxygen than a conventional spaceplane: it functioned as a jet engine during the initial stage of flight, taking in oxygen from the surrounding air; when the air became too thin, it was converted into a rocket, burning oxygen from an on-board supply. The project was developed by British Aerospace and Rolls-Royce but foundered for lack of capital in 1988.

HRG Abbreviation for *hemispherical resonator gyroscope.

H rocket Either of two Japanese-owned launch vehicles, H-1 and H-2, developed for heavier payloads by the *National Space Development Agency (NASDA). 'H' refers to hydrogen, the vehicle's liquid fuel (used with liquid oxygen). The H-1, which launched nine satellites from 1961 to 1991, is no longer used. The two-stage H-2 rocket began launching spacecraft in 1994. A later version, H-2AF, was first launched in 2001; its sixth launch on 29 November 2003 failed to jettison a solid rocket booster and the rocket had to be destroyed.

The first stage of the H-2 is powered by an LE-7 engine that can develop a thrust of 86 000 kg. The second stage is powered by an LE-5A with a 12 000-kg thrust. The rocket, made of aluminium alloy, also has two solid-fuel boosters, one on either side of the first stage, which are jettisoned into the sea when their fuel is used up. NASDA developed a third-generation rocket, the H-2A, which first flew in 2001.

HSM Abbreviation for *high-speed multiplexer.

HST Abbreviation for **Hubble Space Telescope*.

HTV Abbreviation for *H-II Transfer Vehicle.

H-II Transfer Vehicle (HTV) A supply vehicle to the *International Space Station* (*ISS*) built by Japan and launched on its H-IIA rocket. The HTV is an un-manned container about 10 m long that delivers goods to the ISS. After unloading it can be filled with waste which is destroyed when the HTV burns up in the atmosphere.

Hubble, Edwin (Powell) (1889–1953) US astronomer. He discovered the existence of galaxies outside our own, and classified them according to their shape. His theory that the universe is expanding is now generally accepted.

His data on the speed at which galaxies were receding (based on their *redshifts) were used to determine the portion of the universe that we can never come to know, the radius of which is called the **Hubble radius**. Beyond this limit, any matter will be travelling at the speed of light, so communication with it will never be possible. The ratio of the velocity of galactic recession to distance has been named the **Hubble constant**.

Hubble discovered *Cepheid variable stars in the Andromeda galaxy in 1924, proving it to lie far beyond our own Galaxy. In 1925 he introduced the classification of galaxies as spirals, barred spirals, and ellipticals. In 1929 he announced **Hubble's law**, stating that the galaxies are moving apart at a rate that increases with their distance from each other.

Hubble was born in Marshfield, Missouri, and studied at the University of Chicago, Illinois, graduating in 1910, and then at Oxford University, where he gained a degree in law in 1912. He was also an athlete and a heavyweight boxer. He returned to America 1913 and briefly practised law in Louisville, Kentucky, but his real interest was in astronomy, and he went to the Yerkes Observatory, Williams Bay, Wisconsin, as a graduate student in astronomy in 1914. He was awarded a PhD in 1917 for a thesis on the photography of faint nebulae. After serving with the American Expeditionary Force in World War I, he joined the staff at Mount Wilson Observatory, near Pasadena, California, 1919 and carried out research on galactic and extragalactic nebulae. During World War II he was chief of ballistics and director of the Supersonic Wind Tunnel Laboratory at the Aberdeen Proving Ground, Maryland. Afterwards he returned to what was to become the Mount Wilson and Palomar observatories, and was one of the first to use the new 508-cm Hale Telescope that was installed in 1948.

Nearly all Hubble's work related to nebulae, which he was the first to show were extragalactic, that is, located outside our own Galaxy. It has been said that Hubble opened up the observable region of the universe in the same way that the Italian astronomer and physicist Galileo opened up the Solar System in the 17th century, and British astronomers William and John Herschel opened up the Milky Way in the 18th and 19th centuries. He gave an account of some of his researches in *The Realm of the Nebulae* (1936).

Hubble classification A scheme for classifying galaxies according to their shapes, originally devised by US astronomer Edwin *Hubble in the 1920s.

Elliptical galaxies are classed from type E0 to type E7, where the figure denotes the degree of ellipticity. An E0 galaxy appears circular to an observer, while an E7 is highly elliptical (this is based on the apparent shape; the true shape, distorted by foreshortening, may be quite different).

Spiral galaxies are classed as type Sa, Sb, or Sc: Sa is a tightly wound spiral with a large central bulge, Sc is loosely wound with a small bulge, and Sb is in between. Intermediate types are denoted by Sab or Sbc.

Barred spiral galaxies, which have a prominent bar across their centres, are similarly classed as type SBa, SBb, or SBc, with intermediates SBab or SBbc.

Lenticular galaxies, which have no spiral arms, are classed as type S0.

Irregular galaxies, type Irr, can be subdivided into Irr I, which resemble poorly formed spirals, and Irr II which cannot be classified because of disturbance.

The Hubble classification was once believed to reveal an evolutionary sequence (from ellipticals to spirals) but this is now known not to be the case. Our own *Milky Way is a spiral galaxy, classified as type Sb or Sc, but may have a bar.

Hubble constant A measure of the rate at which the universe is expanding, named after US astronomer Edwin Hubble. Observations suggest that galaxies are moving apart at a rate of 50–100 kps for every million *parsecs of distance. This means that the universe, which began at one point according to the *Big Bang theory, is between 10 billion and 20 billion years old. Observations by the **Hubble Space Telescope* in 1996 produced a figure for the constant of 73 kps.

Hubble's law A law that relates a galaxy's distance from us to its speed of recession as the universe expands, formulated in 1929 by US astronomer Edwin Hubble. He found that galaxies are moving apart at speeds that increase in direct proportion to their distance apart. The rate of expansion is known as the Hubble constant.

***Hubble Space Telescope* (*HST*)** A space-based astronomical observing facility, launched in April 1990 and orbiting the Earth at an altitude of 610 km. It consists of a 2.4 m telescope and four complementary scientific instruments, is roughly cylindrical, and is 13 m long and 4 m in diameter, with two large solar panels. The *HST* is a cooperative programme between NASA and the European Space Agency (ESA).

Before the US$2.5 billion *HST* could reach its full potential, a flaw in the shape of its main mirror had to be corrected in 1993 by astronauts aboard the space shuttle *Endeavour*. By October 2003, the telescope had produced about 500 000 images, and had been serviced by astronauts four times. NASA announced in 2004 that it would cancel future servicing of *Hubble*, which will cause it to cease operating in a few years. This was a financial decision, following President George W Bush's announcement in January 2004 that astronauts will be returned to the Moon in 2020.

In December 1995, the *HST* was trained on an 'empty' area of sky near the Plough, now termed the Hubble Deep Field. In 2002, an Advanced Camera for Surveys (ACS) was installed, and in 2004, it revealed the deepest view ever of the

(Image © NASA)
The *Hubble Space Telescope* photographed shortly after its release from the space shuttle Discovery's robot arm, following its third servicing mission (STS-103) in December 1999. The telescope's failed gyroscopes were replaced, its computer upgraded, and other new components installed.

universe, showing galaxies formed shortly (700 million years) after the Big Bang.

http://hubble.stsci.edu/ Everything you could ever want to know about the *Hubble Space Telescope*. This attractive site features a showcase of the very best of Hubble's images and the latest news on the telescope, as well as an illustrated guide to the spacecraft, its instruments, and its optical system.

Hughes Space and Communications Company A US company that has developed and built space systems since 1961. It has produced almost 40% of the world's satellites now in commercial use. With headquarters in Los Angeles, California, it became in 2000 an operating unit of Hughes Electronics Corporation, part of the *Boeing Company. Hughes designed and built the world's first geosynchronous communications satellite, *Syncom*, launched in 1963, and the first geosynchronous meteorological satellite, *ATS-1*, launched in

1966. It is now building the next generation of Tracking and Data-Relay Satellites and Geostationary Operational Environmental Satellites for NASA and the National Oceanic and Atmospheric Administration. The company is also developing a xenon ion rocket propulsion system.

The company also produced the five uncrewed *Surveyor* lunar landers, the first touching down in 1966; the probe on the *Galileo* spacecraft that arrived at Jupiter in 1995; and the radar aboard the *Magellan* spacecraft that reached Venus in 1990. Hughes is now developing an *ion engine as a future propulsion system.

Husband, Rick Douglas (1957–2003) US astronaut. He was part of the space shuttle *Discovery* mission in May 1999 that made the first docking with the *International Space Station*, delivering supplies in readiness for the arrival of the first resident crew. He was selected as an astronaut in 1994, but died when the space shuttle *Columbia* broke up on re-entry in February 2003 after a 16-day mission to the *International Space Station*.

Huygens A European probe, part of the *Cassini* mission to Saturn. *Huygens* was launched attached to *Cassini* on 15 October 1997 and was released on 25 December 2004. On 14 January 2005 it descended to the surface of Saturn's largest moon, Titan, transmitting for nearly two and a half hours during the descent through Titan's atmosphere and then for over an hour on the surface. During its descent, *Huygens* measured wind speeds and atmospheric gases and photographed apparent drainage channels caused by liquid methane and bright islands on darker plains. It landed on a slushy area dotted with boulders composed of water ice, under orange clouds composed of complex hydrocarbons.

Huygens atmospheric structure instrument (HASI) An instrument aboard the *Huygens* probe carried by the *Cassini* spacecraft, launched in 1997 and arriving at Saturn's moon Titan in July 2004. HASI consists of several sensors to measure the physical and electrical make-up of the atmosphere. These include pressure and temperature sensors as well as accelerometers which can determine atmospheric density and wind gusts, or even wave motions on a liquid surface. There is also a permittivity and electromagnetic wave analyser to measure the electron and ion conductivities of the atmosphere and Titan's surface.

hybrid rocket A type of rocket developed in the 1990s that is safer, cleaner, and cheaper than conventional rockets that use solid and liquid propellants. NASA successfully fired a prototype rocket motor in August 1999 and completed its fourth in a series of large-scale hybrid motor tests in 2002 at the Stennis Space Center. The hybrid rocket has solid fuel converted into a rubbery form. An oxidizer is continuously injected into its core, producing hot exhaust gases that vaporize a small layer of fuel, causing a reaction to produce more gases. The rocket can be turned off after ignition using a simple switch. NASA's hybrid rocket motor has 112 500 kg of thrust. NASA's Ames Research Center collaborated with Stanford University to develop a non-toxic, easily handled fuel.

hydrazine fuel A hypergolic rocket propellant composed of monomethyl hydrazine oxidized by nitrogen tetroxide, used on the space shuttle. When the shuttle separates from its external fuel tank at an altitude of about 113 km, it must fire hydrazine fuel from the two rockets of its *orbital manoeuvring system for 2.5 minutes to obtain orbit.

hypergolic fuel A rocket propellant that burns spontaneously when combined with an oxidizer, so that no separate ignition system is required. Hypergolic fuels have been used for Titan and Agena rockets, the *Apollo* lunar module, and the space shuttle.

Hypergolic fuels can be stored at room temperature.

hypoxia An abnormal condition experienced by an astronaut if the body tissues receive less than normal oxygen. This can lead to light-headedness or temporary unconsciousness. NASA has eliminated hypoxia by using pure oxygen in the spacecraft interior.

IAF Abbreviation for *International Astronautical Federation.

IAU Abbreviation for *International Astronomical Union.

ICE Acronym for **International Cometary Explorer.*

ICU Acronym for *instrument control unit.

Ida (asteroid 243) A low-density asteroid (2.9 g/cm^3) with a composition comparable to that of chondrite *meteorites. An asteroid from the Koronis family, it has an irregular size of 58 × 38 km and orbits the Sun between the planets Mars and Jupiter at a distance of around 270 million km. The *Galileo* spacecraft discovered a satellite orbiting Ida—the first natural satellite of an asteroid ever discovered. The satellite, known as Dactyl, is 1.6 × 1.2 km, and is surprisingly round for such a small celestial body. It orbits Ida at a distance of approximately 90 km.

IFOV Abbreviation for *instantaneous field of view.

IGBP Abbreviation for *International Geosphere–Biosphere Programme.

igloo A container in the **Spacelab* cargo bay that provided equipment mounted on *pallets with electrical power, communications, and other services. The igloo was temperature-controlled and pressurized.

IISL Abbreviation for *International Institute of Space Law.

ILC Abbreviation for *International Latex Corporation.

IMAGE Acronym for **Imager for Magnetopause-to-Aurora Global Exploration.*

Imager for Magnetopause-to-Aurora Global Exploration (IMAGE) A NASA satellite, launched on 25 March 2000, which orbits over the Earth's poles to examine how the Earth's magnetosphere responds to changes in the solar wind. *IMAGE* employs a variety of techniques to produce global images of plasma in the inner magnetosphere. Three neutral atom imagers (NAI) provide images at energies from 10 eV to 200 keV while two ultraviolet imagers cover wavelength ranges from 120 to 180 nm (FUV) and 30.4 nm (EUV), and a radio plasma imager (RPI) transmits and receives pulses from 3 kHz to 3 MHz to determine relative motions of the satellite and plasma.

imaging instrument A spacecraft instrument that uses optics such as lenses or mirrors to project an image onto a detector. The image is then converted into digital data and transmitted to ground stations. Science imagers include photometers which measure the intensity of light from a source, spectrometers which measure the intensities of wavelengths, polarimeters which measure the polarization of light reflected from a celestial target, and infrared radiometers which measure the intensity of infrared energy radiated by objects in space. Earlier spacecraft, such as the **Mariner* and **Voyager* probes, carried image detectors that used a vacuum tube called a vidicon. Since the late 1980s spacecraft have employed a *charge-coupled device that is an integrated circuit.

IMP Acronym for *Interplanetary Monitoring Platform.

IMU Abbreviation for *inertia measurement unit.

inclination The angle between the *ecliptic and the plane of the orbit of a planet, asteroid, or comet. In the case of satellites orbiting a planet, it is the angle between the plane of orbit of the satellite and the equator of the planet.

Indian National Satellite System (INSAT) A series of Indian-owned geostationary satellites for simultaneous domestic communications and Earth observation. The four first-generation INSAT vehicles were all built in the USA and launched by NASA or the European Space Agency: *INSAT-1A* was launched in 1982, *INSAT-1B* in 1983, *INSAT-1C* in 1988, and *INSAT-1D* in 1990. The five second-generation vehicles were built in India and launched there: *INSAT-2A* was launched in 1992, *INSAT-2B* in 1993, *INSAT-2C* in 1995, *INSAT-2D* in 1997, and *INSAT-2E* in 1999. The third generation began with *INSAT-3B* in 2000, followed by *INSAT-3C* in 2002, and *INSAT-3A* and *INSAT-3E* in 2003. The satellites have carried a very high-resolution radiometer (VHRR) to observe cloud coverage and cyclone formation, and a high-resolution charge-coupled device (CCD) camera to study cyclones, storms, and climate changes.

Indian Remote-Sensing Satellite (IRS) Any of a series of eight remote-sensing satellites built in India and launched between 1988 and 1999. The early ones were placed in orbit by the USSR (later Russia). *IRS-1A* was launched in 1988, *IRS-1B* in 1991, *IRS-P2* in 1994, *IRS-1C* in 1995, *IRS-P3* in 1996, *IRS-1D* in 1997, and *IRS-P4* in 1999. The latter, also known as *Oceansat*, carries two payloads, an ocean colour monitor (OCM), and a multi-frequency scanning microwave radiometer (MSMR). *IRS-1E* was launched in 1993 but failed to orbit.

Indian Space Research Organization (ISRO) India's space agency. It operates within the Space Commission and Department of Space (DOS) established by the Indian government in 1972. The main launch site is the SHAR Centre, 100 km north of Chennai (formerly Madras). The agency's first satellite was **Aryabhata*, launched in 1975. Subsequent satellites have included the **Rohini* series, first launched in 1979, the *Indian National Satellite System, first launched in 1982, and the *Indian Remote-Sensing Satellites, first launched in 1988. In 2001, ISRO successfully launched its communications satellite *GSAT-1*, when the geosynchronous satellite launch vehicle (GSLV) placed it into orbit. The agency has been planning its first mission to the moon, *Chandrayaan-1*, to carry out high-resolution mapping of the topographic features in three dimensions.

http://www.isro.org/ News of the activities of the Indian Space Research Organization. The site includes information about India's space centre network and research facilities, and well-illustrated pages profiling specific missions.

inertial guidance system A navigation system that makes use of *gyroscopes and accelerometers to monitor and control a vehicle's movements. A computer calculates the vehicle's position and corrects any deviation from the proper course. Inertial navigation is used in aircraft, submarines, spacecraft, and guided missiles.

inertial measurement unit (IMU) An instrument that aids spacecraft navigation by measurements using computer-assisted gyroscopes and accelerometers. Connected to the spacecraft by gimbals (swivelling devices), IMUs establish a fixed base from which the spacecraft's attitude in space can be measured. In the *Gemini* spacecraft, the gyroscopes were mounted at right angles to one another to calculate the vehicle's pitch, yaw, and roll. The accelerometers were mounted in tandem with them to measure the vehicle's reaction to thrusters being fired.

inertial reference device A gyroscope on a spacecraft that is used to provide attitude reference for the attitude and articulation control subsystem (AACS; *see* ATTITUDE CONTROL). The device is employed when celestial references, such as a *star scanner or *star tracker, are not in use. For example, the **Galileo* space probe would only activate its inertial reference device in the event of the loss of its celestial reference during certain manoeuvres. Other spacecraft, such as **Magellan*, made continuous use of the inertial reference device.

Inertial Upper-Stage rocket (IUS) NASA's two-stage rocket used to launch large satellites carried by a space shuttle. Built by the *Boeing Company, the rocket was first deployed from the shuttle in 1983. It is 5 m in length and 3 m in diameter, and weighs approximately 14 625 kg. It has two solid rocket motors, a reaction control system, and redundant systems for guidance, navigation, and communications.

inferior planet A planet (Mercury or Venus) whose orbit lies within that of the Earth, best observed when at its greatest elongation from the Sun, either at eastern elongation in the evening (setting after the Sun) or at western elongation in the morning (rising before the Sun).

in-flight maintenance An early NASA concept that maintenance and repair of spacecraft equipment could be done during a mission by the crew. Although astronauts could diagnose problems, it proved difficult to include a variety of tools and spare parts for each flight. Back-up units were included instead.

***Infrared Astronomical Satellite* (*IRAS*)** A joint US–UK–Dutch satellite launched in 1983 to survey the sky at infrared wavelengths, studying areas of star formation, distant galaxies, and possible embryo planetary systems around other stars. It discovered five new comets in our own Solar System, and operated for ten months.

http://space.gsfc.nasa.gov/astro/iras/iras_home.html Provides information on the wide range of IRAS products, including sky maps, small-field observations of galaxies, and the minor-planet and comet survey. A brief introduction to *IRAS* is provided, as well as links to *IRAS* documents and the project web site at the California Institute of Technology.

infrared astronomy The study of infrared radiation produced by relatively cool gas and dust in space, as in the areas around forming stars. In 1983 the US–Dutch–British *Infrared Astronomical Satellite* (*IRAS*) surveyed almost the entire sky at infrared wavelengths. It found five new comets, thousands of galaxies undergoing bursts of star formation, and the possibility of planetary systems forming around several dozen stars.

Planets and gas clouds emit their light in the far- and mid-infrared regions of the spectrum. The *Infrared Space Observatory* (*ISO*), launched in 1995, observed a broad wavelength (3–200 micrometres) in these regions. The work of these pioneer observations has been extended by the **Spitzer Space Telescope* and, extending observations into the submillimetre range, the **Herschel Space Observatory*.

infrared photo-polarimeter (ISOPHOT) An instrument to detect the amount of infrared radiation emitted by a celestial object. It is part of the European Space Agency's *Infrared Space Observatory. ISOPHOT's 144 detector elements are divided into four arrays and three single detectors in three subsystems (a multi-aperture photo-polarimeter, a photometric camera, and a spectrophotometer).

infrared radiation An electromagnetic radiation of wavelength between about 0.75 micrometres and 1 millimetre—that is, between the limit of the red end of the visible spectrum and the shortest microwaves. All bodies above absolute zero in temperature absorb and radiate infrared radiation. Infrared radiation is used in medical photography and treatment, and in industry, astronomy, and criminology.

The human eye cannot detect infrared, but its effects can be demonstrated. For example, an electric hob operates at high temperatures and only the visible light it gives out can be seen. As it cools down, the visible light is no longer seen; however, the heat (infrared radiation) that continues to be given out can be felt. Infrared absorption spectra are used in chemical analysis, particularly for organic compounds. Objects that radiate infrared radiation can be photographed or made visible in the dark on specially sensitized emulsions.

***Infrared Space Observatory* (*ISO*)** An orbiting telescope with a 60-cm diameter mirror. It was launched on 17 November 1995 by the European Space Agency and spent 18 months in an elongated orbit, at a range from the Earth of 1 000–70 500 km, as far as possible outside the radiation belts that swamp its detectors.

The *ISO* made the first discovery of water vapour from a source beyond the Solar System (in planetary nebula NGC 2027); traced the spiral arms of the Whirlpool galaxy and detected sites of star formation there; and obtained the first comprehensive spectrum of Saturn's atmosphere.

infrared spectrometer (IRS) An instrument flown on spacecraft that uses infrared light to analyse solids and liquids. An IRS was aboard the *Mariner 7* mission to the planet Mars. An IRS launched in 1998 on *Deep Space 1* determined the chemical composition of the asteroid Braille. A composite infrared spectrometer (CIRS) is aboard the *Cassini* probe to measure the temperature, composition, and cloud structures of the planet Saturn (including its rings) and its largest satellite Titan.

ingress–egress platform A platform on a space shuttle's mid-deck used by astronauts to enter and leave the shuttle when it is in its launch position.

ingress–egress side hatch A side opening on the mid-deck of a space shuttle that allows the crew to escape in an emergency during the controlled gliding phase of landing. The side hatch is jettisoned by pyrotechnics and the crew manually deploy an escape pole that goes through and down from the opening. Each crew member attaches his or her parachute harness to a lanyard hook assembly surrounding the pole and leaves through the ingress–egress side hatch, sliding down and off the pole on a trajectory leading below the shuttle's left wing.

ingress flight The inward or return flight of a spacecraft. This includes the space shuttle's re-entry and landing, as well as the *Apollo* flights returning from the Moon in the 1960s and early 1970s. An ingress flight is the opposite of an *egress flight.

initial weight The weight of a rocket stage before its fuel is consumed. As much as 90% of the initial weight can be due to the propellant. The **burnout weight** of a rocket occurs after the fuel consumption.

injection Placing a spacecraft into a particular trajectory. The proper injection often requires manoeuvres after the launch. For example, a space probe to Mars needs a trajectory correction manoeuvre (TCM) about 15 days into the journey. The term is also sometimes used, instead of 'insertion', for placing a space vehicle into an orbit.

Inmarsat Contraction of *International Maritime Satellite Organization.

INPE Abbreviation for *Instituto Naçional de Pesquisas Espaçiais.

INSAT Contraction of *Indian National Satellite System.

insertion (or orbit insertion) Placing an artificial satellite or space probe into orbit, commonly around a planet. On interplanetary missions, navigation and course corrections must occur to ensure that the spacecraft arrives at precisely the correct time and location for insertion into the orbit. When the planet's gravity bends the spacecraft's trajectory, the on-board command sequence manoeuvres the spacecraft into the correct attitude and fires a *retrorocket to decelerate so that the planet pulls the craft into orbit. Later orbital adjustments can be made by *orbit trim manoeuvres, which the *Galileo* probe used for two years around the planet Jupiter, and by *aerobraking.

instantaneous field of view (IFOV) An angular cone of visibility of a sensor on a spacecraft. For example, the IFOV for the advanced very high-resolution radiometer (AVHRR) carried on Polar-Orbiting Environmental Satellites (POEs) of the National Oceanic and Atmospheric Administration, is 1.4 milliradians. The IFOV determines the area of the Earth's surface scanned at any given time from a certain altitude. The area's size is found by multiplying the IFOV by the distance from the sensor to the Earth.

Institute of Space and Astronautical Science (ISAS) Japan's previous main space science institute. Although the development of launch vehicles and satellites has been led by the *National Space Development Agency (NASDA), ISAS conducted space science research by using its own launch vehicles, scientific satellites, planetary probes, and balloons. On 1 October 2003, it was merged with NASDA and the National Aerospace Laboratory of Japan (NAL) to form the Japan Aerospace Exploration Agency.

ISAS was established in 1981 by the reorganization of the former Institute of Space and Aeronautical Science at the University of Tokyo, the centre of Japanese space science since 1964. It remains a university institute, with its main campus in Sagamihara, about 40 km west of Tokyo. Studies of rockets and space science began in the 1950s, under Japanese rocket engineer Hideo Itokawa.

i

Instituto Naçional de Pesquisas Espaçiais (INPE; National Institute for Space Research) Brazil's space agency. An independent part of the Ministry of Science and Technology, it was established in 1971 to replace the Organizing Group for the National Commission on Space Activities formed in 1961. INPE launched Brazil's first satellite (*SCD-1*) in 1993 and its second (*SCD-2*) in 1998. INPE has also been involved in the construction of the *International Space Station*. Its future missions with undetermined launching dates include the China-Brazil Earth Resources Satellite (*CBERS-1*), and a scientific satellite (*SACI-1*).

The agency's research work began in the 1970s, with INPE using the services of foreign meteorological, communications, and Earth-observation satellites. In the late 1970s, the Brazilian government approved INPE's Complete Brazilian Space Mission, and the next decade saw INPE join in the China-Brazil Earth Resources Satellite (*CBERS*) and establish its Integration and Tests Laboratory. Its main research campus is located at São José dos Campos.

Instrumentation Laboratory A laboratory at the Massachusetts Institute of Technology (MIT) that devised the navigation and flight-control computers on board the *Apollo* spacecraft. An on-board computer for *Apollo* was deemed necessary because *Apollo*'s lunar orbit insertion would occur on the far side of the Moon, out of contact with Earth-based computers. The basic design for the computer was completed in 1961 by Milton Trageser of MIT and Robert Chilton of NASA.

The laboratory was established in 1939 by MIT professor Charles Stark Draper. The name was changed to the Charles Stark Draper Laboratory in 1970 and three years later the facility was separated from MIT as a non-profit research and development laboratory with headquarters in Cambridge, Massachusetts. It works in partnership with NASA's Johnson Space Center and Marshall Space Flight Center, having contributed operational support for the space shuttles and *International Space Station*.

instrumentation rack A storage rack on a spaceship for various instruments and equipment, such as computers and radio transmitters and receivers. On the space shuttle, they are located on the mid-deck.

instrument control unit (ICU) A unit on European Space Agency (ESA) spacecraft that receives, decodes, and executes high-level commands for a particular instrument. If a problem occurs, it can automatically take actions to correct the fault. The ICU also autonomously checks the health of the instrument and monitors its parameters. The unit is part of an ESA spacecraft's distributed command and control function.

instrument pointing system (IPS) A precision pointing system for spacecraft's telescopes, other observation instruments, and experiments. It was designed and constructed by the European Space Agency for the **Spacelab* space station. Mounted on a pallet in the shuttle's payload bay, it can provide pointing accuracy for one large instrument or a cluster of smaller ones. The space shuttle can seldom maintain pointing accuracies better than 1/10th of a degree by using its thrusters, whereas the IPS can achieve 1/3 600th of a degree. It is a three-axis gimbal system that has accelerometers to allow it to compensate for motions to the shuttle. The IPS finds predetermined guide stars to determine its correct position.

instrument unit (IU) A vital electronic structure developed for the *Saturn rocket by NASA's *Marshall Space Flight Center. The unit provides guidance, tracking, environmental-control equipment, batteries and power for electronic equipment, and information on the vehicle's performance and environment. The IU is supported by IBM computer programs that guide the vehicle in flight and operate *checkout equipment.

INTEGRAL Contraction of **International Gamma-Ray Astrophysics Laboratory*.

integrated sequence of events (ISOE) A list of expected events during a spacecraft's flight and mission. It is compared with the spacecraft's real-time events to assure that the vehicle is functioning correctly. The times of events come from the spacecraft event file (SEF) drawn up by the *spacecraft team. The list is called 'integrated' because it combines events on the spacecraft with those in the ground stations.

The ISOE is also used by ground stations, such as those of the *deep-space network, to plan command uplinks, usually a week or more in advance of the events.

Intelsat Contraction of *International Telecommunications Satellite Organization.

interdeck access ladder The ladder between the mid-deck and flight deck on a space shuttle. Astronauts use it to enter the flight deck from the mid-deck when the shuttle is in its launch position, and to go from the flight deck to the mid-deck when the shuttle is in its launch and landing positions.

interdeck light shade One of the louvred shades carried in the space shuttle to minimize light leaking between the flight deck and mid-deck during photography inside the shuttle. Velcro is used to attach the shade to the mid-deck ceiling around the access connecting the decks. Photography of the crew's activities is accomplished with three cameras: a 16-mm motion-picture camera, and 35-mm and 70-mm cameras.

interferometry Any of several techniques using interference between multiple beams to obtain high-resolution images or spectra of astronomical objects. The *very long-baseline interferometry system is an example. NASA plans to put radio telescopes, infrared telescopes, and visible-light telescopes in formation in space to produce images of planets around neighbouring stars.

intergalactic space The space existing between the galaxies. It contains intergalactic material, possibly including an enormous amount of *dark matter, and *black holes, as revealed by the **Hubble Space Telescope*, the **Chandra X-Ray Observatory*, and the **Gamma-Ray Observatory*.

Interkosmos A Russian-based cooperative science space programme. It was established by the USSR in 1967 with members of Eastern European and other communist nations; each state financed its own experiments launched on Interkosmos satellites. Cosmonauts were selected from many of these nations for scientific missions. The first Interkosmos satellite was *Vertikal 1*, launched on 28 November 1970.

Following the break-up of the USSR, Western nations have participated in the programme, and Interkosmos joined the International Solar Terrestrial Programme with NASA and the European Space Agency in the 1990s to launch spacecraft to study the *solar wind's influence on the Earth's *magnetosphere. *Interkosmos 26*, launched on 2 March 1994 to study the Sun, included input from scientists in the UK and France.

International Astronautical Federation (IAF) A non-governmental organization that advances and disseminates knowledge about space and the development and application of space assets to benefit humanity. Established in 1951, the IAF in 2004 had 162 members from government organizations, industry, professional associations, and societies in 45 countries. The IAF organizes international symposia, workshops, and events. These include an annual Space Workshop for Developing Nations, organized with the United Nations (UN); space seminars at UN meetings; and an International Astronautical Congress, hosted each year by a different nation. The Congress is organized with the International Academy of Astronautics and the *International Institute of Space Law, which the IAF founded in 1960. Its permanent secretariat is in Paris, France.

International Astronomical Union (IAU) An international astronomy organization founded in 1919. The IAU promotes and coordinates international cooperation in astronomy and other space studies, and is the recognized authority for assigning designations to celestial bodies and their surface features.

The organization's 9 100 members from 67 countries are professional astronomers who are often involved in space missions. Its 12 scientific divisions include those for Space and High-Energy Astrophysics, Planetary Systems Sciences, and Radio Astronomy. There are also 37 commissions that are more specialized, and 83 working and programme groups. Its permanent secretariat is in Paris, France.

http://www.iau.org/ Useful text-only site outlining the organization and activities of the IAU. IAU announcements are posted on the site, and there are links to the divisions and special commissions set up by the Union.

International Cometary Explorer (ICE) A NASA spacecraft that made the first encounter with a comet. *ICE* made a rendezvous with the comet Giacobini–Zinner in September 1985 and passed through its tail, taking measurements. It later went into the vicinity of Halley's Comet in March 1986, becoming the first spacecraft to directly observe two comets.

ICE was launched in 1978 with the name of *International Sun-Earth Explorer 3* (*ISEE-3*) and, in coordination with *ISEE-1* and *ISEE-2*, was intended to monitor the interaction between the *solar wind and the Earth's magnetic field. NASA then decided to change its trajectory to intercept the comet, and renamed the spacecraft accordingly. Its operations were ended in 1997, but its flight continues and the craft will return to the vicinity of the Earth in 2014.

International Gamma-Ray Astrophysics Laboratory (INTEGRAL) A European gamma-ray observatory mission, in conjunction with Russia and the United States, launched on 17 October 2002. It follows a highly eccentric orbit around the Earth so that for most of the time it is above an altitude of 40 000 kilometres, well outside Earth's radiation belts, to avoid background radiation effects.

INTEGRAL undertakes spectroscopy and imaging of gamma-ray sources in the energy range 15 keV to 10 MeV, while an X-ray monitor and an optical camera help identify sources. Its targets include supernovae and the elements formed by them, gamma-ray bursts, and supermassive black holes at the centres of galaxies.

International Geosphere–Biosphere Programme (IGBP) A programme launched in 1986 by the International Council of Scientific Unions to describe and understand the physical, chemical, and biological processes that regulate the Earth's system, as well as the changes occurring and how they are influenced by human actions. IGBP works on six projects centred on the oceans, land, and atmosphere. Scientists from more than 100 countries work on establishing the basis for assessing changes in the Earth's biogeochemical cycles, such as changes controlling the concentration of carbon dioxide and other chemicals in the atmosphere. National committees exist in 72 countries.

The IGBP began the IGBP Data and Information System (IGBP-DIS) to exchange satellite and other data, and to improve access to it. It also established a network of regional research centres under the name of the Global Change System for Analysis, Research, and Training (START) that will add regional perspectives on biogeography and climate.

http://www.igbp.kva.se/cgi-bin/php/frameset.php Presents the aims and work of the International Geosphere–Biosphere Programme (IGBP). There is full information about the activities and meetings of the IGBP, details of science projects, and science highlights. The IGBP's newsletter can be downloaded from the site.

International Institute of Space Law (IISL) An organization for the development of space law and the study of legal and social-science aspects of space exploration. It was established in 1960 by the *International Astronautical Federation (IAF) and is part of the IAF, although it functions autonomously. The IISL's various committees include the Standing Committee on the Status of International Agreements Relating to Activities in Outer Space. It has more than 300 individual and institutional members in 40 countries. Its permanent secretariat is located at the IAF headquarters in Paris, France.

International Maritime Satellite Organization (Inmarsat) An organization that provides satellite communications for ships at sea, for distress and emergency, and for mobile phone customers on the sea, air, and land. It operates services in some 160 countries and on all the oceans, using four main *Inmarsat-3* satellites in geostationary orbits. It supports links for telephone, fax, and data communications to more than 250 000 ships, vehicles, aircraft, and portable terminals. It has been used in Iraq since the 2003 war to link with aid and reconstruction agencies and news media. A Broadband Global Area Network is scheduled to start in 2005, with *Inmarsat-4* being over 100 times more powerful than the *Inmarsat-3*.

***International Microgravity Laboratory* (*IML*)** A *Spacelab* module dedicated to the operation of a suite of international experiments researching the effects of microgravity on materials processing and life sciences. *IML 1* was launched aboard the **Discovery* space shuttle on 22 January 1992 for an eight-day mission with a crew of six. A second *IML* was launched aboard the **Columbia* space shuttle on 8 July 1994, for a 14-day mission with a seven-person crew. Experiments included the observation of the effect of microgravity on crystal growth and chromosome behaviour.

IML was developed by six international space agencies: NASA, the *National Space Development Agency, the *Centre National d'Etudes Spatiales (National Centre for Space Studies), the *European Space Agency, the *Canadian Space Agency, and the Deutsches Zentrum für Luft- und Raumfahrt (German Aerospace Centre).

International Satellite Cloud Climatology Project (ISCCP) A project begun in 1982 to collect and analyse radiance measurements made by satellites, as part of the World Climate Research Programme (WCRP). This data will be used primarily to increase knowledge of the effects of clouds on the radiation balance. It will also improve understanding about the global distribution of clouds, their properties, and variations.

Data is collected by weather satellites operated by several nations. Data collection began in July 1983 and the 20th year of collection was reached in 2003. The first ISCCP regional experiment (FIRE) was the Arctic cloud experiment (ACE) that began in April 1998 to study the impact of Arctic clouds on the radiation exchange between space and the Earth's surface and atmosphere, as well as the influence on surface characteristics on those clouds.

***International Space Station* (*ISS*)** The 430-tonne orbiting *space station being constructed by the USA, Russia, Japan, Canada, Brazil, and Europe (16 nations and around 100 000 workers in all). The final cost was expected to be

(Image © NASA)

One of the first images of the entire *International Space Station* with its solar panels fully deployed. It was taken from the space shuttle *Endeavour*, which had just completed an hour-long undocking manoeuvre after nearly seven days at the space station.

more than US$90 billion. The complete station, orbiting at an altitude of around 378 km, will measure 88.4 m in length, 108.5 m across its solar wings, and 43.6 m in height. It is hoped that teams of up to seven astronauts will live and work for a period of three to six months in an area equivalent to the passenger cabins of two 747 jumbo jets, although a maximum of three crew members per mission is envisaged until at least 2006. The *ISS* is scheduled for completion in 2006, after 45 launches and 160 space walks. It is expected to be used until at least 2016. The first crew arrived at *ISS* in November 2000 under the command of NASA astronaut Bill Shepherd. By 17 September 2001, six habitable modules had been added to the *ISS*.

In November 1998, the *ISS*'s first component was launched from Baikonur Cosmodrome; this was the Russian control module Zarya (Russian 'sunrise'), providing initial propulsion and power. Unity, the six-sided US node (intersection point) that will connect the various modules was bolted on to Zarya the following month. Russia's Zvezda (Russian 'star') module, housing initial living quarters and flight controls, was connected in July 2000. The 'Expedition One' crew of three became the first long-term (four-month) residents on 2 November 2000 and chose their own name for the *ISS*: *Space Station Alpha*. In February 2001, NASA launched the US science laboratory module Destiny, the centrepiece of the station. In April, the *ISS* added the robotic Canadarm and was visited by US industrialist Dennis Tito, the first space tourist, who was reported to have paid around US$20 million to take part in the mission. By October 2001, the *ISS* was an estimated US$4.5 billion over budget. By 2004, it had reached a weight of 186 880 kg with a habitable volume of 425 cu m. Its width was 73 m across solar arrays, the length 44.5 m, and the height 27.5 m. Following the resumption of shuttle flights in 2005, NASA will move 75 600 kg of additional components to the *ISS* to triple the number of science facilities and triple the area of solar arrays.

i Zarya is 12.6 m in length and 4.1 m at its widest point; Zvezda is 13 m long, with a wingspan of 30 m. After Zvezda was connected, it took over propulsion and thruster controls from Zarya, which was then used mostly for its storage capacity and external fuel tanks. Unity serves as a conduit for essential resources running to the living and working areas, and contains more than 50 000 mechanical items, 216 fluid and gas lines, and 121 electrical cables using 9.7 km of wire. Destiny, the first of six planned *ISS* research modules, will serve as the command and control centre. It is 8.5 m long, with a widest point of 4.3 m. In 2004, it housed seven different research facilities.

http://spaceflight.nasa.gov/station/ Breaking news, reports, and articles about the *International Space Station*. There are details of expeditions to the space station and the on-board scientific investigations. A sophisticated calculator and graphical simulator can be used to determine the visibility and path of the space station in the night sky from your location.

International Space University (ISU) An educational institution in Strasbourg, France, that offers courses in space-related disciplines, including a Master of Space Studies (MSS) degree and Professional Development Programmes. The curriculum is geared towards postgraduates, professionals, and those wishing to enter the space sector. Subjects include space mission planning and systems architecture, orbital mechanics, space life sciences, launch vehicles, remote sensing, space and society, and law and policy related to space activities.

International Sun–Earth Explorer (ISEE) A programme consisting of two NASA satellites and one from the European Space Agency (ESA), that contributed to an international magnetospheric study. They carried energetic-ion mass spectrometers (EIMS) for one of the first uses of these instruments to examine ion plasma composition. NASA's *ISEE-1* and ESA's *ISEE-2* spacecraft were launched in October 1977 and re-entered the Earth's atmosphere in September 1987. NASA's *ISEE-3* was launched in August 1978 and was taken out of Earth orbit in 1983 to fly by the comet Giacobini–Zinner in September 1985. To reflect this mission change, *ISEE-3*'s name was changed to the **International Cometary Explorer*.

International Telecommunications Satellite Organization (Intelsat) An organization established in 1964 to operate a worldwide system of communications satellites. In 2001, it became a private company after 37 years as an intergovernmental organization. In 2004, it reached about 200 countries with 29 satellites in orbit. Its headquarters are in Washington, DC. Intelsat satellites are stationed in geostationary orbit over the Atlantic, Pacific, and Indian oceans.

***International Ultraviolet Explorer* (*IUE*)** A joint NASA–European Space Agency orbiting ultraviolet telescope with a 45-cm mirror, launched in 1978 to provide data on ultraviolet sources in space. It was switched off in September 1996, having become the longest-lived astronomical satellite.

interplanetary matter The gas, dust, and charged particles from a variety of sources that occupies the space between the planets. Dust left over from the formation of the Solar System, or from disintegrating comets, orbits the Sun in or near the ecliptic plane and is responsible for scattering sunlight to cause the *zodiacal light (mainly the smaller particles) and for *meteor showers in the Earth's atmosphere (larger particles). The charged particles mostly originate in the *solar wind, but some are cosmic rays from deep space.

***Interplanetary Monitoring Platform* (*IMP*)** A series of spacecraft in NASA's Explorer Program that mainly studied the *solar wind and the Earth's magnetosphere. They were launched over ten years, beginning with *IMP-1* (*IMP-A*) in November 1963 and ending with *IMP-8* (*IMP-J*), launched in October 1973. The latter carried 12 experiments, including a charged-particle measurements experiment (CPME) and a magnetic-field experiment. It also used an electrostatic analyser and solid-state particle telescope to determine the composition and energy of low-energy particles observed during solar flares. The project was closed down in November 2001.

interplanetary space The space existing between the planets of the Solar System. It contains interplanetary dust and small celestial bodies, such as asteroids. Few space probes travel beyond interplanetary space.

Intersputnik A Russian-based international organization of states operating a global satellite telecommunications system. It was established in 1971 by the former USSR to compete with *Intelsat, and in 2001 had 23 member nations. In 1997, Intersputnik and *Lockheed Martin formed a joint satellite-operating venture, Lockheed Martin Intersputnik, and their first satellite, *LMI-1*, was launched in 1999 to provide telecommunications services to the former Soviet states, as well as to European, Asian, and African subscribers. Intersputnik launched two Express-A-series satellites in 2000 to modernize its system, and in 2004 launched its *W3A* satellite to serve Europe, North Africa, and the Middle East, as well as increasing the *European Telecommunications Satellite Organization's (EUTELSAT's) coverage of sub-Saharan Africa. Intersputnik markets and sells EUTELSAT's satellite capacity.

interstellar cirrus The wispy cloud-like structures discovered in the mid-1980s by the **Infrared Astronomical Satellite* and believed to be the remains of dust shells blown into space from cool giant or supergiant stars.

interstellar matter A medium of electrons, ions, atoms, molecules, and dust grains that fills the space between stars in our own and other galaxies. Over 100 different types of molecule exist in gas clouds in our own Galaxy. Most have been detected by their radio emissions, but some have been found by the absorption lines they produce in the spectra of starlight. The most complex molecules, many of them based on carbon, are found in the dense clouds where stars are forming. They may be significant for the origin of life elsewhere in space.

It is only since the mid-20th century that scientists have realized that there is sufficient interstellar matter to have significant effects and that its extent largely determines the form and development of a galaxy. It is most easily observable in the radio region of the spectrum, but was first detected optically. Condensations of such matter are visible as nebulae, while over large parts of the sky interstellar matter dims, reddens, and polarizes the light of distant stars. It also causes a number of characteristic absorption lines in star spectra.

Early radio observations by US radio astronomers Karl Jansky and Grote Reber showed the general extent of interstellar matter; further observations plotted the distribution of its most abundant constituent, neutral hydrogen atoms. Later radio observations located hydroxyl, helium, water, ammonia, and many other molecules, some of them quite complex.

Interstellar matter is not smoothly distributed but occurs in dense and cold clouds. Its fundamental properties are largely determined by the hydrogen component. By mass, helium is 20–30% as abundant as hydrogen. All the other elements together do not amount to more than 3–5%.

Interstellar Probe A proposed NASA mission to send a space probe out of the Solar System to examine the interstellar medium, the ultra-thin gas between the stars. It involves using a solar sail of 200-m radius to accelerate the probe away from the Sun until it leaves the heliosphere, the bubble around the Sun created by the outward flow of the solar wind. In the course of this journey, the probe would explore the boundaries of the heliosphere to reveal how the Sun interacts with its environment, and directly sample the nearby interstellar medium.

http://interstellar.jpl.nasa.gov/ Outlining of a plan for a future interstellar probe. The site has clearly illustrated articles about the interstellar medium, the outer Solar System, and the interaction between the interstellar medium and the solar wind.

intravehicular activity (IVA) The activity by an astronaut within a spacecraft. This includes scientific experiments, maintenance procedures, and giving directions to other crew members during their *extravehicular activity. A study of motions during IVA on the *International Space Station* has been proposed by the Agenzia Spaziale Italiana (ASI; Italian Space Agency). This analysis by the ASI's TE-S2 system, using eight television cameras, would study motion problems caused by microgravity and the adaptation processes.

Io The third-largest moon of the planet Jupiter, 3 643 km in diameter, orbiting in 1.77 days at a distance of 422 000 km. It is the most volcanically active body in the Solar System, covered by hundreds of vents that erupt sulphur (rather than lava), giving Io an orange-coloured surface. Io and Earth are the only two planetary bodies that are undergoing known high-temperature volcanism.

Data gathered by the spacecraft **Galileo* in 1996 indicated that Io has a large metallic core. The *Galileo* probe also detected a 10-megawatt beam of electrons flowing between Jupiter and Io.

In 1997, instruments aboard the spacecraft *Galileo* measured the temperature of Io's volcanoes and detected a minimum temperature of 1 800 K/1 500 °C (in comparison, Earth's hottest volcanoes only reach about 1 600 K/1 300°C).

ionizing radiation Radiation that removes electrons from atoms during its passage, thereby leaving ions in its path. Alpha and beta particles are far more ionizing in their effect than are neutrons or gamma radiation.

IPS Abbreviation for *instrument pointing system.

irregular galaxy A class of galaxy with little structure, which does not conform to any of the standard shapes in the *Hubble classification. The two satellite galaxies of the *Milky Way, the Magellanic Clouds, are both irregulars. Some galaxies previously classified as irregulars are now known to be normal galaxies distorted by tidal effects or undergoing bursts of star formation (*see* STARBURST GALAXY).

IRS Abbreviation for either *Indian Remote-Sensing Satellite or *infrared spectrometer.

Irwin, James Benson (1930–1991) US astronaut. As pilot of the *lunar module during the *Apollo 15* mission in 1971, he drove the first *lunar roving vehicle, spending 18 hours on the Moon, travelling over 25 km. He and David *Scott collected 77 kg of Moon soil and rocks.

Irwin was a test pilot for the US Air Force before being chosen as an astronaut in 1966. He retired from NASA in 1972. After his retirement, he founded the High Flight Foundation, a Christian organization, and served as its president and chair until his death.

ISAS Acronym for *Institute of Space and Astronautical Science.

ISCCP Abbreviation for *International Satellite Cloud Climatology Project.

ISEE Abbreviation for **International Sun–Earth Explorer*.

ISO Abbreviation for **Infrared Space Observatory*.

ISOE Abbreviation for *integrated sequence of events.

isogrid An integral triangular lattice structure, normally made of an aluminium alloy, that is used to reinforce parts of spacecraft. Developed by the *Boeing Company, it is fused to such areas as fuel tanks and payload fairings to provide extra strength.

ISOPHOT Abbreviation for *infrared photo-polarimeter.

ISP Abbreviation for *specific impulse.

ISRO Acronym for *Indian Space Research Organization.

ISS Abbreviation for **International Space Station.*

ISU Abbreviation for *International Space University.

Italian Space Agency English name for the *Agenzia Spaziale Italiana.

IU Abbreviation for *instrument unit.

IUS Abbreviation for *inertial upper-stage rocket.

IVA Abbreviation for *intravehicular activity.

Ivins, Marsha Sue (1951–) US astronaut. She logged more than 1 318 hours in space on five shuttle flights. These include the *Atlantis* mission (January 1997) to dock with the Russian space station *Mir*, and a further *Atlantis* mission (February 2001) to construct and supply the *International Space Station* (*ISS*). She has since been assigned to three branches involving the *ISS*, shuttle, and advanced projects. Ivins joined the Johnson Space Center as an engineer in 1974 and was selected as an astronaut in 1984.

J-2 engine A rocket engine produced in the USA by the Rocketdyne Propulsion and Power Division of the *Boeing Company. It was used in the Saturn V launch vehicle for the *Apollo* programme, with five J-2 engines on the second stage and one on the third stage. The engine is more than 3.4 m long and has a thrust of 101 250 kg. It uses liquid-hydrogen and liquid-oxygen propellants.

James Webb Space Telescope (JWST) An orbiting infrared observatory that will take the place of the *Hubble Space Telescope*. It will study the Universe as it appeared when galaxies were forming during the first billion years or so after the Big Bang. As light from such great distances is subjected to very high *redshift, such observations are best performed in the infrared. The *JWST* will be capable of detecting radiation at wavelengths from 0.6 to 28 microns. In addition, it will be able to see objects 400 times fainter than those visible to the largest ground-based infrared telescopes or the current generation of space-based infrared telescopes, and with a resolution comparable to the *Hubble Space Telescope (HST)*.

To achieve these objectives, *JWST* will have a beryllium mirror 6.5 m in diameter consisting of 18 hexagonal segments. The mirror will be folded for launch, as will the large sunshield, the size of a tennis court, that will shade it from the Sun's rays once in space. It will be equipped with an infrared camera and spectrometer. The *JWST* will be positioned at the L2 *Lagrangian point of the Earth's orbit, 1.5 million km from Earth on the side away from the Sun.

The telescope is named after James E. Webb, a former NASA administrator. It is being built by NASA with contributions from ESA and Canada and is scheduled for launch in 2012.

http://www.jwst.nasa.gov/home/ Official NASA site chronicling the development of the telescope and its instruments, with a summary of the questions in cosmology it is designed to address.

jansky The unit of radiation received from outer space, used in radio astronomy. It is equal to 10^{-26} watts per square metre per hertz, and is named after US engineer Karl Jansky.

Japan Aerospace Exploration Agency (JAXA) The Japanese national space agency formed in October 2003 from the merger of three previous organizations: the Institute of Space and Astronautical Science (ISAS), which was devoted to space and planetary research; the National Aerospace Laboratory of Japan (NAL), which focused on aviation research and development; and the National Space Development Agency of Japan (NASDA), which was responsible for the development of launch vehicles, satellites and Japanese contributions to the *International Space Station*.

(Image © NASA)
Artist's view of the *Japanese Experiment Module* (*JEM*), one of five international research laboratories attached to the *International Space Station*. The cylindrical sections are the pressurized module (PM). The square units in the foreground below *JEM*'s robot arm are the exposed facility (EF).

***Japanese Experiment Module* (*JEM*; or *Kibo*)** A Japanese facility forming part of the *International Space Station* (*ISS*). *Kibo*'s components are scheduled for launch by the *National Space Development Agency (NASDA) in 2006. *Kibo* is Japan's first crewed spacecraft, designed to accommodate four astronauts to conduct experiments over a long duration.

The major part of *Kibo* is a pressurized module (PM), which is 11.2 m long and 4.4 m in diameter. It is furnished with ten racks of equipment to be used mainly for microgravity experiments. The other experimental component is the exposed facility (EF) located outside the *ISS*, enabling direct use of the space environment. Both components include an experiment logistics module (ELM) attached as a storage area. The ELMs are detachable, allowing their return to Earth via space shuttle where they will be filled with cargo for another launch. The final unit of *Kibo* is a remote manipulator system (JEMRMS), to move experiment payloads from the PM to other areas.

Jason A spacecraft mission between NASA and France's *Centre National d'Etudes Spatiales launched on 7 December 2001 to monitor the Earth's oceans for a period of five years. The mission is recording ocean circulation, measuring global sea-level changes, monitoring special natural events, such as El Niño, and exploring the links between the oceans and atmosphere. *Jason*, with a mass of 500 kg, orbits at an altitude of 1 336 km and carries such instruments as an altimeter, radiometer, and laser retroflector array. It works together with the **TOPEX/Poseidon* satellite to discover ties between the oceans and the atmosphere. The *Jet Propulsion Laboratory is in charge of mission operations.

JAXA Abbreviation for *Japan Aerospace Exploration Agency.

JEM Acronym for **Japanese Experiment Module*.

Jemison, Mae Carol (1956–) US astronaut and physician. She made history as the first black woman in space, on the first Japanese **Spacelab* mission, aboard the space shuttle *Endeavour* in September 1992. Jemison was selected as an astronaut in 1987 and left NASA in 1993.

Jernigan, Tamara Elizabeth (1959–) US astronaut. Her space shuttle flights include the *Discovery* mission in May 1999 for the first docking with the *International Space Station* (*ISS*). She has logged more than 1 512 hours in space and nearly eight hours of extravehicular activity. She next served as the lead astronaut for *ISS* external maintenance before leaving NASA. Jernigan served as a research scientist at NASA's Ames Research Center from 1981 to 1985, when she was selected as an astronaut.

jet A narrow luminous feature seen protruding from a star or galaxy and representing a rapid outflow of material. *See* ACTIVE GALAXY.

jet propulsion A method of propulsion in which an object is propelled in one direction by a jet, or stream of gases, moving in the other.

The **ramjet** is used for some types of missiles. At twice the speed of sound (Mach 2), pressure in the forward-facing intake of a jet engine is seven times that of the outside air, a compression ratio that rapidly mounts with increased speed (to Mach 8), with the result that no compressor or turbine is needed. The ramjet consists of an open-ended, barrel-shaped tube, burning fuel in its widest section. It is cheap, light, and easily made. However, fuel consumption is high and it needs rocket-boosting to its operational speed.

Jet Propulsion Laboratory (JPL) A NASA installation at Pasadena, California, operated by the *California Institute of Technology. Established in 1944, it is the command centre for NASA's deep-space probes such as the **Voyager*, **Magellan*, and **Galileo* missions, with which it communicates via the *deep-space network. It is the leading US centre for robotic exploration of the Solar System, with its spacecraft having visited all the planets except Pluto. It is managing the *Mars Exploration Rover mission.

http://www.jpl.nasa.gov/ Frequently updated site with news and articles on Solar System exploration. Pages describe new technologies and the laboratory's Earth-observation programme; there are activities for younger students, and the acclaimed Solar System Simulator. If you've ever wondered how Saturn would look from the vicinity of Titan, or Earth from Pluto, the Solar System Simulator is for you—you can quickly and easily generate your own views of the Solar System.

Jett, Brent Ward, Jr (1958–) US astronaut. His three space shuttle flights include the *Atlantis* mission in January 1997 to dock with the Russian space station *Mir* for the second exchange of astronauts. In November 2000, he flew with the *Endeavour* mission to assemble the *International Space Station.* He has logged more than 669 hours in space. His 2004 shuttle flight was postponed after the 2003 *Columbia* space shuttle disaster. Jett was selected as an astronaut in 1992.

jettison The act of releasing a depleted or unneeded part of a spacecraft into space. A space shuttle jettisons its solid rocket boosters and external fuel tank before achieving orbit. Space probes landing on a planet or moon jettison their *aeroshells in order to carry out their exploratory missions.

Jiuquan Satellite Launch Centre An alternative name for *Shuang Cheng Tzu.

Johnson Space Center (JSC) A NASA field centre at Houston, Texas, home of the *mission control team for crewed space missions. Established in 1961, it is also NASA's main centre for the design and development of spacecraft, and the location for the **Astronaut Selection Office** and the training of astronauts. By 2004, it had trained 295 US astronauts and 50 from other countries.

http://www.jsc.nasa.gov/ Large NASA site with information on all aspects of crewed space exploration. There are pages dedicated to the history of the Johnson Space Center (JSC) and crewed space flight as well as the latest news of space shuttle missions and work aboard the *International Space Station.* The JSC site hosts a vast online image archive, with more than 250 000 press releases and Earth-observation images.

Jones, Thomas David (1955–) US astronaut. By 2001, he had logged 1 272 hours in space on four flights, including the *Atlantis* space shuttle mission in February 2001 to deliver the Destiny laboratory module to the *International Space Station.* During the mission, he performed three space walks, totalling more than 19 hours, to assist its installation. He has since left NASA. Jones worked on developing NASA's space probe missions before being selected as an astronaut in 1990.

JPL Abbreviation for *Jet Propulsion Laboratory.

JSC Abbreviation for *Johnson Space Center.

Juno rocket An early NASA rocket. Juno 1 launched the first successful US satellite, *Explorer 1*, in January 1958. The rocket was created by adding a fourth-stage Sergeant engine to a three-stage Jupiter-C rocket. In 1961, the power of the Juno 5 rocket was increased by the installation of a cluster of engines, and it was renamed the *Saturn rocket.

Jupiter The fifth planet from the Sun and the largest in the Solar System, with a mass equal to 70% of all the other planets combined and 318 times larger than that of the Earth. Its main feature is the Great Red Spot, a cloud of rising gases, 14 000 km wide and 30 000 km long, revolving anticlockwise.

mean distance from the Sun 778 million km
equatorial diameter 142 980 km
rotation period (equatorial) 9 hours 51 minutes
year 11.86 Earth years

atmosphere consists of clouds of white ammonia crystals, drawn out into belts by the planet's high speed of rotation (the fastest of any planet). Darker orange and brown clouds at lower levels may contain sulphur, as well as simple organic compounds. Temperatures range from −140 °C in the upper atmosphere to as much as 24 000 °C near the core. This is the result of heat left over from Jupiter's formation, and it is this that drives the turbulent weather patterns of the planet. The Great Red Spot was first observed in 1664. Its top is higher than the surrounding clouds; its colour is thought to be due to red phosphorus. The Southern Equatorial Belt in which the Great Red Spot occurs is subject to unexplained fluctuation. In 1989 it sustained a dramatic and sudden fading. Jupiter's strong magnetic field gives rise to a large surrounding magnetic 'shell', or magnetosphere, from which bursts of radio waves are detected. Jupiter's faint rings are made up of dust from its moons, particularly the four inner moons

surface largely composed of hydrogen and helium, which under the high pressure and temperature of the interior behave not as a gas but as a supercritical fluid. Under even more extreme conditions, at a depth of 30 000 km, hydrogen transforms into a metallic liquid. Jupiter probably has a molten rock core whose mass is 15 to 20 times greater than that of the Earth.

In 1995, the **Galileo* probe revealed Jupiter's atmosphere to consist of 0.2% water, less than previously estimated

satellites Jupiter has over 60 known moons, although most are only small. The four largest moons, Io, Europa, Ganymede, and Callisto, are the *Galilean satellites, discovered in 1610 by Galileo Galilei (Ganymede, which is larger than Mercury, is the largest moon in the Solar System). In 1979 the US **Voyager* probes discovered a faint ring of dust around Jupiter's equator 30 000–145 000 km above the cloud tops

The comet Shoemaker-Levy 9 crashed into Jupiter in July 1994. Impact zones were visible for several months.

http://www.solarviews.com/eng/jupiter.htm Full details of the planet and its moons, including a chronology of exploration, various views of the planet and its moons, and links to other planets.

JWST Abbreviation for **James Webb Space Telescope.*

Kagoshima Space Centre A Japanese launch site in Uchinoura, Kagoshima Prefecture, founded in February 1962. It was renamed *Uchinoura Space Centre (USC) when the Japan Aerospace Exploration Agency (JAXA) was formed in October 2003.

Kapustin Yar A small launch complex in Russia, last used for a space launch in 1987. Located about 100 km southeast of the city of Volgograd, it was the launch site of small satellites and sounding rockets. Space activity at the site was frequent in the 1960s but averaged only one launch a year throughout the 1970s and 1980s.

Kavandi, Janet Lynn (1959–) US astronaut. She flew aboard the space shuttle *Discovery* in June 1998, the final mission to dock with the Russian space station *Mir*. In February 2000, she flew on *Endeavour*, as part of the *Shuttle Radar Topography Mission. In July 2001, she flew aboard *Atlantis* to install a new airlock module on the *International Space Station*. She has logged 33 days in space. Kavandi joined the Boeing Company in 1984, working on such space projects as the *Inertial Upper-Stage rocket, before being selected as an astronaut in 1994.

KC-135 aircraft A modified Boeing jet transport aircraft used by NASA for weightless training. The four-engine plane flies in parabolas, which produces 25 to 30 seconds of weightlessness over the top of the course. At this time, astronauts (who have named the plane the 'Vomit Comet') practice using different types of space shuttle equipment, as well as eating and drinking. Each week there are normally four flights of 40 parabolas each, with training sessions lasting between one and two hours. The KC-135 is also used for zero-gravity experiments.

http://jsc-aircraft-ops.jsc.nasa.gov/kc135/index.html Describes the KC-135 aircraft used for microgravity research and astronaut training. As well as technical and safety information, there are diagrams, photographs, and links to other sites with more information about the aircraft.

Kelly, James McNeal (1964–) US astronaut. He was the pilot on the March 2001 flight of the space shuttle *Discovery*, which transported the second resident crew to the *International Space Station*. His second flight was as pilot of *Discovery* in July–August 2005 on the first mission since the Columbia disaster of 2003. He has logged more than 307 hours in space. Kelly was selected as an astronaut in 1996.

Kelly, Mark Edward (1964–) US astronaut. He was selected as an astronaut in 1996 and his first space shuttle flight was as the pilot of *Endeavour* in December 2001 to the **International Space Station*, which included the exchange of resident crews. He is the twin of the US astronaut Scott Kelly.

Kelly, Scott Joseph (1964–) US astronaut. He was the pilot of the *Discovery* space shuttle mission that upgraded the **Hubble Space Telescope* in December 1999. His flight in 2004 was postponed following the 2003 *Columbia* space shuttle disaster. Kelly was selected as an astronaut in 1996. He is the twin of the US astronaut Mark Kelly.

Kennedy, John F(itzgerald) (1917–63) The 35th president of the USA 1961–3. An enthusiastic supporter of space exploration, he called space 'our great New Frontier'. Kennedy initiated the *space race in May 1961 by declaring that the USA should land a person on the Moon within a decade.

Kennedy Space Center (KSC) A NASA launch site on Merritt Island, near *Cape Canaveral, Florida, used for **Apollo* project and *space shuttle launches. It was established in 1962 and celebrated its 40th anniversary in 2002. The first *Apollo* flight to land on the Moon (1969) and **Skylab*, the first orbiting laboratory (1973), were launched from the site. In 2003, it launched the two spacecraft of the *Mars Exploration Rover mission.

The centre, named after US president John F Kennedy, is dominated by the *Vehicle Assembly Building, 160 m tall, used for assembling *Saturn rockets and space shuttles. The centre is in charge of the checkout, launch, and landing of shuttles. On 12 August 2003, the deputy director, James W Kennedy, succeeded Roy D Bridges, to become the eighth director of KSC.

http://www.ksc.nasa.gov/ Latest news and information about space shuttle and rocket launches, progress on the *International Space Station*, development of spaceport technology, and other activities at the Kennedy Space Center. There are plentiful textual and multimedia resources, including a space flight archive profiling crewed space programmes and an excellent history of space flight.

Kennelly–Heaviside layer Former term for the *E-layer of the ionosphere.

Kepler A NASA satellite designed to detect Earth-sized planets around other stars. *Kepler* will continuously monitor 100 000 main-sequence stars in a 10-degree-square patch of sky in the constellation Cygnus for four years. It carries a 0.95-m diameter telescope with a highly sensitive photometer that will detect changes in the stars' brightness due to planets passing in front of them. To detect an Earth-size planet, the photometer must be able to sense a drop in brightness of only 0.01 of a percent. *Kepler* is scheduled for launch in 2008.

Kepler's laws Three laws of planetary motion formulated in 1609 and 1619 by German mathematician and astronomer Johannes Kepler: (1) the orbit of each planet is an ellipse with the Sun at one of the foci; (2) the radius vector of each planet sweeps out equal areas in equal times; (3) the squares of the periods of the planets are proportional to the cubes of their mean distances from the Sun.

Kepler derived the laws after exhaustive analysis of numerous observations of the planets, especially Mars, made by Danish astronomer Tycho Brahe without telescopic aid. British physicist and mathematician Isaac Newton later showed that Kepler's laws were a consequence of the theory of universal gravitation.

Kerwin, Joseph Peter (1932–) US astronaut and physician. He was the first US physician in space, when he flew on *Skylab 2* (*see* SKYLAB) in May 1973 as the science-pilot of the first crewed mission to the *Skylab* space station. He studied the crew's reaction to microgravity and carried out an extravehicular activity. Kerwin was selected as an astronuat in 1965 and resigned from NASA in 1987.

Kevlar A polymer that is thin but five times stronger than steel. Created in 1965 by scientists at the DuPont company, it is used to reinforce space shuttle suits and in gloves worn by astronauts, being fine enough to allow the picking up of items as thin as a coin.

Khrunichev Space Centre (KhSC; officially Khrunichev State Research and Production Space Centre) A Russian company that builds spacecraft, including the *Proton rocket, various satellites, and the Zvezda module for the **International Space Station* (*ISS*). KhSC is an active partner with Western nations, including a project with *Lockheed Martin to produce the simpler, low-cost Angara rocket. The more advanced Angara A4B was scheduled to be launched in 2005.

Located in Moscow, KhSC is a former defence enterprise. Since 1995, KhSC has also used its space technologies to produce medical equipment, such as pressure chambers to treat respiratory and cardiovascular diseases. In 1997, it joined with Motorola for the Iridium system of global satellites for mobile telephones.

Khrushchev, Nikita Sergeyevich (1894–1971) Soviet politician, secretary general of the Communist Party 1953–64, premier 1958–64. His policy of competition with capitalism was successful in the space programme, which launched the world's first satellite, **Sputnik*, and the first person in space, Yuri *Gagarin.

KhSC Abbreviation for *Khrunichev Space Centre.

Kibo Alternative name for the **Japanese Experiment Module*.

kick stage Any stage of a *multistage rocket used to provide extra velocity to propel a spacecraft into its designated trajectory.

Komarov, Vladimir (1927–67) Soviet cosmonaut. Selected in 1959 as one of the first cosmonauts, he flew as commander of the three-person *Voskhod 1* (*see VOSKHOD*) mission in October 1964, conducting scientific and medical research. He piloted the trouble-plagued *Soyuz 1* (*see SOYUZ*) orbital flight that ended in his death in April 1967 when a parachute failed to open and it plummeted to the ground with Komarov inside.

Kondakova, Elena Vladimirovna (1957–) Russian cosmonaut. She was flight engineer aboard the Russian space station *Mir* from October 1994 to March 1995. In May 1997, she was a mission specialist on the space shuttle *Atlantis* when it docked with *Mir*. Konakova was a researcher for the *Energiya Rocket and Space Complex from 1980 and was selected as an astronaut in 1989.

Koptev, Yuri Nikolayevich (1940–) The director general of *Rosaviakosmos from its founding in 1992 until 2004. He oversaw the deorbiting of the *Mir* space station voyage, and the controversial visit of US industrialist and space tourist Dennis Tito to the *International Space Station*. He is credited with the recovery of the declining Russian space programme to pre-1993 levels. In 2004, he became chief of Rosoboronexport, Russia's state arms exporter.

Korolev, Sergei Pavlovich (1906–66) The Russian designer of the first Soviet intercontinental missile, used to launch the first **Sputnik* satellite in 1957, and of the **Vostok* spacecraft in which Yuri *Gagarin made the world's first space flight in 1961.

Korolev and his research team built the first Soviet liquid-fuel rocket, launched in 1933. His innovations in rocket and space technology include ballistic missiles, rockets for geophysical research, launch vehicles, and crewed spacecraft. Korolev was also responsible for the **Voskhod* spaceship, from which the first space walks were made.

Korolev was a member of the Institute for Jet Research from its foundation in 1933, becoming head of the Rocket Vehicle Department in 1934, and worked as an engine designer 1924–46. Later he was appointed head of the large team of scientists who developed high-powered rocket systems.

Korolev published his first paper on jet propulsion in 1934. By 1939 he had designed and launched the Soviet 212 guided wing rocket. This was followed by the RIP-318-1 rocket glider, which made its first piloted flight in 1940.

http://www.hq.nasa.gov/office/pao/History/sputnik/korolev.html
Biography of the Soviet rocket designer from NASA, the institution he spent his career attempting to outdo. There is a photograph of Korolev, an account of his technical achievements, and details of the imprisonment he suffered at the hands of Stalin.

Koronas Alternative spelling of **Coronas*.

Korzun, Valeri Grigorievich (1953–) Russian cosmonaut. He spent 197 days aboard the Russian space station *Mir* from August 1996 to March 1997. During his residency he performed two space walks totalling more than 12 hours. He was a member of the fifth resident crew on the **International Space Station*, from June to December 2002, taking two space walks to install panels and a frame. Korzun graduated from the Gagarin Air Force Academy in 1987, and was selected as a cosmonaut in the same year.

Kourou The second-largest town of French Guiana, northwest of Cayenne, site of the *Centre Spatial Guyanais (Guianese Space Centre) of the European Space Agency (ESA); population (1996 est) 20 000 (20% of the total population of French Guiana). ESA launches have included *SMART-1* on 27 September 2003, a flight to orbit the Moon, and **Rosetta* on 2 March 2004, a mission to land on a comet.

Situated near the Equator, it is an ideal site for launches of satellites into *geostationary orbit.

Kregel, Kevin Richard (1956–) US astronaut. His four space flights include piloting the *Discovery* space shuttle mission in July 1995 to deploy the sixth and final *Tracking and Data-Relay Satellite, and in February 2000 commanding *Endeavour* on the *Shuttle Radar Topography Mission. He was then assigned to the Space Launch Initiative Project at the Johnson Space Center before leaving NASA. Kregel joined NASA in 1990 as an aerospace engineer and instructor pilot, and was selected as an astronaut in 1992.

Krikalev, Sergei Konstantinovich (1957–) Russian cosmonaut. His space flights include two long residencies as a flight engineer on the Russian space station *Mir*. His first mission on the station lasted from November 1988 to April 1989 and the second from May 1991 to March 1992. He was also a member of the first resident crew aboard the *International Space Station*, from October 2000 to March 2001. Krikalev logged more than one year and five months in space, with seven space walks. He joined the *Energiya Rocket and Space Complex in 1981, testing space-flight equipment, and was selected as a cosmonaut in 1985.

Kuiper belt A ring of small, icy bodies orbiting the Sun beyond the outermost planet. The Kuiper belt, named after Dutch-born US astronomer Gerard Kuiper, who proposed its existence in 1951, is thought to be the source of comets that orbit the Sun with periods of less than 200 years. The first member of the Kuiper belt was seen in 1992. In 1995 the first comet-sized objects were discovered; previously the only objects found had diameters of at least 100 km (comets generally have diameters of less than 10 km).

LAGEOS Contraction of **Laser Geodynamics Satellite*.

Lagrangian point One of five locations in space between two bodies where the centrifugal and gravitational forces of the two bodies neutralize each other; a third, less massive body located at any one of these points will be held in equilibrium with respect to the other two. Three of the points, L1–L3, lie on a line joining the two large bodies. The other two points, L4 and L5, which are the most stable, lie on either side of this line. For example, there is a Lagrangian point between the Earth and the Sun 1.5 million km from the Earth. Here the Sun's gravity and the Earth's gravity are equal, meaning that a spacecraft can orbit the point as if it were a planet. The points were discoverd by the 18th-century Italian-born French mathematician, Joseph Lagrange.

lander A space probe that lands on a planet or moon. Among the best-known landers are the **Mars Pathfinder*, the *Mars Exploration Rovers*, the lunar **Surveyor* series, and the **Venera* missions to Venus. Landers are designed to withstand harsh conditions long enough to allow them to transmit significant data back to Earth stations.

landing site An area in which a spacecraft lands on the Earth, another planet, or a moon. On Earth, the prime landing sites for a space shuttle are Edwards Air Force Base (EAFB) in California and the Kennedy Space Center (KSC) in Florida, from where the craft is also launched. The EAFB, which experiences better weather conditions and has concrete and dry lake-bed runways, is preferred when a payload makes the shuttle heavier than normal. Landing at the KSC, however, saves around US$1 million as well as up to seven days of processing, because the shuttle does not have to be returned to Florida. The Pacific Ocean was chosen as the *Apollo* landing site and is often used for deactivated, uncrewed satellites. The Russian space station *Mir* was also brought down into the Pacific Ocean, on 23 March 2001.

Landsat A series of US satellites launched from 1972 that monitor the Earth's resources and the effects of pollution, deforestation, and desertification.

Landsat 1, originally called *ERTS-A* (*Earth Resources Technology Satellite-A*), orbited 1972–5, acquiring more than 300 000 images of the Earth's surface. *Landsat 4*, launched in June 1982, contained a new kind of technology, the thematic mapper (TM). The *National Oceanic and Atmospheric Administration took over the running of the Landsat programme from NASA in 1983.

Langley Research Center A NASA facility in Hampton, Virginia, used for aeronautics research. Langley was established in 1917 as the USA's first civil aeronautics laboratory. It conducts programmes in atmospheric sciences and develops technology for advanced space transportation systems, such as the X-33 reusable launch vehicles (due to replace the space shuttle, but cancelled in 2001). The centre also does research in such areas as remote sensing, spacecraft structures, materials, and design. Its design expertise has been used for planetary probes, including the **Mars Pathfinder*. Its Wind Tunnel Enterprise has 28 facilities. In 2004, Langley had 3 800 civil service and contract employees.

http://www.larc.nasa.gov/ Details the centre's activities in aeronautics, Earth science, space technology, and structures and materials. In addition to news articles on current research, users can obtain technical reports and fact sheets. The site is well-organized and easy to navigate.

***Laser Geodynamics Satellite* (*LAGEOS*)** A NASA series of passive *satellites whose data are used to measure the motion of the Earth's tectonic plates, its gravitational field, and the wobble in its rotation. They can also provide a more accurate measurement of the length of an Earth day. The satellites have aluminium shells covered with retroreflectors, which reflect laser beams transmitted from ground stations around the world.

LAGEOS-1 was launched in May 1976 into a circular orbit 5 800 km above the Earth. *LAGEOS-2* was built by the Agenzia Spaziale Italiana (Italian Space Agency), as a joint project with NASA. It was launched in October 1992 into an orbit 5 900 km above the Earth. A third multinational satellite is planned. Both satellites should remain in orbit for 8.4 million years, and *LAGEOS-1* contains a plaque for future generations. It depicts three maps of the Earth and its continents: as they looked 268 million years ago, as they appear today, and as they will appear at the end of the spacecraft's life.

***Laser Interferometer Space Antenna* (*LISA*)** A joint ESA/NASA mission to detect gravitational waves from objects such as black holes. LISA will consist of three spacecraft flying in triangular formation 5 million km apart, 20 degrees (50 million km) behind the Earth in its orbit. Together, they act as a Michelson interferometer to measure the distortion of space caused by passing gravitational waves. Lasers in each spacecraft will measure minute changes in the separation distances of free-floating solid blocks, known as 'proof masses', within each spacecraft. A gravitational wave passing through the spacecraft will change the separation between the proof-masses to vibrate. These tiny movements are observed by means of laser interferometry. *LISA* is due for launch in 2014. Technical challenges involved will be addressed by a precursor mission known as **LISA Pathfinder*.

laser ring gyroscope A spacecraft gyroscope that can detect the *Doppler effect that is initiated in light beams when the gyro is rotated. It does this using *interferometry, producing wave interference to measure wavelength displacement. An advantage of the laser ring gyroscope is that it does not have moving parts that can wear out.

launch The act of propelling a space vehicle through the atmosphere and into space. This normally involves rockets, a satellite or space probe, and the payloads on board the spacecraft. Although a launch begins with lift-off, months are spent on pre-launch activities.

http://www.fas.org/spp/guide/usa/launch/ Part of the World Space Guide, this site includes statistics and a history of the USA's launch systems. Each spacecraft description is accompanied by a simple diagram or photograph.

launch and early orbit phase (LEOP) A designation used by the European Space Agency for the launch and first few orbital revolutions of a spacecraft. The launch phase begins 10 hours before launch and ends 18 minutes and 25 seconds after lift-off as the satellite separates from its launch vehicle. Before a satellite goes into orbit, its on-board computer initiates an acquisition sequence of automatic operations, pointing the payload instruments to the Earth and otherwise preparing the spacecraft for orbit. The early orbit phase during the first few hours comprises operations on board and at ground stations to make the spacecraft fully operational.

launch azimuth The direction taken by a space vehicle as it is launched. The launch azimuth is given by the angle between the vehicle's trajectory (mapped onto the horizon) and a vertical line from the north point of the horizon. If the vehicle is heading due east, the launch azimuth is 90°. The Kennedy Space Center allows launch azimuths between 35° (to the northeast) and 120° (to the southeast), to avoid flying over populated areas.

launch control centre (LCC) NASA's four-storey building at the *Kennedy Space Center that houses the team that oversees all aspects of a spacecraft's *checkout and launch operation. The automatic *launch processing system is located on the first floor, and the *firing rooms are on the second. The LCC is attached to the *Vehicle Assembly Building and is 115 m long, 55 m wide, and 23.5 m high.

launch/entry suit (LES) The pressure suit worn by astronauts during launch and re-entry. The suit is coloured orange for detection during an ocean rescue, is partially pressurized, and has an oxygen supply, parachute, life raft, and additional survival equipment.

launch escape system A system used to rescue astronauts during a launch emergency. It consists of a small rocket or rockets on a tower above the command module. If the launch fails, the rockets fire to lift the spacecraft clear. The *Mercury* and *Apollo* spacecraft had one escape rocket on a 3-m high steel frame. An abort handle to the left of the commander's seat would activate the rocket's 25 200-kg thrust. During a successful launch, the tower was jettisoned at the altitude of 97 km, which was too high for its use. A space shuttle has no launch escape system: the weight of any rocket that could pull the shuttle out and away from a malfunctioning solid rocket booster or external fuel tank would be prohibitive.

launch pad A prepared area on which a spacecraft stands when it is fuelled and launched. The pad includes such support facilities as the service tower, *flame deflector, and safety equipment. Each launch pad has a designation, such as Pad 39A at the *Kennedy Space Center from which space shuttles have been launched.

(Image © NASA)
The first space shuttle *STS-1* awaits lift-off. This timed exposure shows clearly the intricate fixed and rotating service structures, as well as the launch platform itself.

launch phase A timetable of operations during the launching of a spacecraft. At NASA, it is part of the *assembly, test, and launch operations. Its launch phase is carried out by personnel at the *Space Flight Operations Facility in the *Jet Propulsion Laboratory, and by engineers and controllers involved in the upper-stage vehicle. This phase is followed by the *cruise phase.

launch processing system (LPS) An automated system at the Kennedy Space Center, Florida, that performs a checkout of the space shuttle for launch, handles other launch operations, and conducts the countdown. It is located in the *launch control centre. The LPS interfaces with the shuttle's systems and main engines, the solid rocket boosters, and the external tank. Computer programs perform the checkout and, if unsatisfactory, provide data that isolate the fault.

launch site The geographical area and its launch complex from which spacecraft are launched. Each complex includes *launch pads and a control centre. The best location for a launch site is the Equator, which gives spacecraft a full *free ride due to the Earth's rotation. However, the only major facility near the Equator is the European Space Agency's *Centre Spatial Guyanais at Kourou in French Guiana.

Other sites, among more than a dozen in the world, include the *Kennedy Space Center at Cape Canaveral, Florida, the *Baikonur Cosmodrome in Kazakhstan, and the *Uchinoura Space Centre in Japan.

launch umbilical tower (LUT) Former name for *umbilical tower.

launch vehicle The rocket or system of rockets used to launch spacecraft. 'Launch vehicle' is an inclusive term used in space exploration, because *booster rockets are sometimes added, and a *multistage rocket can be composed of two or three different rockets.

http://www.esa.int/export/esaLA/index.html Provides details of ESA rockets, launch programme, and development. There is a description of the European spaceport of Kourou and lots of pictures of the installation in the site's image gallery.

launch weight Alternative term for *gross lift-off weight.

launch window The time span during which a spacecraft can be launched safely and best accomplish its mission trajectory. This can vary from a few hours to a period of several weeks.

The launch window depends on such factors as the weather, the Earth's rotation and orbital motion, and the motion of the destination planet or rendezvous target, for example the *International Space Station*. Launch windows for missions to the planet Mars occur approximately only every 26 months.

Lawrence, Wendy Barrien (1959–) US astronaut. Two of her four space shuttle flights docked with the Russian space station *Mir*: the *Atlantis* mission in September 1997 and the *Discovery* flight in June 1998 (the final shuttle–*Mir* docking). In July–August 2003 she flew aboard *Discovery* on the first space shuttle mission since the *Columbia* disaster of 2003. Lawrence was selected as an astronaut in 1992.

LCC Abbreviation for *launch control centre.

LCG Abbreviation for *liquid cooling garment.

LDEF Abbreviation for **Long-Duration Exposure Facility*.

LECP Abbreviation for *low-energy charged-particle detector.

Lee, Mark Charles (1952–) US astronaut. His space shuttle missions include the *Atlantis* mission in May 1989 to deploy the **Magellan* spacecraft. In February 1997, he flew with the *Discovery* mission to upgrade and repair the **Hubble Space Telescope*, when he conducted three space walks totalling more than 19 hours. Lee was selected as an astronaut in 1984. He retired from NASA and the US Air Force in 2001.

Leestma, David Cornell (1949–) US astronaut. His three space shuttle flights include the *Challenger* mission of October 1984 that demonstrated the feasibility of satellite refuelling. Leestma was selected as an astronaut in 1980, and in 1998 became deputy director of engineering at NASA's Johnson Space Center (JSC). In 2001, he became JSC project manager for the Space Launch Initiative and is now assistant programme manager for the Orbital Space Plane.

Lenoir, William Benjamin (1939–) US astronaut. He was aboard the first operational flight of the space shuttle *Columbia* in November 1982. Lenoir was involved in satellite experiments as an assistant professor at the *Massachusetts Institute of Technology from 1965, before his selection as a scientist-astronaut in 1967. He resigned from NASA in 1984, returning 1989–92 as associate administrator for space flight.

lenticular galaxy A lens-shaped galaxy with a large central bulge and flat disc but no discernible spiral arms.

LEO Abbreviation for *low Earth orbit.

Leonardo An uncrewed *multi-purpose logistics module (MPLM) built by the Italian Space Agency to serve the **International Space Station (ISS)*. Owned by NASA, it is designed to carry laboratory racks filled with equipment, experiments, and supplies to and from the station in the space shuttle. *Leonardo*'s first launch was in March 2001 on the space shuttle *Discovery* when it carried 16 racks filled with equipment and supplies to outfit the US Destiny Laboratory Module.

Leonardo is named for Leonardo da Vinci, the Italian inventor, scientist, and artist. The Italian Space Agency also built two other MPLMs, called *Raffaello* and *Donatello*.

Leonov, Aleksei Arkhipovich (1934–) Soviet cosmonaut. On 18 March 1965 he became the first person to walk in space, from *Voskhod 2* (*see* VOSKHOD). He was also one of the cosmonauts on the **Apollo–Soyuz* Test Project in 1975.

Leonov was commander of the cosmonaut team 1976–82 and was deputy director of the Gagarin Cosmonaut Training Centre until his retirement in 1991.

Born in Listvyanka, Siberia, Leonov served in the Soviet Air Force in East Germany before being selected as a cosmonaut in 1959.

LEOP Abbreviation for *launch and early orbit phase.

LES Abbreviation for *launch/entry suit.

Lewis Research Center Former name of the *Glenn Research Center.

LGA Abbreviation for *low-gain antenna.

libration point Another name for a *Lagrangian point.

lidar An instrument used on spacecraft to detect and observe the Earth's cloud patterns and aerosol pollution in the atmosphere. A lidar uses a laser beam and measures its back scatter (scattered reflection of the beam) to create the distant images.

life-support system (LSS) A general term for a system that provides an astronaut with the necessary conditions for survival in space. These include communications and the adequate control of oxygen, temperature, pressure, humidity, and waste removal. Life-support systems are used to regulate the atmosphere of a spacecraft and for *extravehicular activity (EVA). During EVAs, the system used is a portable device, also known as an **extravehicular mobility unit** (EMU), which is mounted as a backpack on the astronaut's spacesuit.

http://www.hq.nasa.gov/office/pao/History/alsj/plss.html Technical information, diagrams, and photographs of the *Apollo* spacesuit and the portable life-support system.

lift-off The action of a rocket rising off its launch pad. Lift-off occurs when the thrust of the rocket's engines exceeds the force of gravity. A space shuttle's engines begin to ignite three seconds before lift-off, which occurs when the eight bolts holding the shuttle assembly to the pad are blown off by explosive charges.

lift-to-drag ratio (L/D) The figure obtained by dividing the lift by the drag. Lift is the force acting on a body, perpendicular to the direction of motion. Drag is the retarding force acting upon a body, parallel to the direction of motion. The ratio is an important part of aerodynamics, and is used to calculate the path of spacecraft and aircraft as they travel through the Earth's atmosphere.

light The *electromagnetic waves in the visible range, having a wavelength between about 400 nanometres in the extreme violet and about 770 nanometres in the extreme red. Light is considered to exhibit both particle and wave properties, and the fundamental particle, or quantum, of light is called the photon. The speed of light (and of all electromagnetic radiation) in a vacuum is 299 792.5 km/s and is a universal constant denoted by *c*.

light second A unit of length, equal to the distance travelled by a beam of light in a vacuum in one second. It is equal to 2.99792458×10^8 m *See* LIGHT YEAR.

light year The distance travelled by a beam of light in a vacuum in one year. It is equal to approximately 9.4605×10^{12} km.

limb sounding The process of directing a satellite's radiometer horizontally to collect radiation from the upper layer of the Earth's atmosphere over a long distance but thin width. In this way, observations can be made of radiation in minute concentrations.

Lind, Don Leslie (1930–) US astronaut. He flew with the first operational **Spacelab* mission on the space shuttle *Challenger* in April 1985. Lind joined NASA's Goddard Space Flight Center as a space physicist in 1964 and was selected as an astronaut in 1966. He left NASA in 1986.

Lindsey, Steven Wayne (1960–) US astronaut. He was the pilot of two space shuttle flights, including the *Discovery* mission in October 1998 to deploy the *Spartan 201 spacecraft. In July 2001, he commanded *Atlantis* on a mission to install a new airlock module on the *International Space Station*. He has logged more than 896 hours in space. His later flight to the *ISS* was postponed following the 2003 *Columbia* space shuttle disaster. Lindsey was selected as an astronaut in 1994.

Linenger, Jerry Michael (1955–) US astronaut and physician. In January 1997, he flew on the space shuttle *Atlantis* to reach the Russian space station *Mir*. With two Russian cosmonauts, he spent more than 132 days on the station, the longest duration time in space for a male NASA astronaut. After a series of problems on the mission, including a serious fire, he returned to Earth on *Atlantis* in May 1997. Linenger was selected as an astronaut in 1992. He retired from NASA and the US Navy in 1998.

linknet A soft fishnet Dacron (synthetic polyester) layer used in *spacesuits to keep the pressurized fabric from stretching into a rigid form that hampers mobility. Linknet, which is lightweight but strong, was introduced in 1956 for US Air Force pilots. NASA's *Gemini* astronauts were the first to have spacesuits equipped with this restraint layer.

link-up The connecting of two space vehicles or their parts in space. Link-ups include docking during the **Apollo–Soyuz* Test Project, and the connecting of modules for the **International Space Station*.

Linnehan, Richard Michael (1957–) US astronaut and veterinary surgeon. He was a mission specialist aboard the Life and Microgravity **Spacelab*, launched in June 1996 on the space shuttle *Columbia*. In April 1998, he was payload commander on the *Columbia* mission to study the effects of *microgravity on the brain and nervous system. In March 2002, he flew on *Columbia* to service the *Hubble Space Telescope*, making three space walks totalling more than 21 hours. Linnehan was selected as an astronaut in 1992.

liquid cooling garment (LCG) A body layer worn by astronauts. It has water circulating from the portable life-support system (PLSS) and over the astronaut's body through small polyvinyl tubing woven into nylon spandex, a stretchy polyurethane fabric. The water then carries body heat back to the PLSS. The cooling virtually eliminates perspiration. It was devised by NASA and D R Burton of the Royal Aircraft Establishment in Farnborough, England, for the extravehicular mobility unit used by the *Apollo* astronauts on the Moon.

liquid fuel Rocket fuel in liquid form. It is a gas, normally hydrogen, that has been cooled to a liquid by a cryogenic process. The liquid fuel is combined in the combustion chamber with a liquid oxidizer, usually liquid oxygen, to convert it back into a pressurized gas. This is called a **bipropellant system**.

Russian scientist Konstantin *Tsiolkovsky designed a rocket in 1903 using liquid hydrogen and liquid oxygen and US rocket pioneer Robert *Goddard introduced petrol as a fuel in 1930. Germany in the 1940s used alcohol for their wartime V-2 rockets; the USA in the 1950s chose kerosene for the first Atlas military rockets (*see* ATLAS ROCKET) and, in the next decade, liquid hydrogen for the upper-stage Centaur engine; and the Soviets in the 1960s tested nitric acid for their intercontinental ballistic missiles.

liquid oxygen (LOX) The rocket propellant oxidizer, normally combined with liquid hydrogen. Spacecraft must carry an oxidizer as well as fuel in order to be able to burn the fuel, there being no oxygen in space. The oxygen and hydrogen (LOX–LH) are in a cryogenic state whose super-coldness turns them into liquids; LOX is chilled to −148 °C. It then expands by a factor of 900 when turned into gas on combustion.

US rocket pioneer Robert *Goddard used liquid oxygen mixed with petrol for his first liquid-fuelled rocket engine, and Germany's V-2 World War II rocket was propelled by liquid oxygen and alcohol.

LISA Acronym for **Laser Interferometer Space Antenna*.

LISA Pathfinder The joint ESA/NASA precursor mission to test technology for detecting gravitational waves which will be used in the subsequent **Laser Interferometer Space Antenna* (*LISA*) mission. For *LISA* to work properly scientists must be able to guarantee that a solid body can float freely in space completely undisturbed, its trajectory always constant. They will also have to be able to control the spacecraft position with an accuracy of a few millionths of a millimetre. *LISA Pathfinder* will contain two cube-shaped 'proof' masses housed in separate vacuum cans. The displacement of the cubes with respect to their housing is measured in three dimensions and the position of the surrounding spacecraft is adjusted so that it remains centred on the proof mass. The launch of *LISA Pathfinder* is planned for 2009. *LISA Pathfinder* will be placed at the L1 *Lagrangian point of the Earth's orbit.

Little Joe rocket An early NASA booster-rocket cluster made up of four modified Sergeant rockets, used to test *Mercury* spacecraft systems, including the launch escape system.

LMLSTP Abbreviation for *Lunar–Mars Life-Support Test Project.

Local Group A cluster of about 30 galaxies that includes our own, the Milky Way. Like other groups of galaxies, the Local Group is held together by the gravitational attraction between its members, and does not expand with the expanding universe. Its two largest galaxies are the Milky Way and the Andromeda galaxy; most of the others are small and faint.

Lockheed Martin (formerly Lockheed) A US aircraft manufacturer, the world's largest military contractor. Its Space Systems Company has built or contributed to major spacecraft. It designed and built the *Hubble Space Telescope*, provided the eight solar arrays for the *International Space Station*, and built the *Terra*, *Gravity Probe-B*, and *IMAGE* spacecrafts. The Lockheed company was founded in 1916 by two brothers, Allan and Malcolm Loughead (they later changed the spelling of their name), who had built their first seaplane in 1913, with headquarters in Burbank, California. Lockheed built the Vega plane in 1926 (later used by Amelia Earhart in her solo transatlantic flight), the first fully pressurized aircraft, the *XC-35*, in 1937, the P-38 Lightning fighter of World War II, and the TriStar passenger plane of the 1960s. It has been heavily involved in space exploration, building nearly 1 000 spacecraft for the military, NASA, and the commercial sector. In the late 1990s, it developed the F-22, an advanced tactical surveillance (ATS) and fighter aircraft. A STOL (short takeoff and landing) type, it combined stealth with unprecedented agility.

The P-38 Lightning shot down more Japanese aircraft than any other US fighter in the Pacific campaign. In 1974, the company was implicated in a scandal with the Japanese government, in which the then premier, Kakuei Tanake, was found guilty of accepting bribes from the Lockheed corporation and forced to resign.

After a merger in 1995 of Lockheed with the Martin Marietta Corporation, the company became Lockheed Martin Aeronautical Systems, the largest defence company in the world.

lock-on In space communications, tuning to a signal. A spacecraft's receiver can lock on to an uplink from a ground station, and a ground receiver, such as one in NASA's deep-space network, locks on to a downlink. In telemetry, after a ground receiver has locked on to a downlink, its system is also said to lock on to data when it reliably recognizes the decoded data frames.

LOI Acronym for *lunar orbit insertion.

Lola cloth A fireproof material used in cosmonaut spacesuits. Lola cloth does not burn, but extinguishes fire automatically. The Soviets developed it in the 1970s for use in the pure-oxygen atmosphere of the *Apollo* spacecraft in the **Apollo–Soyuz* Test Project. By contrast, NASA astronauts involved in the project wore spacesuits whose fire-retardant cloth burned slowly.

Lonchakov, Yuri Valentinovich (1965–) Russian cosmonaut. His first flight was aboard the space shuttle *Endeavour* in April 2001. The mission carried the second resident crew to the *International Space Station*, as well as the robotic arm, *Canadarm 2. Lonchakov was selected as a cosmonaut in 1997.

***Long-Duration Exposure Facility* (*LDEF*)** A NASA satellite deployed in April 1984 by the space shuttle *Challenger* to collect long-term data on the space environment and its effects on space systems and operations. It was retrieved in January 1990 by the space shuttle *Columbia*. *LDEF* carried 57 experiments, which involved more than 200 principal investigators from 7 NASA centres, 33 companies, 21 universities, 9 US Department of Defense laboratories, and 8 foreign countries. The results, kept in the LDEF Archive System at the *Langley Research Center, have led to research in such fields as radiation, meteoroids and debris, and contamination.

http://www-curator. jsc.nasa.gov/curator/seh/ldef/index.htm Describes the aims and operation of the LDEF. There is in-depth documentation of the spacecraft experiments and their results. Among the site's resources are pictures of the LDEF, its test trays, and damage caused by micro-meteorite impacts.

Long March rocket A Chinese rocket series whose Long March 1 launched the nation's first satellite on 24 April 1970 and put China's first man into space in October 2003. The two-stage Long March rocket is a modified intercontinental ballistic missile manufactured by the Shanghai Xinxin Machine Factory. The Long March 2C, in use from the early 1990s, can lift a payload of 2 199 kg into low Earth orbit. It stands 35 m high and has a range of up to 5 000 km.

Lopez-Alegria, Michael Eladio (1958–) US astronaut. He was on the *Discovery* mission in October 2000 to expand the *International Space Station (ISS)*. During the mission he carried out two space walks totalling more than 14 hours. Lopez-Alegria flew on the *Endeavour* to the *ISS* from 23 November to 7 December 2002, exchanging resident crews, and he made three space walks of nearly 20 hours to install the P1 Truss. He has logged 42 days in space with five space walks of nearly 35 hours. Lopez-Alegria was selected as an astronaut in 1992.

LOR Abbreviation for *lunar-orbit rendezvous mode.

Loria, Gus (born Christopher Joseph Loria) (1960–) US astronaut. He was based with NASA's *mission control centre as an ascent and entry CapCom (spacecraft communicator) awaiting assignment as pilot of a space shuttle flight. Loria worked on the *X-33* prototype of a reusable vehicle, intended to replace the space shuttle, at the *Dryden Flight Research Center from July 1994 to June 1995 before being selected as an astronaut in 1996. He was assigned to the *Endeavour* flight of November 2002 to the *International Space Station* but had to withdraw because of an injury. In 2002 and 2003, he was chief of flight testing for the Orbital Space Plane programme, and went on to become a NASA Fellow at the Kennedy School of Government, Harvard University, studying for a masters degree in public administration.

LOS Acronym for *loss of signal.

loss of signal (LOS) A temporary or permanent loss of a spacecraft's signal to a ground station. This is usually brief, as when the vehicle rotates during a manoeuvre or when a space probe enters *occultation as it orbits around the back of a planet or other celestial body. Permanent LOS occurs during a failure, as experienced by the **Mars Polar Lander* probe.

Lounge, John Michael (1946–) US astronaut. His space shuttle flights include the *Discovery* mission in September 1988 to deploy the *Tracking and Data-Relay Satellite. He joined NASA's Johnson Space Center in 1978, becoming a principal engineer for satellites launched from space shuttles. He was selected as an astronaut in 1980 and left NASA in 1991.

Lousma, Jack Robert (1936–) US astronaut. He was pilot of the two-month *Skylab 3* (*see* SKYLAB) mission in July 1973, and spent 11 hours on two space walks. In March 1982, he commanded the *Columbia* space shuttle mission, the first to use the *Canadarm. His time in space totalled 1 619 hours. Lousma was selected as an astronaut in 1966 and left NASA in 1983.

louvre (or louver) One of several angled slats that open and close to help regulate the temperature in a spacecraft. They are activated by a type of thermostat. When temperatures on board a spacecraft are low, the louvres close to reflect and retain heat; when temperatures are high, they open to release the heat into space. Their action helps save the electrical power used to heat or cool the spacecraft.

Lovell, James Arthur, Jr (1928–) US astronaut. He piloted *Gemini 7* in 1965, commanded *Gemini 12* in 1966, and orbited the Moon on *Apollo 8* (the first crewed lunar flight) in 1968. Lovell was also commander of the ill-fated *Apollo 13* mission that had to abort a Moon landing when an oxygen tank ruptured in the *service module.

Born in Cleveland, Ohio, Lovell was awarded a Bachelor of Science (BSc) degree by the US Naval Academy, then became a naval test pilot. He was selected as an astronaut in 1962. He retired in 1973. After his retirement, he established Lovell Communications.

Low, (George) David (1956–) US astronaut. His three space shuttle flights include the first flight of **Spacelab* on board the shuttle *Endeavour* in June 1993. *Microgravity experiments were conducted during the mission and Low, who was the payload commander, carried out nearly six hours of extravehicular activity. He logged 714 hours in space, leaving NASA in 1996 to join Orbital Sciences Corporation. He worked as an engineer in NASA's Jet Propulsion Laboratory from 1980 until 1984, when he was selected as an astronaut.

low Earth orbit (LEO) An orbit at an altitude of between 160 km and 1 500 km, usually used by spacecraft in polar or highly inclined orbits. (If the Earth was the size of a beachball, a LEO satellite would circle 1.3 cm above it.) LEO satellites complete an orbit of the Earth in between 90 minutes and 2 hours. The **Hubble Space Telescope* orbits at a LEO altitude of 600 km.

The low altitude reduces the time of communications with the Earth and makes LEO satellites accessible to television dishes, mobile phones, and palm computers. They are also used for remote sensing, meteorology, surveillance, and other missions. However, their closeness to the Earth's surface requires a speed of about 28 150 kph to avoid gravity's pull. This rapid orbit means more sophisticated Earth aerials are needed for tracking and communications.

low-energy charged-particle detector (LECP) A spacecraft instrument that measures the energy, composition, and angular distributions of charged particles in interplanetary space and within planetary systems. LECPs have been carried by such space probes as **Ulysses* and the two **Voyager* spacecraft.

lower torso assembly (LTA) A lower part of the space shuttle *extravehicular activity spacesuit. It resembles a pair of trousers and includes integrated boots and a waist ring to attach it to the *hard upper torso.

low-gain antenna (LGA) A small spacecraft antenna that provides a low amplification of radio frequency signals. An LGA is capable of transmitting and receiving in almost any direction, and does not require highly accurate pointing. Sometimes an LGA is mounted above the spacecraft's *high-gain antenna (HGA), as on **Galileo* and **Cassini*. LGAs can be used with space probes to the nearer planets, as with **Magellan* investigating Venus, but probes travelling further, such as the **Voyager* probes, must use high-gain antennae.

low-inclination orbit A spacecraft orbit that has a small *inclination. Such spacecraft are normally launched near the Equator, such as NASA's *High-Energy Transient Explorer 2*, launched on 9 October 2000 from the Kwajalein Missile Range in the central Pacific Ocean. Satellites in *geosynchronous orbits have a low inclination.

LOX Abbreviation for *liquid oxygen.

LPS Abbreviation for *launch processing system.

LRL Abbreviation for *Lunar Receiving Laboratory.

LRV Abbreviation for *lunar roving vehicle.

LSS Abbreviation for *life-support system.

Lu, Edward Tsang (1963–) US astronaut. He flew on the *Atlantis* space shuttle mission in May 1997 that docked with the Russian space station *Mir* and the *Atlantis* mission in September 2000 to prepare the *International Space Station* for its first resident crew. He was then a member of the seventh resident crew from 25 April to 27 October 2003, becoming the first US flight engineer on a *Soyuz* spacecraft and the first US astronaut to travel both ways to the *ISS* on that spacecraft. Lu was selected as an astronaut in 1994.

Lucid, Shannon Matilda Wells (1943–) US astronaut. She flew five space shuttle missions 1985–96, including the *Atlantis* mission in 1989 to deploy the space probe **Galileo*. Her last flight, launched in March 1996, delivered her to the *Mir* space station for a mission during which she clocked 188 days in space, the most time by a female and by any US astronaut on a single mission. Her total time in space of 223 days is also a US record. Lucid was selected as a mission specialist astronaut in 1978.

luminosity (or brightness) The amount of light emitted by a star, measured in *magnitudes. The apparent brightness of an object decreases in proportion to the square of its distance from the observer. The luminosity of a star or other body can be expressed in relation to that of the Sun.

***Luna* (or *Lunik*)** A series of 24 crewless Soviet probes launched towards the Moon between 1959 and 1976. *Luna 1* missed the Moon in January 1959. *Luna 2* was the first object from Earth to hit the Moon, in September 1959. *Luna 3* passed behind the Moon a month later, and took the first pictures of the Moon's far side. *Luna 9* was the first spacecraft to make a soft landing on the Moon, in February 1966, transmitting photographs of the surface to the Earth. *Luna 10* was the first craft to orbit the Moon. *Luna 17* (November 1970) and *Luna 21* (January 1973) landed *Lunokhod roving vehicles, which were driven over the Moon's surface by remote control from Earth. *Luna 16* (September 1970), *Luna 20* (February 1972), and *Luna 24* (August 1976) returned soil samples to Earth.

LunaCorp A private US company that promotes future space travel through the development of technologies and finances. LunaCorp has built and tested a prototype lunar rover, Nomad, developed by the Robotics Institute at Carnegie

Mellon University in Pittsburgh, Pennsylvania. In June 2000, the company joined with US company Radio Shack to announce a US$130 million programme, Lunar Icebreaker, to send Nomad to the Moon to search for ice at the poles. The rover is solar-powered and also runs on batteries. LunaCorp believes revenue could come from the public paying to operate the rover on the Moon by remote control. It also has plans for a SuperSat spacecraft with broadband communications to be assembled on the *International Space Station* and sent into orbit around the Moon.

Former astronaut Buzz *Aldrin is an advisor to the company, which is located in Fairfax, Virginia.

Lunar-A A Japanese probe to study the Moon's surface and interior. *Lunar-A* will carry two 13-kg penetrometers, 0.9 m long and 0.14 m wide, which will impact the Moon at about 300 m/s, burrowing 1–3 m into the surface. One penetrator will be targeted at the near side, in the region of the *Apollo 12* and *14* landing sites, while the other will be dropped onto the far side. The seismometers will monitor moonquakes and the flow of heat from the Moon's interior for a year, giving information on the Moon's internal structure and possibly its origin. Meanwhile, the *Lunar-A* orbiter will photograph the Moon's surface with a resolution of 30 m. Launch is expected in or after 2006.

lunar base A permanent human occupation of a site on the Moon. This is not the current stated goal of any nation, although US president George Bush made a long-range commitment on 20 July 1989 that the USA would go 'back to the Moon, back to the future, and this time back to stay'. NASA acknowledged that planning a future lunar base would be helped by data gathered by its **Lunar Prospector* mission to map the lunar surface and search for ice deposits. Some private organizations, such as the international *Artemis* project and the Apollo Society of Honolulu, Hawaii, have the stated goal of establishing a private base on the Moon. A lunar base is one of the proposed milestones in the **Aurora* programme to put humans on Mars.

lunar dust The dust on the surface of the Moon caused by meteoroid and cosmic-ray bombardment of surface rocks. Lunar dust proved problematic for the *Apollo* astronauts (1969–72), since the dust thrown up by the lunar rover or other activity settled on their spacesuits, tools, and scientific experiments.

lunar gravity The gravitational pull of the Moon's surface. Lunar gravity is only about one-sixth that of the Earth. A person who weighs 45 kg on the Earth would only weigh 7.7 kg on the surface of the Moon. Since the Moon is less massive than the Earth, it has a smaller gravitational effect.

Lunar–Mars Life-Support Test Project (LMLSTP) A series of tests by NASA at the Johnson Space Center to support the concept that a human life-support system can be devised for space travel to supply food, water, and oxygen indefinitely by using regenerative or recycling technologies.

The first test, then known as the Early Human Testing Initiative (EHTI Phase I), was conducted in August 1995 in a closed 3-m variable pressure growth chamber (VPGC) to show that a crop of wheat could revitalize air for 15 days. The second test, LMLSTP Phase II, took place in June 1996 in a 6-m chamber, the life-support systems integration facility, that recycled air and water for a crew of four. LMLSTP Phase IIA repeated this for 60 days from 13 January to 14 March 1997 using hardware simulating the *International Space Station*. The final test, LMLSTP Phase III, which began on 19 September 1997, had the crew complete 91 days in the chamber, the longest duration for a closed-chamber test in the USA. The LMLSTP tests were followed by the *Bioregenerative Planetary Life-Support Systems Test Complex (BIO-Plex).

http://advlifesupport.jsc.nasa.gov/lmlstp.html Describes each phase of the Lunar–Mars Life-Support Test Project. There are in-depth articles about project systems, photographs and diagrams showing its layout, crew diaries, and pictures of the crew during the trials.

lunar module (LM) The module used by *Apollo* astronauts to land on the Moon. Carrying two astronauts, it disengaged from the *command and service modules (CSM) orbiting the Moon. Its descent stage had retrorockets to slow it down and four legs whose pads contained probes to register a landing. The ascent stage had an engine to launch the astronauts back to a rendezvous with the CSM.

The LM was built by the Grumman Aircraft Engineering Corporation (later *Grumman Aerospace) and stood 7 m tall on the Moon's surface. Since it only flew in space and was designed for the Moon's weak gravity, the LM was not streamlined and was so delicate it would have collapsed on the Earth's surface. The first LM to land on the Moon in 1969 for the *Apollo 11* mission was named the **Eagle*, inspiring astronaut Neil *Armstrong to announce, 'The *Eagle* has landed.'

Lunar Orbiter Any of a series of five US photographic reconnaissance satellites put into orbit around the Moon 1966–7 to map landing sites for the *Apollo* programme. A total of 1950 high-resolution images covering more than 99% of the lunar surface were transmitted back to Earth, some showing features as small as a metre in size. Precise tracking of the orbital motions of the satellites allowed scientists to map the gravitational field of the Moon.

lunar orbit insertion (LOI) The placement of a spacecraft into orbit around the Moon. *Apollo 11* used a retroburn (*see* RETROROCKET) to enter an elliptical LOI and, after one orbit, fired another retroburn to circularize the orbit. NASA's **Lunar Prospector* probe, launched in 1998, changed its spin axis before encountering the Moon in order to prepare for LOI. At LOI, it fired its thrusters to begin an initial 12-hour orbit and within a few days settled into a circular polar orbit 100 km above the lunar surface.

lunar-orbit rendezvous mode (LOR) A method used by NASA to land the *Apollo* astronaunts on the Moon by separating their *lunar module (LM) from the *command module (CM) in orbit around the Moon, and then returning the LM to rendezvous with the CM. LOR was chosen instead of the other options of *Earth-orbit rendezvous mode (EOR) and the *direct ascent mode.

Lunar Prospector A spin-stabilized spacecraft (1.4 × 1.2 m) sent into orbit on 7 January 1998 as a NASA Discovery Mission to study the Moon. It was launched on an Athena 2 launch vehicle from the Kennedy Space Center.

Mission objectives were to 'prospect' the lunar crust and atmosphere for potential resources (minerals, water ice, and gases) and to map the Moon's gravitational and magnetic fields and core. It was deliberately crashed into a crater near the Moon's south pole on 31 July 1999 in an unsuccessful attempt to discover traces of water.

http://spacelink.nasa.gov/NASA.Projects/Space.Science/Solar.System/Lunar.Prospector/ Describes the spacecraft, instrument payload, launch vehicle, and mission objectives. There are artists' views and photographs of *Prospector* and video-clip presentations by the science team. The comprehensive Frequently Asked Questions list covers the *Lunar Prospector* spacecraft, the mission, and the Moon in general.

Lunar Receiving Laboratory (LRL) An isolation facility created by NASA in Houston, Texas, for the *Apollo 11* astronauts, the first crew to return from the Moon. Immediately upon being picked up on 25 July 1969 by the USS *Hornet* rescue vessel, Neil *Armstrong, Michael *Collins, and Buzz *Aldrin were given *biological isolation garments and quarantined in a modified house trailer on the ship for three days. They were then taken to Houston to be quarantined behind a locked door in the LRL until 10 August 1969. The facility included an exercise room, good food, and a bar. During this time they debriefed NASA personnel, who observed them and the white mice that were put in the room to indicate any early infection from lunar microbes. The LRL also had a large laboratory for testing the Moon rocks brought back by the mission for toxicity and potential biohazards.

***Lunar Reconnaissance Orbiter* (*LRO*))** A NASA Moon-orbiter probe scheduled for launch in October 2008. While mapping the lunar surface from a polar orbit at an altitude of 30–50 km *LRO* will look for potential sites for future manned landings. It will study the permanently shadowed regions near the Moon's poles in search of possible possible deposits of ice on or just under the surface. Its instruments will include a mapping camera with resolution of one metre or better, a laser altimeter to measure slopes, a radiometer to map temperatures, and a neutron detector to search for evidence of water ice. *LRO* is the first mission in NASA's Robotic Lunar Exploration Program.

lunar roving vehicle (LRV; or Moon buggy) A small vehicle, resembling a dune buggy, used on the surface of the Moon by the astronauts of *Apollo 15* (1971), *Apollo 16* (1972), and *Apollo 17* (1972). Battery powered and weighing 209 kg on Earth, the LRV was built to withstand extreme temperatures and a vacuum atmosphere. It had an antenna shaped like an inverted umbrella mounted on top to send television pictures and sound to Earth; the *Apollo 15* camera also provided the first pictures of a lift-off from the Moon. On *Apollo 17*, the crew of the last mission to visit the Moon covered 34 km in the two-seat, four-wheel vehicle, travelling over the rough surface to collect rocks.

The Soviet Union sent two uncrewed **Lunokhod* roving vehicles to the Moon in 1970 and 1973 to take photographs and scoop up soil that was returned to Earth.

http://nssdc.gsfc.nasa.gov/planetary/lunar/apollo_lrv.html Simple page with a description of the lunar rover and the part it played in the *Apollo* landings. There are some very good pictures of the rover on the Moon. Links are provided to several other lunar rover sites and to the very technical *Lunar Rover Operations Handbook.*

Lunik Alternative name for the **Luna* probes.

Lunokhod Either of two crewless roving vehicles sent to the Moon by the USSR in the 1970s. Both of the solar-powered eight-wheeled rovers carried television cameras, soil-testing equipment, a laser reflector, and several other scientific instruments. *Lunokhod 1* was taken to the Moon by the *Luna 17* spacecraft, landing in the Sea of Showers on 17 November 1970. It travelled 10.5 km before being shut down on 4 October 1971. *Lunokhod 2*, carried on *Luna 21*, landed in the Sea of Serenity on 15 January 1973 and travelled 37 km before ceasing to function on 4 June 1973.

McArthur, William Surles, Jr (1951–) US astronaut. His three space shuttle flights included the *Discovery* mission in October 2000 to expand the *International Space Station*, during which he logged more than 13 hours on two space walks. He was assigned to NASA's Johnson Space Center in 1987 as an engineer for the space shuttle, and was selected as an astronaut in 1990.

McAuliffe, (Sharon) Christa (born Sharon Christa Corrigan) (1948–86) US teacher and civilian astronaut. She became the first teacher selected for the NASA Teacher in Space programme. In 1986 she died along with the rest of the crew on *Challenger* when the space shuttle exploded soon after its launch.

Born in Framingham, Massachusetts, she started teaching in 1970. She impressed NASA officials with a course she had developed entitled 'The American Woman', which gained her access to the NASA programme.

McBride, Jon Andrew (1943–) US astronaut. He was pilot of the *Challenger* space shuttle mission in October 1984 to conduct observations of the Earth, and undertook an extravehicular activity to demonstrate the possibility of satellite refuelling. He was selected as an astronaut in 1978 and retired from NASA in 1989.

McCandless, Bruce, II (1937–) US astronaut. Selected in 1966 for a potential *Apollo* mission, McCandless did not travel to space until February 1984 as a mission specialist on the space shuttle *Challenger*. He helped develop the *manned manoeuvring unit, which he used to become the first person to fly as an independent satellite during a space walk on the mission. His other space shuttle flight was in *Discovery*, launched in April 1990 to deploy the **Hubble Space Telescope*.

McCool, William Cameron (1961–2003) US astronaut. He was selected as an astronaut in 1996, and worked on upgrading the space shuttle cockpit. He died when the space shuttle *Columbia* broke up on re-entry in February 2003 after a 16-day mission to the *International Space Station*.

McCulley, Michael James (1943–) US astronaut. He was pilot of the *Atlantis* space shuttle mission in October 1989 that deployed the **Galileo* spacecraft. McCulley was selected as an astronaut in 1984, retiring from NASA in 1990 to join the Lockheed Space Operations Company as deputy director of the Kennedy Space Center launch site. He then became vice-president of the United Space Alliance.

McDivitt, James Alton (1929–) US astronaut. He piloted the *Gemini 4* flight in June 1965, during which Edward *White made the first US space walk. In March 1969, he commanded *Apollo 9* in Earth orbit to test the spacecraft's lunar-landing hardware. McDivitt was selected as an astronaut in 1962. From 1969 he was manager of the **Apollo* project, and he left NASA in 1972.

McDonnell Douglas A US aircraft manufacturer that has contributed to many space missions, including the **Mercury* project, building its capsule, and the **Gemini* project. It produced Saturn V third-stage rockets and converted one into the space station **Skylab*, launched on 14 May 1973. The corporation, whose headquarters are in St Louis, Missouri, merged with the *Boeing Company in 1997.

On 30 August 1984, Charles Walker became the first astronaut to represent a private company when he flew on the shuttle **Discovery* in charge of McDonnell Douglas's electrophoresis operations in space (EOS) experiment, which investigated ways to process materials in microgravity.

The company was established in 1920 as the David Douglas Company. It became the Douglas Company in 1921, Douglas Aircraft Company in 1928, and McDonnell Douglas after a merger with the McDonnell aircraft company in 1967. It merged with the Boeing Company in 1997.

Mach number The ratio of the speed of a body to the speed of sound in the medium through which the body travels. In the Earth's atmosphere, Mach 1 is reached when a body (such as an aircraft or spacecraft) 'passes the sound barrier', at a velocity of 331 m per second (1 192 kph) at sea level. A *space shuttle reaches Mach 15 (about 17 700 kph an hour) 6.5 minutes after launch.

The Mach number is named after Austrian physicist Ernst Mach.

m

MacLean, Steven G (1954–) Canadian astronaut. Selected as an astronaut by the Canadian Space Agency in 1983, he flew with the space shuttle *Columbia* in October 1992, in charge of Canadian experiments. In 1993 he became a science advisor for the *International Space Station*, and in 1994 he was named as acting director-general of the Canadian Astronaut Programme. His 2004 flight was postponed following the *Columbia* tragedy the previous year. In 1996 MacLean was selected to train as a NASA mission specialist for a future shuttle flight, and was subsequently transferred to the Robotics Branch of NASA's Astronaut Office.

McMonagle, Donald Ray (1952–) US astronaut. He made three space shuttle flights, including the *Atlantis* mission in November 1994, which he commanded, to carry the third *Atmospheric Laboratory for Applications and Science. He became manager of launch integration at the Kennedy Space Center in 1997. McMonagle was selected as an astronaut in 1987. In 1996, NASA assigned him the task of establishing a new Extravehicular Activity Project Office, which also involved him in upgrading spacesuit design. He logged 605 hours in space before leaving NASA.

McNair, Ronald E (1950–86) US astronaut and physicist. He served on the space shuttle *Challenger* mission in February 1984 that marked the first flight of the *manned manoeuvring unit and the first use of the *Canadarm, which he operated. McNair died in the *Challenger* tragedy in 1986.

Born in Lake City, South Carolina, McNair was awarded a PhD in Physics by the Massachusetts Institute of Technology in 1976. He did research at the Hughes Research Laboratories in Malibu, California, on satellite-to-satellite space communications. He was selected as an astronaut in 1978.

Magellan The NASA space probe to the planet *Venus, launched from the space shuttle *Atlantis* on 4 May 1989. The probe went into orbit around Venus in August 1990 to make a detailed map by radar of 99% of the planet. It revealed the existence of volcanoes, impact craters, and mountains, on the planet's surface.

In October 1994, *Magellan* was purposely destroyed, entering the atmosphere around Venus, where it burned up.

http://www.jpl.nasa.gov/magellan/ Details of the NASA *Magellan* project that sent a probe to Venus. It includes a full mission overview, technical details about the planet, many images, and an animated view of Venus.

magnetic field The region around a permanent magnet, or around a conductor carrying an electric current, in which a force acts on a moving charge or on a magnet placed in the field. The force cannot be seen; only the effects it produces are visible. The field can be represented by lines of force parallel at each point to the direction of a small compass needle placed on them at that point. These invisible lines of force are called the magnetic field or the flux lines. A magnetic field's magnitude is given by the magnetic flux density (the number of flux lines per unit area), expressed in teslas.

magnetosphere The volume of space, surrounding a planet, in which the planet's magnetic field has a significant influence. The Earth's magnetosphere extends 64 000 km towards the Sun, but many times this distance on the side away from the Sun. That of Jupiter is much larger, and, if it were visible, would appear from the Earth to have roughly the same extent as the full Moon. The Russian-led space missions *Coronas-I* (launched in 1994) and *Coronas-F* (launched in 2001) were designed to investigate the magnetosphere of the Sun.

The Earth's magnetosphere is not symmetrical. The extension away from the Sun is called the **magnetotail**. The outer edge of the magnetosphere is the **magnetopause**. Beyond this is a turbulent region, the **magnetosheath**, where the *solar wind is deflected around the magnetosphere. Inside the magnetosphere, atomic particles follow the Earth's lines of magnetic force. The magnetosphere contains the *Van Allen radiation belts.

magnitude A measure of the brightness of a star or other celestial object. The larger the number denoting the magnitude, the fainter the object. Zero or first magnitude indicates some of the brightest stars. Still brighter are those of negative magnitude, such as Sirius, whose magnitude is −1.46.

Apparent magnitude is the brightness of an object as seen from the Earth; **absolute magnitude** is the brightness at a standard distance of 10 parsecs (32.616 light years).

Each magnitude step is equal to a brightness difference of 2.512 times. Thus a star of magnitude 1 is $(2.512)^5$ or 100 times brighter than a sixth-magnitude star just visible to the naked eye. The apparent magnitude of the Sun is −26.8, its absolute magnitude +4.8.

main engine cut-off A command to stop the throttled engines, signalled by computer then called out by the commander. This operation occurs nearly nine minutes into the flight and 16 seconds before the empty external *fuel tank is jettisoned.

Malenchenko, Yuri Ivanovich (1961–) Russian cosmonaut. He was commander of the crew that stayed on the Russian space station *Mir* July–November 1994. In September 2000, he flew with the space shuttle *Atlantis* mission to prepare the *International Space Station* (*ISS*) for its first resident crew. During the mission he undertook a six-hour space walk to connect cables. He was commander of the eighth resident crew on the *ISS* from 25 April to 27 October 2003. He has logged 321 days in space, with three space walks totalling 18 hours. Malenchenko was selected as a cosmonaut in 1993.

mandatory programme A programme of activities of the European Space Agency (ESA) that requires the participation of all member nations. This includes the space science programme and the agency's general budget. Such activities are funded by contributions from all ESA members according to each country's gross national product. For example, the **Hipparcos* satellite launched in August 1989 was designated a mandatory programme in 1980.

The alternative to the mandatory programme is the *optional programme. The two types of programme were created when ministers in charge of space from ten European nations met in 1973 in Brussels, Belgium, to establish the ESA.

manned manoeuvring unit (MMU; informally called space bike) A jet backpack worn by an astronaut in order to be able to move accurately through space during an *extravehicular activity. The MMU, which connects to the *primary life-support system (PLSS), provides propulsion by means of nitrogen jets. The 24 thrusters are operated by hand controls on the end of arm rests, and the jets allow an astronaut to move forwards, backwards, sideways, up, and down. The device weighs about 139.5 kg on Earth. Astronauts retrieve and service nearby satellite payloads by using the MMU.

***Manned Orbital Laboratory* (*MOL*)** A joint NASA and US Air Force programme in the 1960s that failed to meet its objectives. The programme was initiated to test the long-term residence of astronauts in space and to develop Earth observation, including military surveillance (of the USSR in particular). The *MOL* was to use a modified *Gemini* capsule and a cylindrical module to house a laboratory measuring 12.5 m in length. A polar orbit was planned for five flights of 30 days each, with the first given a scheduled launch date of 1968 with a crew that would conduct 25 experiments. However, the programme was hindered by a lack of funds during the protracted Vietnam War. The only *MOL* flight was a successful test of the modified *Gemini* capsule in November 1966.

manoeuvre A controlled movement of a spacecraft to change its attitude, trajectory, or orbit. The manoeuvre is normally carried out via computer commands to the craft's thrusters.

man-rated Of a spacecraft, reliable enough to carry astronauts. During the early days of NASA, military rockets had to be upgraded in order to be man-rated. The Atlas rocket, among other changes, was fitted with an *abort sensing and implementation system and a more reliable type of engine valve.

Mao 1 Alternative name for **China 1*.

MAP Abbreviation for *Multifoil Microabrasion Package.

m

mare (plural: maria) A dark lowland plain on the Moon. The name comes from the Latin word for 'sea', because these areas were once wrongly thought to be water.

Mariner A series of US space probes that explored the planets Mercury, Venus, and Mars 1962–75.

Mariner 1 (to Venus) had a failed launch. *Mariner 2* (1962) made the first fly-by of Venus, at 34 000 km, confirmed the existence of *solar wind, and measured Venusian temperature. *Mariner 3* (launched 4 November 1964) did not achieve its intended trajectory to Mars. *Mariner 4* (launched 25 November 1964) passed Mars at a distance of 9 800 km, and took photographs revealing a dry, cratered surface. *Mariner 5* (1967) passed Venus at 4 000 km and measured Venusian temperature, atmosphere, mass, and diameter. *Mariner 6* and *Mariner 7* (1969) photographed Mars's equator and southern hemisphere respectively, and also measured temperature, atmospheric pressure and composition, and diameter. *Mariner 8* (to Mars, in 1971) had a failed launch. *Mariner 9* (1971) mapped the entire Martian surface, and photographed Mars's moons. Its photographs revealed the changing of the polar caps, and the extent of volcanism, canyons, and features, which suggested that there might once have been water on Mars. *Mariner 10* (1974–5) took close-up photographs of Mercury and Venus, and measured temperature, radiation, and magnetic fields.

Mariner 11 and *Mariner 12* were renamed *Voyager 1* and *Voyager 2* (launched 1977) (*see* VOYAGER).

http://nssdc.gsfc.nasa.gov/planetary/mars/mariner.html Profiling the *Mariner 4*, *Mariner 6*, *Mariner 7*, and *Mariner 9* missions to Mars. The site includes introductory text, technical information about the spacecraft and their experiments, and access to the Mariner imagery.

Mars The fourth planet from the Sun. It is much smaller than Venus or Earth, with a mass 0.11 that of Earth. Mars is slightly pear-shaped, with a low, level northern hemisphere, which is comparatively uncratered and geologically 'young', and a heavily cratered 'ancient' southern hemisphere.

mean distance from the Sun 227.9 million km
equatorial diameter 6 792 km
rotation period 24 hours 37 minutes
year 687 Earth days
atmosphere 95% carbon dioxide, 3% nitrogen, 1.5% argon, and 0.15% oxygen. Red atmospheric dust from the surface whipped up by winds of up to 450 kph accounts for the light pink sky. The surface pressure is less than 1% of the Earth's atmospheric pressure at sea level
surface the landscape is a dusty, red, eroded lava plain. Mars has white polar caps (water ice and frozen carbon dioxide) that advance and retreat with the seasons
satellites two small satellites: *Phobos and Deimos

There are four enormous volcanoes near the equator, of which the largest is Olympus Mons 21 km high, with a base 600 km across, and a crater 80 km wide. To the east of the four volcanoes lies a high plateau cut by a system of valleys, Valles Marineris, some 4 000 km long, up to 200 km wide and 6 km deep; these features are apparently caused by faulting and wind erosion. Recorded temperatures vary from −100 °C to 0 °C.

Mars may approach Earth to within 54.7 million km. The first human-made object to orbit another planet was *Mariner 9*, launched in 1971 (*see* MARINER). The two **Viking* probes, which landed in 1976, provided more information and analysed the soil in an unsuccessful search for life.

In December 1996, NASA launched the **Mars Pathfinder*, which made a successful landing on Mars in July 1997 on a flood plain called Ares Vallis. Its 0.3-m rover, **Sojourner*, explored the Martian landscape until September 1997.

The **Mars Global Surveyor*, launched on 7 November 1996, entered Martian orbit in September 1997. Its data revealed that Mars's magnetic field is a mere 800th that of the Earth. In February 1999, the spacecraft established its orbit for mapping the surface of the planet. In 2001, it located two regions on the surface rich in haematite, providing further evidence for the existence of water at some time in the planet's history.

The NASA probe **Mars Odyssey* was launched in April 2001, and discovered water in the form of subsurface ice at Mars's south pole in March 2002.

The European Space Agency's **Mars Express* went into orbit around Mars in 2003 and discovered methane in the atmosphere and water ice at the poles. It carried a lander for soil collection, **Beagle 2*, which descended to the surface on 25 December 2003 but communications were immediately lost.

NASA's twin *Mars Exploration rovers, named Spirit and Opportunity, successfully landed on 3 January and 24 January 2004 respectively. They found traces of previous water. US president George W Bush announced in January 2004 plans to send and return astronauts to Mars in the 21st century.

http://www-mars.lmd.jussieu.fr/ This Web interface allows the user to access climate data about the planet Mars, produced by numerical simulations of the Martian atmosphere using a computerized weather forecasting model.

http://mars.jpl.nasa.gov/ Well-presented NASA site with comprehensive information on current and future missions to Mars. There are fascinating and well-written accounts of Pathfinder and Global Surveyor, and large numbers of images of the planet.

http://www.solarviews.com/eng/mars.htm Detailed description of the planet Mars, including statistics and information about its surface, volcanoes, satellites, and clouds, supported by a good selection of images.

http://www.sciencemonster.com/virtualmars View a three-dimensional image of Mars. Clicking on a land feature centres your view at that spot and allows you to rotate the planet. Click on the button below the image to toggle the names of surface features and landing sites, including the *Pathfinder* and *Viking 2* landing sites. Night and day areas are recalculated every five minutes to adjust to rotation.

http://www.spaceref.com/mars/ Regularly updated with links to breaking stories of Mars research and events. There are links to other major Mars sites, and to pages that deal with certain key topics such as Mars meteorites and missions to Mars.

Mars A series of largely unsuccessful Soviet space probes launched towards the planet *Mars between 1962 and 1996. Contact with *Mars 1*, launched in November 1962, was lost en route. *Mars 2* crashed on Mars on 27 November 1971, becoming the first spacecraft from Earth to reach the planet. *Mars 3* landed safely five days later and transmitted video pictures for 20 seconds before communication was lost. *Mars 4* and *Mars 5*, launched in July 1973, were intended to be orbiters but only *Mars 5* achieved orbit. *Mars 6* and *Mars 7* were launched in August 1973, but contact was lost with *Mars 6* on its final descent and *Mars 7* failed to enter orbit. *Mars 96* crashed to Earth soon after its launch in November 1996.

Mars Climate Orbiter A NASA Mars probe launched in December 1998 to investigate the planet's cycles of water, carbon dioxide, and dust. Communications with the spacecraft were lost in September 1999. An investigation found that the spacecraft's manoeuvres to place it into a Mars orbit failed because one NASA spacecraft team had used imperial measurements and the other team had used metric ones.

Mars Exploration Rover Mission A NASA mission launched on two spacecraft on 10 June and 7 July 2003 and arriving at Mars on 4 January and 25 January, 2004. The craft released two rovers, Spirit and Opportunity, weighing 150 kg, on different regions of the planet to search for evidence of water from the planet's past. Each was intended to travel approximately 100 m over the surface of the planet per Martian day (24 hours 40 minutes). The rovers were designed with much greater mobility than the *Sojourner rover of the 1997 **Mars Pathfinder* mission. The 2003 rovers descended by parachute, and inflated air bags bounced them onto the planet's surface. Their instruments included a panoramic camera, a microscopic imager, three spectrometers, and a rock abrasion tool (RAT) to expose new rock surfaces. In March, Opportunity discovered evidence that its landing site on the flat Meridiani Planum was previously covered in water and Spirit on the opposite side of the planet found that water had once existed in the Gusev Crater.

m

Mars Express A European Space Agency mission to Mars launched on 2 June 2003 and which went into polar orbit around the planet on 25 December 2003. It discovered in early 2004 evidence of methane and previous water. The spacecraft has obtained high-resolution images of the surface and studied the atmosphere and subsurface, as well as their interaction. The first data from *Mars Express* in January 2004 were a direct measurement of water present in the Martian ice caps. The science payload includes seven experiments, using such measures as subsurface radar, neutral and charged particle sensors, and spectrometers. Mounted on top of the *Mars Express* orbiter was the **Beagle 2* lander to study the surface, but communications failed after it touched down on 25 December 2003. *Mars Express* is the first fully European mission to any planet.

http://sci.esa.int/marsexpress/ Full details and news of the European *Mars Express* mission. The scientific goals of the mission are outlined and the orbiter's sensor payload described. Among the site's resources are a news archive and image gallery. There is also a Webcam.

***Mars Global Surveyor* (*MGS*)** A US spacecraft, launched in November 1996, that went into orbit around *Mars in September 1997 to conduct a detailed photographic survey of the planet, commencing in March 1998. The spacecraft used a previously untried technique called **aerobraking** to turn its initially highly elongated orbit into a 400 km circular orbit by dipping into the outer atmosphere of the planet. The *Global Surveyor* established its correct orbit for mapping the surface of Mars in February 1999, a year later than planned.

In June 2000, US astronomers announced that photographs from the *Mars Global Surveyor* showed channels that seemed to have been formed by large amounts of water seeping to the surface and causing landslides. Around 65 000 images taken in the previous year were examined, and were thought to be evidence of a ground-water supply. In 2001, the *Global Surveyor* located two regions on the surface rich in haematite, providing further evidence for the existence of water at some time in the planet's history. On 31 January 2001, *Mars Global Surveyor* completed its primary mission and then went into an extended mission phase, continuing to transmit data back to Earth.

http://mars.jpl.nasa.gov/mgs/ Regularly updated with images from the cameras of *Mars Global Surveyor* (*MGS*), this site is now home to a huge collection of high-resolution images of the planet. There are also results from the thermal emission spectrometer and Mars Orbiter laser altimeter. A graphical simulation shows Mars's position with respect to the Earth, and the position of the *MGS* spacecraft relative to Mars.

Marshall Space Flight Center (MSFC) A NASA field centre at Huntsville, Alabama, established in 1960, where the series of *Saturn rockets and the *space shuttle engines were developed. It also manages missions, such as the *Gravity Probe B*, as well as the propulsion for the space shuttle and various payloads, including *microgravity experiments. The Destiny science laboratory for the **International Space Station* was built by the *Boeing Company using the centre's facilities.

The centre's first director was German rocket engineer Wernher *von Braun.

http://www.msfc.nasa.gov/ Outlining the activities of the centre, this well-designed site has pages introducing their science projects and describing the spacecraft missions with which it is involved. Animations and video clips accompany many of its news releases. The Marshall image exchange allows easy access to image galleries of its space flight and science projects past and present.

Mars Observer A NASA space probe launched in 1992 to orbit Mars and survey the planet, its atmosphere, and the polar caps over two years. The US$980 million project miscarried, however, when the probe stopped transmitting in August 1993 after a suspected explosion in its orbital insertion engine.

Mars Odyssey NASA space probe launched on 7 April 2001, arriving at Mars on 24 October 2001 for a four-year mission. The main instruments were a thermal-emission imaging system (THEMIS) to determine the distribution of minerals and rock types; a gamma-ray spectrometer (GRS) to search for 20 chemical elements, especially hydrogen in the form of water ice in the shallow subsurface; and the Mars radiation environment experiment (MARIE) to study the radiation environment. Its images in December 2003 indicated that Mars may be coming out of an ice age. The spacecraft also acted as a communications relay for the two rovers that landed in January 2004 on the *Mars Exploration Rover Mission. *Odyssey* achieved the milestone of being in orbit around Mars for a full Martian year of 687 Earth days, on 11 January 2004.

Mars Pathfinder A NASA spacecraft launched on 4 December 1996, landing on the planet *Mars on 4 July 1997. It carried a small six-wheeled roving vehicle called *Sojourner to examine rock and soil samples around the landing site. *Mars Pathfinder* was the first to use air bags instead of retrorockets to cushion the landing.

Pathfinder sent its last data transmission on 27 September 1997. It sent over 23 million bits of information back to Earth, including over 16 500 images from the landing site, 15 chemical analyses of the surrounding soil, and previously unknown data on the Martian wind and weather.

http://mars.jpl.nasa.gov/MPF/index1.html Features galleries of both *Pathfinder* (lander) and Sojourner (rover) images. The scientific findings directory has chapters on geology, mineralogy, and meteorology, and interpretation of the *Pathfinder* data and images. There are high-resolution pictures and images viewable with red/green glasses.

Mars Polar Lander A NASA probe launched in January 1999 to search for evidence of water or ice beneath the surface of the planet Mars. The probe was lost in December 1999 when it was scheduled to land close to the planet's south pole. The spacecraft's disappearance was blamed on a faulty sensor that switched off the spacecraft's landing rockets too soon, causing it to crash. It was NASA's second consecutive Mars mission to fail in just over two months, following the disappearance of the **Mars Climate Orbiter* in September.

Mars Reconnaissance Orbiter A NASA space probe, launched 12 August 2005, whose mission is to orbit Mars and conduct the most comprehensive inspection of the planet ever made. It will image Martian landscapes with a resolution of 0.2–0.3 m. Other instruments will use radar to probe underground layers for water and ice, determine the composition and origins of surface minerals, and track changes in atmospheric water and dust. A spectrometer will identify types of water-related minerals. The orbiter will lay the groundwork for future Mars surface missions, such as the *Phoenix* lander to launch in 2007 and the Mars Science Laboratory rover to launch in 2009.

Mars Science Laboratory A proposed NASA mission, for launch in 2009 or later, to land a roving science laboratory on Mars. The probe would pave the way for a future sample-return mission.

mascon Contraction of *mass concentration*. Any of a number of areas on the Moon where gravity is stronger, particularly under the major mare basins, evidently due to localized area of denser rock. Mascons cause slight perturbations to the motions of craft orbiting the Moon.

maser Acronym for microwave amplification by stimulated emission of radiation, A high-frequency microwave amplifier or oscillator. The *deep-space network uses masers as low-noise amplifiers of signals received from artificial satellites.

mass The quantity of matter in a body as measured by its inertia, including all the particles of which the body is made up. Mass determines the acceleration produced in a body by a given force acting on it, the acceleration being inversely proportional to the mass of the body. The mass also determines the force exerted on a body by gravity on Earth, although this attraction varies slightly from place to place (the mass itself will remain the same). In the SI system, the base unit of mass is the kilogram.

At a given place, equal masses experience equal gravitational forces, which are known as the weights of the bodies. Masses may, therefore, be compared by comparing the weights of bodies at the same place. The standard unit of mass to which all other masses are compared is a platinum-iridium cylinder of 1 kg, which is kept at the International Bureau of Weights and Measures in Sèvres, France.

Massachusetts Institute of Technology (MIT) A US university that is a leading institution in space technology. Its developments include spacecraft, such as the **Rossi X-Ray Timing Explorer* launched in 1995 to probe the Milky Way and other areas of space; systems, such as the **High-Energy Transient Explorer 2* satellite; and instruments, including the *all-sky monitor (ASM) and a new flight data system. MIT was also involved in 1996 in experiments to grow animal tissue on the Russian space station *Mir*.

The university's Space Systems Laboratory (SSL) works on technologies for small spacecraft and the *International Space Station* (*ISS*). There is also the Space Propulsion Laboratory (SPL), Man Vehicle Laboratory (MVL), Aeronautical Systems Laboratory, and a Center for Space Research (CSR). Many former MIT students have become astronauts, including five in the *Apollo* programme—Buzz Aldrin, Charles Duke, Edgar Mitchell, Russell Schweickart, and David Scott—and William Shepherd, who commanded the first *ISS* crew, Expedition 1, in October 2000.

mass driver An electromagnetic accelerator designed to accelerate payloads or objects without rockets. Its projected design will allow it to operate on the surface of the Moon and other celestial bodies. In weak gravity, objects could be accelerated to *escape velocity. The mass driver is essentially a catapult tube comprising a magnetic object that runs along the tunnel, which is lined with electromagnets, but there is no standard design.

Potentially, it would be capable of transporting lunar material to construct a settlement on the Moon. Its design would also allow the machine to land on an asteroid heading dangerously towards the Earth; it would then break off pieces of the asteroid and hurl them into space. The idea was developed at the *Massachusetts Institute of Technology.

mass ratio The mass of a vehicle (usually a rocket) at lift-off, divided by the mass of the empty vehicle without propellants. Mass ratio is often used as an indicator of a rocket's performance. The greater the mass ratio, the more propellant the rocket can carry, resulting in a better lifting performance from the vehicle. The ratio can be increased by the use of more lightweight materials on the rocket and improvements to the method of introducing the propellants to the combustion chamber. The choice of propellants is also a factor: some are much heavier than others. Typical mass ratios range from 5:1 for boosters using pressure-fed liquid hydrogen propellant, to 17:1 for pumped hydrogen peroxide.

mass spectrometer The apparatus for analysing chemical composition by separating ions by their mass; the ions may be elements, isotopes, or molecular compounds. Positive ions (charged particles) of a substance are separated by an electromagnetic system, designed to focus particles of equal mass to a point where they can be detected. This permits accurate measurement of the relative concentrations of the various ionic masses present. A mass spectrometer can be used to identify compounds or to measure the relative abundance of compounds in a sample.

Mastracchio, Richard Alan (1960–) US astronaut. He was a mission specialist on the *Atlantis* space shuttle flight in September 2000 to prepare the **International Space Station* for the arrival of the first resident crew. Mastracchio joined the *Johnson Space Center in 1987, and became an engineer for NASA in 1990. His 2004 flight was postponed following the 2003 *Columbia* space shuttle disaster. He has logged over 283 hours in space. He was selected as an astronaut in 1996.

mate–demate facility A structure that lifts a space shuttle on or off a Boeing 747 aircraft. The 747 is used to fly the shuttle from its landing site at Edwards Air Force Base, California, to the launch site at the Kennedy Space Center in Florida. There is a mate–demate facility at both locations.

material science An astronautical term for creating things in space. This includes growing crystals, making purer versions of chemicals, and placing medicines in nearly perfect capsule spheres. Crystals grow more perfectly without gravity, and the experiments contribute towards manufacturing computer chips and semiconductors. Growing protein crystals in space also enables deeper understanding of their structure, which contributes towards the development of drugs.

The **Microgravity Science Laboratory* on the space shuttle *Columbia* launched in October 1995 was especially used for material-science experiments, including growing crystals for 16 days.

Mathilde (or Asteroid 253) An asteroid photographed by NASA's **NEAR Shoemaker* spacecraft during a 25-minute fly-by in June 1997 at a distance of 1 200 km. Mathilde is a rocky, carbon-rich asteroid, one of the darkest objects in the Solar System, reflecting only 3.6% of the sunlight it receives. It is approximately 50 × 50 × 70 km in size, one-millionth the mass of the Moon, heavily cratered (five craters are larger than 20 km), and rotates only every 17.4 days. *NEAR Shoemaker* took 534 images of Mathilde, including some in colour, using its sensitive multispectral camera.

The data from Mathilde was the first science return from NASA's *Discovery Program.

Mattingly, Thomas Kenneth, II (1936–) US astronaut. He was the command module pilot on the *Apollo 16* flight, the fifth crewed Moon-landing mission, during which he carried out more than an hour of extravehicular activity. He was spacecraft commander for two space shuttle flights: on the final orbital test flight of the space shuttle *Columbia* (June 1982), and on the *Discovery* mission (January 1985) that deployed an *Inertial Upper Stage rocket. Mattingly was selected as an astronaut in 1966, and left NASA in 1985.

MAXIM Pathfinder A development flight for NASA's proposed *MicroArcsecond X-Ray Imaging Mission (MAXIM), expected to be launched after 2015. The MAXIM Pathfinder mission will consist of an optics spacecraft initially flying 200 km away from a detector spacecraft. In this configuration MAXIM Pathfinder will have a resolution of 100 microarcseconds, 5 000 times better than that of the *_Chandra X-ray Observatory_. After a year or so of observations, the optics spacecraft will split into six sections arranged in a ring around a central hub, with the detector spacecraft moved to a distance of 20 000 km. This will increase resolution by up to a hundredfold, but the eventual full-scale MAXIM will have a resolution 10 times better still.

max Q (or maximum dynamic pressure) The greatest force of the Earth's atmosphere on a rocket during its launch. During lift-off, a spacecraft will shake due to the combination of max Q and the effects of approaching Mach 1 (the speed of sound; *see* MACH NUMBER).

MCC Abbreviation for *mission control centre.

MCT Abbreviation for *mission control team.

Meade, Carl Joseph (1950–) US astronaut. His three space shuttle flights include the *Discovery* mission in September 1994 when he tested a self-rescue jetpack (a backpack with thruster propulsion to be used to rescue a crew member who becomes untethered), performing NASA's first untethered extravehicular activity for ten years. Meade was a US Air Force test pilot before being selected as an astronaut in 1985. He left NASA and the air force in 1996 to join the Lockheed Martin aircraft corporation as the deputy project manager for the *X-33* reusable launch vehicle.

Mean Local Solar Time (MLST) The average *local time of the descending node of a spacecraft's *Sun-synchronous orbit. Around the Earth, the time will vary by up to 15 minutes because of the eccentricity of the Earth's orbit around the Sun.

mechanical devices subsystem (DEV) A spacecraft subsystem that deploys mechanical equipment, such as louvres and pyrotechnic devices, after a launch. Once the equipment is in motion, it cannot be stopped or controlled by feedback or other means.

MECO Acronym for *main engine cut-off.

medium Earth orbit (MEO) An orbit having an *apogee of more than 3 000 km but less than 30 000 km. Some sources may give different altitude figures. Communications and navigation satellites are often placed in MEO.

medium-gain antenna (MGA) A spacecraft antenna that provides a medium amplification of radio frequency signals, between those produced by *low-gain antennae (LGA) and *high-gain antennae (HGA). MGA area coverage also lies between that provided by LGA and HGA. The *_Magellan_ space probe used an MGA when its HGA failed to point towards the Earth during some manoeuvres.

medium-resolution imaging spectrometer (MERIS) A remote-sensing instrument launched by the European Space Agency on its **Envisat*. The primary mission of MERIS is to observe the Earth's oceans and coastal zones, but it will also investigate the atmosphere and land areas. This will increase the understanding of the climate system and global climate changes. The spectrometer uses five cameras that cover a 1 150-km band of the Earth's surface. Measurements of the Sun's reflected radiation from the Earth are taken in the visible and near-infrared part of the spectrum. The ocean data will measure algae, plankton blooms, suspended matter (including sediments), dissolved organic material, and coastal erosion. The atmospheric studies will provide information on clouds, water vapour, and aerosols, and the land observations will focus on vegetation.

Melnick, Bruce Edward (1949–) US astronaut. He flew with the *Discovery* space shuttle flight that deployed the **Ulysses* spacecraft in October 1990. He also flew on the first *Endeavour* space shuttle mission in May 1992. He left NASA that year to become director of Process Improvement Technology at Lockheed Space Operations and from 1994 to 1996 served as vice-president and director for Shuttle Engineering at the company, then renamed United States Alliance. He was selected as an astronaut in 1987.

Melroy, Pamela Ann (1961–) US astronaut. She was pilot of the *Discovery* space shuttle mission to expand the **International Space Station (ISS)* in October 2000 and flew the *Atlantis* flight to the *ISS* in October 2002 where a truss was installed. She was selected as an astronaut in 1994.

MEO Acronym for *medium Earth orbit.

Merbold, Ulf (1941–) German astronaut, the first non-US citizen to fly on a space shuttle and the first European Space Agency (ESA) astronaut aboard the Russian space station *Mir*. His shuttle flight on *Columbia* with *Spacelab 1* (*see* SPACELAB), in November–December 1983, included 72 scientific experiments. In January 1992, Merbold was the ESA *payload specialist on the **International Microgravity Laboratory* aboard *Discovery*, with 55 scientific experiments. He was a research astronaut on *Mir* in October–November 1994.

Merbold was selected as an astronaut by the ESA in 1977. He was appointed head of the Astronaut Office of the Deutsches Zentrum für Luft- und Raumfahrt (DLR; German Aerospace Centre) in 1987. He was subsequently involved in the development of ESA's contribution to the *International Space Station*.

Mercury The closest planet to the Sun. Its mass is 0.056 that of Earth. On its sunward side the surface temperature reaches over 400 °C, but on the 'night' side it falls to 170 °C.

mean distance from the Sun 58 million km
equatorial diameter 4880 km
rotation period 59 Earth days
year 88 Earth days
atmosphere Mercury's small mass and high daytime temperature mean that it is impossible for an atmosphere to be retained

surface composed of silicate rock often in the form of lava flows. In 1974 the US space probe *Mariner 10* showed that Mercury's surface is cratered by meteorite impacts (see illus. on p. 208)

satellites none

NASA's *Mariner 10* probe, launched on 3 November 1973, arrived at Mercury on 29 March 1974, and provided the first close-up images of the planet. In August 2004 the **Mercury Surface, Space Environment, Geochemistry and Ranging* (*MESSENGER*) probe was launched to the planet.

Mercury's largest known feature is the Caloris Basin, 1 400 km wide. There are also cliffs hundreds of kilometres long and up to 4 km high, thought to have been formed by the cooling of the planet billions of years ago. Inside is an iron core three-quarters of the planet's diameter, which produces a magnetic field 1% the strength of the Earth's.

http://www.solarviews.com/solar/eng/mercury.htm Detailed description of the planet Mercury. It includes statistics and information about the planet, along with a chronology of its exploration supported by a good selection of images.

***Mercury* project** A US project 1961–3 to put a human in space using the one-seat *Mercury* spacecraft.

The first two *Mercury* flights, on *Redstone rockets, were short suborbital voyages to the edge of space and back. The flights were known by the double name of *Mercury–Redstone*, with Alan *Shepard flying on *Mercury–Redstone 3* to become the first US astronaut in space on 5 May 1961. The orbital flights were launched by *Atlas rockets. The *Mercury–Atlas 6* flight was made on 20 February 1962 by John *Glenn, who became the first US astronaut to orbit the Earth.

***Mercury Surface, Space Environment, Geochemistry, and Ranging* (*MESSENGER*)** A NASA mission to the planet Mercury, launched on 3 August 2004. The spacecraft will fly past Mercury three times, in January and October 2008 and September 2009, before going into near-polar orbit around the planet in March 2011. During the flybys *MESSENGER* will map nearly the entire planet in colour, including most of the areas unseen by *Mariner 10* in 1974, the only previous probe there, as well as measuring the composition of the surface, atmosphere, and magnetosphere. Once in orbit *MESSENGER* will investigate the geology and composition of Mercury's Caloris basin, an immense impact scar. Instruments aboard the probe include the Mercury Dual Imaging System (MDIS), a Gamma-Ray and Neutron Spectrometer (GRNS), X-ray Spectrometer (XRS), Mercury Laser Altimeter (MLA), Atmospheric and Surface Composition Spectrometer (MASCS) and Energetic Particle and Plasma Spectrometer (EPPS). A magnetometer is on the end of a 3.6-m boom.

http://messenger.jhuapl.edu Introduces the *MESSENGER* mission to the planet Mercury. There are articles describing the spacecraft, the programme's scientific objective, and the mission design. The site's press room has news releases and spacecraft animations.

MERIS Contraction of *medium-resolution imaging spectrometer.

MESSENGER Contraction of **Mercury Surface, Space Environment, Geochemistry, and Ranging.*

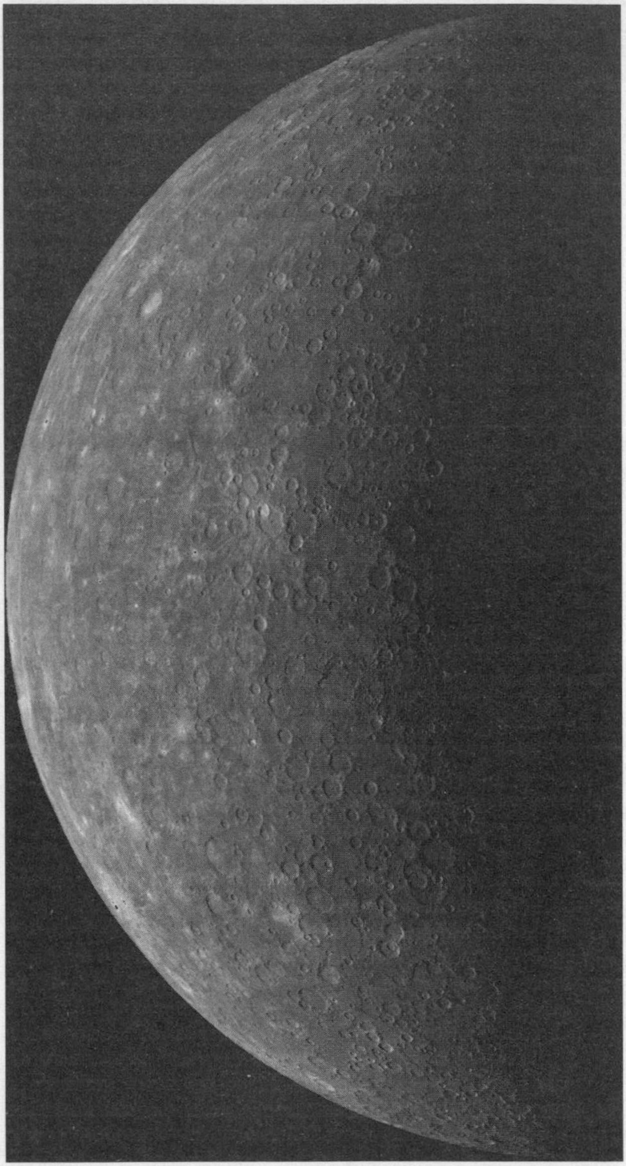

(Image © NASA)

The surface of Mercury constructed from a montage of images taken by the *Mariner 10* space probe in 1974 and 1975.

Messier catalogue A catalogue of 103 galaxies, nebulae, and star clusters (the Messier objects) published in 1781 by French astronomer Charles Messier. Catalogue entries are denoted by the prefix 'M'. Well-known examples include M31 (the Andromeda galaxy), M42 (the Orion nebula), and M45 (the Pleiades star cluster).

Messier compiled the catalogue to identify fuzzy objects that could be mistaken for *comets. The list was later extended to 109.

MET Acronym for either *mission elapsed time or *modularized equipment transporter.

meteor A flash of light in the sky, popularly known as a **shooting** or **falling star**, caused by a particle of dust, a *meteoroid, entering the atmosphere at speeds up to 70 kps and burning up by friction at a height of around 100 km. On any clear night, several **sporadic meteors** can be seen each hour.

Several times each year, the Earth encounters swarms of dust shed by comets, which give rise to a **meteor shower**. This appears to radiate from one particular point in the sky, after which the shower is named; the **Perseid** meteor shower in August appears in the constellation Perseus. The **Leonids** shoot out from the constellation Leo and are caused by dust from the comet Tempel-Tuttle, which orbits the Sun every 33 years. The Leonid shower reaches its peak when the comet is closest to the Sun.

A brilliant meteor is termed a **fireball**. Most meteoroids are smaller than grains of sand. The Earth receives an estimated 16 000 tonnes of meteoric material every year.

http://www.seds.org/billa/tnp/meteorites.html Informative collection of facts about meteorites: how they are formed, classification, and what happens when one hits Earth. The site includes images of a selection of meteorites.

meteor-burst communications A technique for sending messages by bouncing radio waves off the trails of *meteors. High-speed computer-controlled equipment is used to sense the presence of a meteor and to broadcast a signal during the short time that the meteor races across the sky.

The system, first suggested in the late 1920s, remained impracticable until data-compression techniques were developed, enabling messages to be sent in automatic high-speed bursts each time a meteor trail appeared. There are usually enough meteor trails in the sky at any time to permit continuous transmission of a message. The technique offers a communications link that is difficult to jam, undisturbed by storms on the Sun, and would not be affected by nuclear war.

meteorite A fragment of one planetary body that lands on another planetary body. Most meteorites found on Earth are thought to be fragments of asteroids or comets, but some originate from the Moon and Mars. Most meteorites are stony (chondrites and achondrites), although some are made of iron and a few are mixtures of stone and iron (pallasites).

http://www.barringercrater.com/ Attractive site all about the Barringer Crater in Arizona. The site documents the history of the crater's study, and provides background on meteorites in general. The site includes meteorite news, a bulletin board, and activities such as an interactive impact simulator.

http://www.jpl.nasa.gov/snc/ Information about and images of the various meteorites from Mars that have been discovered on Earth, including the latest news on research that debunks NASA's initial claim that a meteorite contained evidence of Martian life. You can research the news archives, and there is information on future missions to Mars.

meteoritics The branch of geology dealing with the study of *meteorites, their composition, texture, formation, and origin, and applying these observations to a more thorough understanding of the evolution of the Solar System and planets, including Earth.

meteoroid A small natural object in interplanetary space. Meteoroids are smaller than *asteroids, ranging from the size of a pebble up, and move through space at high speeds. There is no official distinction between meteoroids and asteroids, although the term 'asteroid' is generally reserved for objects larger than 1.6 km in diameter.

Meteoroids are believed to result from the fragmentation of asteroids after collisions. Some meteoroids strike the Earth's atmosphere, and their fiery trails are called **meteors**. If they fall to Earth, they are named **meteorites**.

A meteoroid could puncture a spacecraft or spacesuit. The **Skylab* space station carried a meteoroid shield, and a protective layer of felt was added to the spacesuits of astronauts on the **Gemini* project.

m

meteorological satellite (METSAT) A geostationary Earth satellite that observes weather patterns and is used for weather forecasting. NASA's **Tiros* was the first such satellite, in 1959. A worldwide METSAT system now exists, including a series of **Meteosat* craft operated by the European Meteorological Satellite Organization (EUMETSAT). The first craft in the Meteosat series was launched by the European Space Agency in 1977.

Meteosat A series of weather satellites launched by the European Space Agency (ESA) into *geostationary orbit 36 000 km above the coast of Africa, from where detailed images of weather systems are transmitted to ground stations. *Meteosat 7* was launched in 1997. ESA's upgraded Meteosat Second Generation (*MSG-1*) was first launched in August 2002. The satellites are now operated by Eurnetsat.

METSAT Contraction of *meteorological satellite.

MGA Abbreviation for *medium-gain antenna.

MGLAB Abbreviation for *Microgravity Laboratory of Japan.

Michelson interferometer for passive atmospheric sounding (MIPAS) A spectrometer used on spacecraft to detect and measure emissions of the trace gases that play a major role in atmospheric chemistry. MIPAS, which operates in the near- to mid-infrared part of the spectrum, studies the chemical composition, dynamics, and radiation of the middle atmosphere. It has flown on the European Space Agency's **Envisat*.

MicroArcsecond X-Ray Imaging Mission (MAXIM) A proposed NASA mission to image black holes using the technique of X-ray interferometry.

MAXIM will consist of up to 32 separate X-ray reflecting mirrors arranged in a circle hundreds of metres wide, feeding a detector some 500 km behind them. MAXIM would have a resolution of 0.1 microarcseconds, sufficient to resolve the event horizon around black holes. To test the techniques involved, a precursor mission called *MAXIM Pathfinder is planned.

microgravity The virtual lack of gravity experienced by astronauts in space. The term is more accurate than 'zero gravity' or 'weightlessness', both of which imply a complete absence of gravity. The environment of microgravity is excellent for experiments concerning chemicals and materials, as conducted in the **Microgravity Science Laboratory*. The phenomenon also causes *space sickness.

http://microgravity.msfc.nasa.gov/ News of NASA's microgravity research programme. Articles introduce gravitational principles and outline microgravity research in various disciplines, including biotechnology and material science. Programme reports are made available and the results of experiments are clearly explained.

http://esapub.esrin.esa.it/onstation/onstation.htm Current and archived pages of the online magazine *On Station*. Journal articles are clearly written and accompanied by photographs. They contain announcements of events and meetings, as well as updates on both flight- and ground-based microgravity research. The site also includes links to archive issues of *Microgravity News*.

Microgravity Laboratory of Japan (MGLAB) A commercial facility at Toki City, Japan, providing a drop tower for microgravity research. The free-fall tube is 100 m deep, which creates a microgravity atmosphere for 4.5 seconds. Each of the three payload capsules is 2.28 m long and can accommodate a payload weight of 400 kg. An experiment cycle normally takes 90 minutes.

***Microgravity Science Laboratory* (*MSL-1*)** The laboratory flown on the space shuttle *Columbia* 1–17 July 1997. Its 16 days of scientific experiments were crucial to long-term investigations planned for the **International Space Station*. Besides experiments in *material science, the *MSL-1* crew members carried out research on protein crystal growth and on combustion, which may improve fuel efficiency and reduce the pollution of internal combustion engines.

On *MSL-1*, scientists used Germany's TEMPUS electromagnetic levitation furnace to examine the supercooling of metals and alloys. The process involves taking a material below its freezing temperature but keeping it in a liquid state (because microgravity keeps it from freezing to a surface). Supercooling can lead to the creation of new forms of metals and alloys.

The original flight of *MSL-1* was aborted on 8 April 1997 because of a fuel cell problem. However, NASA decided the mission would fly with the same crew and payload as planned, the first time this has occurred with a shuttle.

micrometeoroid impact The collision of a micrometeoroid with a spacecraft. Micrometeoroids are small particles of dust weighing no more than a millionth of a gram. However, the particles can be dangerous, especially in great numbers, since they usually advance at high velocities. The **Voyager* probes were hit thousands of times by micro-meteoroids the size of smoke particles during their journey past the planets Jupiter, Saturn, Uranus, and Neptune. Spacecraft can be protected from these impacts by multi-layer, thermal insulation blankets made of strong fabrics such as *Kevlar.

Microvariability and Oscillations of Stars (MOST) A small Canadian Space Agency satellite launched on 30 June 2003 carrying a 0.15-m telescope to monitor the tiny brightness variations in stars caused by surface oscillations and also to look for reflected light from any planets that may be orbiting them.

microwave limb sounder (MLS) A remote-sensing instrument that measures microwave thermal emission from the limb (edge) of the Earth's atmosphere in order to sense vertical profiles of atmospheric gases, temperature, and pressure. The first MLS was launched in 1991 on NASA's **Upper Atmosphere Research Satellite* to investigate ozone depletion. The 280-kg instrument has a sensor, spectrometer, and power supply. An advanced MLS, developed for the 2004 *Aura* mission, will help increase our understanding of global climate changes. MLS experiments are also carried on balloons and aircraft.

microwave radiometer (MWR) An instrument flown on spacecraft for atmospheric studies. The MWR on the European Space Agency's **Envisat* is used to measure atmospheric humidity, as well as soil moisture, and ice characterization. It was derived from radiometers used on the *European Remote-Sensing Satellite*.

m

Midcourse Space Experiment (MSX) A test satellite of the US Ballistic Missile Defense Organization (BMDO) launched on 24 April 1996. Its primary purpose was to identify and track ballistic missiles during their mid-course flight phase. Its instruments were also used to observe the aurorae and celestial sources at wavelengths from infrared to ultraviolet, against which missiles must be tracked.

Milky Way A faint band of light crossing the night sky, consisting of stars in the plane of our Galaxy. The name Milky Way is often used for the Galaxy itself. It is a spiral galaxy, 100 000 light years in diameter and 2 000 light years thick, containing at least 100 billion *stars. The Sun is in one of its spiral arms, about 25 000 light years from the centre, not far from its central plane.

The densest parts of the Milky Way, towards the Galaxy's centre, lie in the constellation Sagittarius. In places, the Milky Way is interrupted by lanes of dark dust that obscure light from the stars beyond, such as the Coalsack *nebula in Crux (the Southern Cross). It is because of these that the Milky Way is irregular in width and appears to be divided into two between Centaurus and Cygnus.

The Milky Way passes through the constellations of Cassiopeia, Perseus, Auriga, Orion, Canis Major, Puppis, Vela, Carina, Crux, Centaurus, Norma, Scorpius, Sagittarius, Scutum, Aquila, and Cygnus.

miniworkstation A NASA term for the tool belt worn by an astronaut. The most important item is the battery-powered **pistol grip tool** (PGT), a type of powerful ratchet wrench. The PGT was used during maintenance work on the **Hubble Space Telescope*.

minor planet A name sometimes given to the larger members of the *asteroid belt.

Minotaur rocket A four-stage US rocket for launching small satellites. Part of the motor came from decommissioned Minuteman military missiles. The first Minotaur was launched in early 2000, and had placed 12 satellites in orbit by 2001. These included some belonging to the US Air Force and science satellites for universities. A classifed Air Force launch was planned for 2005.

MIPAS Acronym for *Michelson interferometer for passive atmospheric sounding.

***Mir* (Russian 'peace' or 'world')** Russian space station, the core of which was launched on 20 February 1986. It was permanently occupied until 1999, and then purposely brought down on 23 March 2001 to crash into the Pacific Ocean. During its life, *Mir* travelled more than 3 billion km, and was home to 104 cosmonauts.

The world's first modular space station, *Mir* weighed 135 tonnes and its core, weighing 20.9 tonnes, had four compartments for work, living, the engines, and docking. *Mir* evolved from, and somewhat resembled, the earlier series of **Salyut* space stations, but carried several improvements. Instead of one docking port there were six, four of which eventually had scientific and technical modules attached to them. The first was the 11-tonne *Kvant* (Russian 'quantum') astrophysics module in 1987. *Mir* had expanded to six modules by the early 1990s. The space shuttle *Atlantis* docked on 27 June 1995, exchanging

m

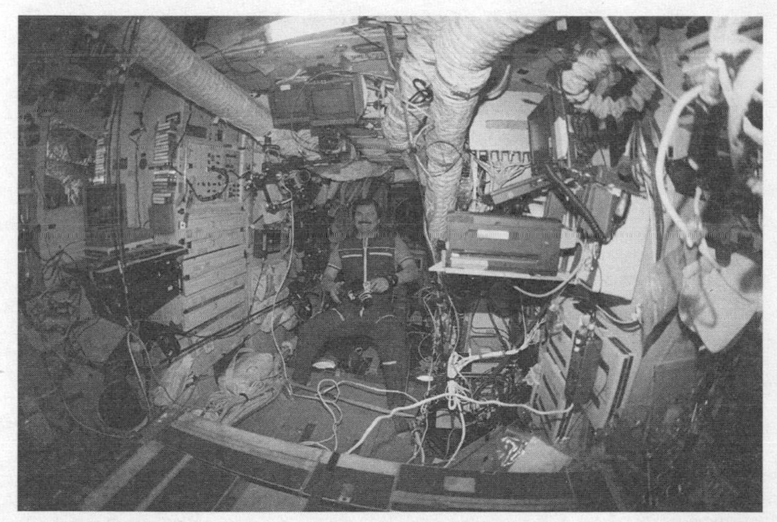

(Image © NASA)

With shuttle mission STS-91, US astronauts made their final visit to the *Mir* space station. Pictured here is cosmonaut Nikolai Budarin (flight engineer) in *Mir*'s base block, the station's central module.

crew members in the first of nine joint US–Russian missions to the station. A small wheat crop, the first successful cultivation of a plant from seed in space, was harvested on 6 December 1996.

During its orbiting life, *Mir* suffered more than 1 500 problems, including a fire in 1997, but all these were overcome by its crew. The most serious was a collision with an uncrewed cargo ship in June 1997, which damaged a solar panel and temporarily depressurized one of the modules.

The station's many residents included Valeri *Poliakov, who established a space-endurance record of 14 months from January 1994 to March 1995; Helen *Sharman, the first Briton in space, in May 1991; and Japanese journalist Toyohiro Akiyama, the first passenger in space, in December 1990.

http://spaceflight.nasa.gov/history/shuttle-mir/ Official site of the shuttle–*Mir* joint operations. There is extensive information ranging from the technical to human interest stories from crew members. The site also includes some low- and high-resolution images of the spacecraft, the crew at work, and the Earth.

mission The goal of a space flight. Typical missions include Earth observations, satellite deployment, planetary probes, and military surveillance. The spacecraft itself is also sometimes called the mission.

http://www.space.com/missionlaunches/index.html Well-designed site with comprehensive news coverage of crewed and uncrewed space flights worldwide. A launch log lists all major launches that have taken place since 1999.

mission control centre (MCC) The area from which a spacecraft mission is controlled. At NASA, the MCC for crewed space flights is at the *Johnson Space Center, and the MCC for space probes is at the *Jet Propulsion Laboratory. Russia's MCC in *Kaliningrad is operated by the *Energiya Rocket and Space Complex.

mission control team (MCT) The NASA ground team that operates a spacecraft. The team usually consists of around 20 members who control transmissions to a spacecraft and receive *telemetry from it. The team monitors a spacecraft's performance in real time and confirms the proper execution of transmitted sequences. MCT also schedules the resources and coordination for the ground stations and helps determine the amount of tracking time needed for the mission.

Mission Demonstration Test Satellite A series of Japanese testbed satellites to develop technology for future missions. The first *Mission Demonstration Test Satellite* (*MDS-1*), also known as *Tsubasa*, was launched on 4 February 2002 into a highly elliptical orbit (a geostationary transfer orbit) which took it through the Earth's Van Allen radiation belts. During its 20-month flight it gathered data on radiation in space for the development of radiation-resistant components.

Mission Design Section A NASA working group at the *Jet Propulsion Laboratory that devises the overall design for a space mission. After a project is funded, the section divides tasks among other groups, such as the Spacecraft Systems Engineering Section, the Navigation Systems Section, and the Mission Execution and Automation Section.

mission elapsed time (MET) The amount of time recorded from the instant a spacecraft is launched. The MET at launch is zero, and the elapsed time of a flight is then measured in normal units. If the time of a mission is recorded as 3/06:20:51, this means 3 days, 6 hours, 20 minutes, and 51 seconds. MET is recorded by clocks on board a spacecraft (which also carries *Universal Time Coordinated clocks). The planned events of a mission, such as deploying a satellite, are timed by MET because it remains the same whenever a spacecraft is launched. If normal Greenwich Mean Time (GMT) time were used, a launch delay would require changing all of the GMT times for events.

mission operations and data analysis (MO & DA) Alternative term for *operations phase.

mission specialist A space shuttle astronaut whose responsibilities include scientific experiments and payload operations, shuttle systems, crew activity planning, and consumables. Mission specialists assist the *payload specialists on the flight, deploy satellites from the cargo bay, perform *extravehicular activities, and operate the *Canadarm.

mission station An area of a space shuttle that contains the controls and displays for managing the vehicle's routine 'housekeeping', including the systems connecting to its payloads. The mission station is located on the right-hand side of the flight deck, slightly back of the pilot's seat.

mission status report An official report on the progress of a space mission. These regular reports are released to the press and public. They normally come from the office responsible for a mission, such as NASA's mission control centre at the Johnson Space Center or its media relations office at the Jet Propulsion Laboratory.

MIT Abbreviation for *Massachusetts Institute of Technology.

Mitchell, Edgar Dean (1930–) US astronaut. He was the *lunar module pilot of the *Apollo 14* mission in January 1971, commanded by Alan *Shepard, landing on the Moon to collect rocks, take soil samples, and set up a nuclear-powered science station. Mitchell was selected as an astronaut in 1966, and retired from NASA in 1972.

mixture ratio The weight ratio of oxidizer to fuel in the *combustion chamber of a rocket engine. For example, the mixture ratio for the space shuttle's main engine is 6:1, meaning six parts of oxidizer (liquid oxygen) to one part of fuel (liquid hydrogen).

In 1962, NASA's Marshall Space Flight Center developed control circuits and valves for a Saturn 1B launch vehicle that could change the mixture ratio during the engine's operation. Extra thrust could be created by changing the ratio from 4:5 to a maximum of 5:5.

MLI Abbreviation for *multilayer insulation.

MLS Abbreviation for *microwave limb sounder.

MLST Abbreviation for *Mean Local Solar Time.

MMS Abbreviation for *multi-mission modular spacecraft.

MMU Abbreviation for *manned manoeuvring unit.

modularized equipment transporter (MET) A two-wheeled vehicle used on the Moon by the *Apollo 14* astronauts, Alan *Shepard and Edgar *Mitchell in 1971. It was nicknamed 'the rickshaw'. The MET was tubular shaped and was 218 cm long, 99 cm wide, and 81 cm high. It had inflated tyres, two legs for stability at rest, and was pulled by a single handle. Its Earth weight was 12 kg, and it could carry 63 kg. The MET aided surface exploration on the Moon's surface by serving as a carrier for geology hand tools, four cameras, film, bags for rock and soil samples, a work table, and a lunar surface penetrometer to determine the hardness of the lunar surface.

modular locker A standard storage locker on a space shuttle and space stations. The lockers are used for a wide variety of items. On a shuttle, the mid-deck modular lockers hold such items as medical kits, dosimeters (to measure radiation levels), and wet-trash (human waste) containers; scientific experiments are also sometimes mounted in these lockers.

module A component of a spacecraft. Modules are designed in coordination. For the *Apollo* Moon landings, the craft comprised a *command module (for working, eating, sleeping), *service module (electricity generators, oxygen supplies, manoeuvring rocket), and *lunar module (to land and return the astronauts).

m

Mohri, Mamoru (1948–) A Japanese astronaut with the *National Space Development Agency (NASDA). He flew on the space shuttle *Endeavour* in September 1992 with the US–Japanese **Spacelab* mission. He also flew on *Endeavour* in February 2000 as part of the *Shuttle Radar Topography Mission. Mohri logged 459 hours in space. He was selected as an astronaut by NASDA in 1985.

MOL Acronym for **Manned Orbital Laboratory*.

momentum desaturation (desat; or momentum unload) A manoeuvre to remove the excess momentum that builds up in a spacecraft's *momentum wheels. Frequent desats are usually required. The manoeuvre applies torque (force causing rotation) to the spacecraft by using its *thrusters. The **Hubble Space Telescope*, which would be contaminated by thruster exhaust, achieves desat by using on-board magnets that interact with the Earth's magnetic field to produce torque.

momentum wheel The massive electrically-powered reaction wheels used to stabilize a spacecraft's attitude. They are mounted on board in three axes at right angles. The spacecraft can be rotated in one direction by spinning the relevant wheel in the opposite direction. If the wheel is slowed down, the vehicle will rotate back. Momentum wheels add weight to a spacecraft but give it a steadier attitude than *thrusters can provide.

MON Abbreviation for *monitor system.

monitor system (MON) A system used by NASA's *deep-space network (DSN) to report on the operation and performance of the DSN itself. The data is collected from the DSN's seven systems, such as the *telemetry system and *tracking system, and used both at the Jet Propulsion Laboratory to manage and advise DSN operations, and in each *deep-space communications complex (DSCC) to control its own activities.

monopropellant A rocket propellant that has the fuel and oxidizer already mixed together. A chemical reaction, normally caused by a catalyst, produces the energy. Monopropellants can be solid or liquid. *Hydrozine, a liquid fuel, is used in a catalytic decomposition engine, flowing over a heated catalyst to decompose and create a temperature of about 927°C.

Moon A natural satellite of Earth, 3 476 km in diameter, with a mass 0.012 (approximately one-eightieth) that of Earth.

Its surface gravity is only 0.16 (one-sixth) that of Earth. Its average distance from Earth is 384 400 km, and it orbits in a west-to-east direction every 27.32 days (the **sidereal month**). It spins on its axis with one side permanently turned towards Earth. The Moon has no atmosphere and no liquid water, although there may be subsurface ice at the poles.

phases The Moon is illuminated by sunlight, and goes through a cycle of phases of shadow, waxing from **new** (dark) via **first quarter** (half Moon) to **full**, and waning back again to new every 29.53 days (the **synodic month**, also known as a **lunation**). On its sunlit side, temperatures reach 110 °C, but during the two-week lunar night the surface temperature drops to −170 °C.

origins The origin of the Moon is still open to debate. Earlier theories that it split from the Earth, that it was a separate body captured by Earth's gravity, or that it formed in orbit around Earth are now discounted in favour of the view that it was formed from debris thrown off when a body the size of Mars struck Earth shortly after our own planet had formed.

research About 70% of the far side of the Moon was photographed from the Soviet *Luna 3* in October 1959. Much of our information about the Moon has been derived from this and other photographs and measurements taken by US and Soviet Moon probes, from geological samples brought back by US *Apollo* astronauts and by Soviet *Luna* probes, and from experiments set up by US astronauts 1969–72. The US probe *Lunar Prospector*, launched in January 1998, examined the composition of the lunar crust, recorded gamma rays, and mapped the lunar magnetic field. It also discovered signs of ice at the Moon's poles in 1998. In January 2004, US president George W Bush announced a plan to put astronauts back on the Moon by 2020.

composition The Moon is rocky, with a surface heavily scarred by *meteorite impacts that have formed craters up to 240 km across. Seismic observations indicate that the Moon's surface extends downwards for tens of kilometres; below this crust is a solid mantle about 1 100 km thick, and below that a silicate core, part of which may be molten. Rocks brought back by astronauts show that the Moon is 4.6 billion years old, the same age as Earth. It is made up of the same chemical elements as Earth, but in different proportions, and differs from Earth in that most of the Moon's surface features were formed within the first billion years of its history when it was hit repeatedly by meteorites.

(Image © NASA)
A full Moon photographed from the *Apollo 11* spacecraft as it made the three-day, 386 000-km journey back to Earth. This photograph was taken about 18 520 km from the Moon.

The youngest craters are surrounded by bright rays of ejected rock. The largest scars have been filled by dark lava to produce the lowland plains called seas, or **maria** (plural of *mare). These dark patches form the so-called 'man-in-the-Moon' pattern. Inside some craters that are permanently in shadow is up to 300 million tonnes/330 million tons of ice existing as a thin layer of crystals.

The US lunar probe *Clementine* discovered an enormous depression on the far side of the Moon in 1994. The South Pole–Aitken Basin is 2 500 km across and 13 km deep, making it the largest such feature on the Moon.

http://www.fourmilab.ch/earthview/ Graphical simulations of the Moon and Earth. The Moon (or locations on the Moon) can be viewed from different locations on Earth, from the Sun, or as a map showing day and night across the lunar surface.

http://www.lpi.usra.edu/expmoon/future/future.html Directory of lunar science missions planned by various international space agencies. There are also links to sites that discuss proposals for further crewed exploration of the Moon.

http://www.inconstantmoon.com/inconstant.htm Packed with information, maps, lunar tables, and many other useful items. The site features a lunar tour, with photographs showing features to look out for as the Moon changes phase.

http://www.solarviews.com/solar/eng/moon.htm Detailed description of the Moon. It includes statistics and information about the surface, eclipses, and phases of the Moon, along with details of the *Apollo* landing missions. The site is supported by a good selection of images.

moon Any natural *satellite that orbits a planet. Mercury and Venus are the only planets in the Solar System that do not have moons.

Moon probe A crewless spacecraft used to investigate the Moon. Early probes flew past the Moon or crash-landed on it, but later ones achieved soft landings or went into orbit. Soviet probes included the **Luna* series. US probes (**Ranger* space probes, **Surveyor*, **Lunar Orbiter*) prepared the way for the **Apollo* project crewed flights.

The first space probe to hit the Moon was the Soviet *Luna 2* on 13 September 1959, and *Luna 9* performed the first soft landing on 3 February 1966. The first successful US Moon probe was *Ranger 7*, which took close-up photographs before it hit the Moon on 31 July 1964. *Surveyor 1*, on 2 June 1966, was the first US probe to soft-land on the lunar surface. Between 1966 and 1967 a series of five *Lunar Orbiters* photographed the entire Moon in detail, in preparation for the *Apollo* landings 1969–72.

In March 1990, Japan put a satellite, **Hagoromo*, into orbit around the Moon. NASA's **Clementine* probe discovered an enormous crater on the far side of the Moon in 1994, and its **Lunar Prospector* discovered ice on the Moon in 1998.

Moon Treaty (officially Agreement Governing the Activities of States on the Moon and Other Celestial Bodies) A *space law treaty drafted in 1979 by the United Nations to ban military activity and national claims to the Moon and other celestial bodies. The treaty was never ratified by the nations involved in space exploration.

The treaty stated that any nation could conduct scientific investigations and establish stations on the Moon and planets, but should not contaminate the environments. It added that 'The exploration and use of the Moon shall be the province of all mankind and shall be carried out for the benefit and in the interests of all countries, irrespective of their degree of economic or scientific development.'

Moon walk An extravehicular activity on the Moon to explore and test its surface. Moon walks were carried out by 12 US astronauts on 6 *Apollo* spacecraft missions from 1968 to 1972. The astronauts were Neil Armstrong and Buzz Aldrin on *Apollo 11*, Pete Conrad and Alan Bean on *Apollo 12*, Alan Shepard and Edgar Mitchell on *Apollo 14*, David Scott and James Irwin on *Apollo 15*, John Young and Charles Duke on *Apollo 16*, and Eugene Cernan and Jack Schmitt on *Apollo 17*.

Morelos Satellite System (MSS) A Mexican communication satellite system for which two geosynchronous satellites were launched in 1985 by Hughes Communications International. Each satellite had 22 *transponders (transmitter-receivers). The two main users were the Mexican communications company Televisa and DGT (Direcion General de Telecomunicaciones), a telecom agency.

Morukov, Boris Vladimirovich (1950–) Russian cosmonaut and physician. His first space flight was with NASA's space shuttle *Atlantis* mission in September 2000 to prepare the *International Space Station* for its first resident crew. He logged 11 days in space. Morukov provided medical support for crewed space flights for more than 20 years before he began his training as a cosmonaut in 1990.

MOST Acronym for **Microvariability and Oscillations of Stars* (*MOST*).

MPLM Abbreviation for *Multi-Purpose Logistics Module.

MSC Abbreviation for *Munich space chair.

MSFC Abbreviation for *Marshall Space Flight Center.

MSS Abbreviation for *Morelos Satellite System.

MSSL Abbreviation for *Mullard Space Science Laboratory.

MSX Abbreviation for **Midcourse Space Experiment*.

Mueller, George E (1939–) The head of NASA's *Apollo* project from 1963 to 1969. He was also responsible for the *Gemini* and Saturn programmes. He originated the **Skylab* space station and has been called the 'Father of the Space Shuttle'. After leaving NASA in 1969, Mueller was senior vice-president of the General Dynamics Corporation 1969–71 and held executive positions in other corporations before becoming chief executive officer of Kistler Aerospace Company in 1995. He was awarded the 1970 President's National Medal of Science and has three NASA Distinguished Service Medals.

Mukai, Chiaki (1952–) The first female Japanese astronaut in space, and the first Japanese to fly twice. A physician, she was a payload specialist on the *Columbia* space shuttle 8–23 June 1994, on which medical and microgravity experiments were conducted. She also flew on the *Discovery* flight in 29 October–7 November 1998, which included investigations on space flight and the ageing process.

Mullane, Richard Michael (1945–) US astronaut. His three space shuttle flights include the first *Discovery* mission in August 1984 to deploy three satellites. Mullane was chosen as an astronaut in 1978 and retired from NASA in 1990.

Mullard Space Science Laboratory (MSSL) The UK's largest university space physics institute, located at Holmbury St Mary, Surrey. It is part of the Department of Space and Climate Physics, University College London, with scientific research groups in astrophysics, solar physics, plasma physics, and climate physics. MSSL has designed and built instruments for some 40 spacecraft and 250 rockets. Its instruments have flown on missions including *Cassini, *Exosat, *Röntgen Satellite, *Rosetta, *Solar and Heliospheric Laboratory *Spacelab*, and *X-MM Newton*. It was also involved in the atmospheric measurements of the *Mars Express*.

http://www.mssl.ucl.ac.uk/pages/ Presents the research activities and facilities at the Mullard Space Science Laboratory (MSSL). The site has details of the many space missions to which the MSSL has contributed and is updated with the latest science news from the laboratory.

multi-angle imaging A technique used by *Earth-observation satellites to acquire images from several angles simultaneously. NASA's Multi-Angle Imaging Spectroradiometer (MISR), launched on the *Terra* satellite in December 1999, acquires Earth images at nine angles using nine cameras pointing forwards, backwards, and downwards along its flight path. Its images of the Pine Island Glacier in Antarctica taken in March 2001 revealed differences in the surface texture and could distinguish between rough crevasses and smooth ice.

Multifoil Microabrasion Package (MAP) An experiment to count impacts made by meteoroid and space debris while in space. MAP was flown on NASA's *Long-Duration Exposure Facility* satellite that was launched in April 1984 by the space shuttle *Challenger*. It was retrieved in January 1990 by *Columbia*. MAP consisted of thin foil layers that were pointed in five directions. During their 69 months in space, the foils received 2 342 perforations. Most came from 'unbound particles' of extraterrestrial origin rather than 'bound particles' from the Earth orbital area.

multilayer insulation (MLI) Thin layers of polyester or polyamide, used to provide thermal insulation for spacecraft. MLI protects the vehicle from overheating by reflecting sunlight, but it retains enough heat inside. These layers have been coated with a thin film of aluminium or gold. To minimize the conductance between layers, they are often separated by such materials as glass-fibre paper, foam, or plastic and silk netting.

multi-mission modular spacecraft (MMS) A spacecraft designed to facilitate different types of scientific observations and experiments. It provides all the necessary utilities for payloads, the modules of which can be attached via a platform equipment deck. The MMS, which can be serviced while orbiting, was developed at NASA's *Goddard Space Flight Center for the *Solar Max* spacecraft (1980). Later examples have included the *Upper-Atmosphere Research Satellite* and the *Extreme-Ultraviolet Explorer*.

multiple-docking adapter A space-station component that allows more than one visiting spacecraft to dock at the same time. *Skylab* was fitted with such an adapter. The Russian space station *Mir* was fitted with a multiple-docking adapter with five ports, a large improvement on the Russian *Salyut* space station. The adapter on *Mir* allowed *Soyuz* spacecraft and supply vehicles to attach at various points, and extra modules could also be added.

multiplexing The simultaneous transmission of different data from a spacecraft using a single channel. The data stream is separated into frames that carry codes for different information, such as temperatures, pressures, and the state of on-board computers. The Jupiter probe **Cassini* had more than 13 000 different measurements. Ground stations can then recognize the codes and isolate a specific measurement.

multi-probe mission A spacecraft mission that sends more than one probe into the atmosphere or onto the surface of a planet or a moon. The *Pioneer Venus* multi-probe mission in 1978 deployed four atmospheric probes that sent data to Earth during their 55-minute descents through Venus's atmosphere. The spacecraft's large probe was released on 15 November and three small probes on 20 November. The *Mars Exploration Rover mission of 2003 sent twin rovers to the Martian surface in January 2004.

Multi-Purpose Logistics Module (MPLM) These are any of three re-usable pressurized modules used to transport cargo to and from the *International Space Station (ISS)*.

The MPLMs are cylinders approximately 6.4 m long and 4.6 m in diameter. Each MPLM is carried in the space shuttle cargo bay and then docked to the *ISS* by the robotic arm of the shuttle or the *ISS*. After exchange of cargo, the MPLM is detached from the station and re-positioned in the shuttle's cargo bay for return to Earth. The three MPLMs were built by the Italian Space Agency in exchange for Italian access to research time on the *ISS*. They are named **Leonardo*, **Raffaello*, and **Donatello* after famous Italians of the past. *Leonardo* was the first to be launched, on shuttle mission STS-102 in March 2001. *Raffaello* had its first flight in April 2001 aboard STS-100. *Donatello* had not flown by the time of the space shuttle accident in February 2003.

multistage rocket A rocket launch vehicle made up of several rocket stages (often three) joined end to end. The bottom, or first, stage fires first, boosting the vehicle to high speed, then it falls away. The second stage fires, thrusting the now lighter vehicle even faster. The remaining stages fire and fall away in turn, ending with the upper stage, boosting the vehicle's payload (cargo) to an orbital speed that can reach 28 000 kph.

Munich space chair (MSC) An innovative chair for use in space, designed by architecture students at the Munich Technical University in Munich, Germany. It has a body-restraint system by which an astronaut straps the chair to his or her body at three points, up to the hip. An adjustable table can also be attached to the chair. An MSC was installed on the Russian space station *Mir* in 1995 and the design will also be used in the *International Space Station*.

Musgrave, (Franklin) Story (1935–) US astronaut and physician. He logged more than 1280 hours in space on six flights, including the first mission of the *Challenger* space shuttle in April 1983. He also flew with the first servicing and repair mission to the **Hubble Space Telescope* on the space shuttle *Endeavour* in December 1993. Musgrave was selected as an astronaut in 1967 and retired from NASA in 1997.

NACA Acronym for *National Advisory Committee for Aeronautics.

nadir The point on the celestial sphere vertically below the observer and hence diametrically opposite the **zenith**.

Nagel, Steven Ray (1946–) US astronaut. His four space shuttle flights include two German **Spacelab* missions: on *Challenger* in October 1985 and on *Columbia* in April 1993. Nagel was selected as an astronaut in 1979. He left NASA's Astronaut Office in 1995 to become deputy director of the Johnson Space Center. In 1996 he became a NASA research pilot.

NASA Acronym for *National Aeronautics and Space Administration

NASA Communications Network (NASCOM) The NASA network that provides communications support for its Tracking and Data-Relay Satellite System (TDRSS), which manages *Tracking and Data-Relay Satellites. NASCOM interconnects customers and other groups that support the TDRSS, such as the Flight Dynamics Facility (FDF) and Network Control Center (NCC). NASCOM provides data, computer, television, teletype, and voice services.

NASCOM Contraction of *NASA Communications Network.

NASDA Contraction of *National Space Development Agency.

NASM Abbreviation for *National Air and Space Museum.

National Advisory Committee for Aeronautics (NACA) A predecessor of NASA. It was established in 1915 to 'supervise and direct the scientific study of the problems of flight, with a view to their practical solutions'. By 1946, NACA had almost 7 000 employees at six centres throughout the USA. The organization's research and testing programmes resulted in the Bell X-1 rocket plane breaking the sound barrier on 14 October 1947. NACA subsequently developed the X-15 rocket plane and ideas for crewed spacecraft, but was replaced by NASA in 1958.

National Aeronautics and Space Act 1958 A US Congressional Act that established *NASA as a replacement for the *National Advisory Committee for Aeronautics (NACA). The final version of the Act was greatly influenced by the Senate's special committee on space and astronautics, chaired by the Senate's majority leader, Lyndon B *Johnson. US president Dwight D *Eisenhower signed the Act into law on 29 July 1958. It did not divide responsibilities between NASA and the US military, but Eisenhower's intention was for NASA to manage the crewed space programme.

National Aeronautics and Space Administration (NASA) The US government agency for space flight and aeronautical research, founded in 1958 by the National Aeronautics and Space Act. Its headquarters are in

Washington, DC, and its main installations include the *Kennedy Space Center on Merritt Island in Florida, the *Johnson Space Center in Houston, Texas, the *Jet Propulsion Laboratory in Pasadena, California, the *Goddard Space Flight Center in Beltsville, Maryland, and the *Marshall Space Flight Center in Huntsville, Alabama. NASA's early planetary and lunar programmes included the *Pioneer* probes, from 1958, which gathered data for the later crewed missions, and the **Apollo* project, which took the first astronauts to the Moon in *Apollo 11* on 16–24 July 1969.

NASA launched the first space shuttle in 1981. In the early 1990s, the agency moved towards lower-budget missions, such as the **NEAR Shoemaker* craft and the **Lunar Prospector*. It also established a *New Millennium Program to identify, develop, and fly advanced technologies at lower costs. The programme's first launch was **Deep Space 1* in 1998, and its Space Technology 6 (ST6) series has been developing new technologies for future flights. A notable recent success was the *Mars Exploration Rover Mission which in January 2004 landed two rovers on that planet, leading to the discovery of previous water. However, the break-up of the *Columbia* shuttle on 1 February 2003, killing all seven astronauts, led to all shuttles being grounded until 2005, as well as an extensive investigation and upgrade of safety. In January 2004, US president George W Bush announced plans to send astronauts back to the Moon in 2020 and later to Mars.

Its other recent major project, in partnership with 15 other nations, is the US$60 billion **International Space Station*, scheduled for completion in 2006. The first crew arrived at *ISS* in November 2000 under the command of NASA astronaut Bill Shepherd; by September 2001, six habitable modules had been added to the *ISS*.

Other NASA installations are located in Virginia (*Langley Research Center and *Wallops Flight Facility); California (*Ames Research Center and the Flight Research Center); and Ohio (*Lewis Research Center). The Office of Manned Space Flight is responsible for space missions with crews and for the space station and space shuttle programs. The Office of Space Science and Applications deals with the scientific exploration of space. The Office of Advanced Research and Technology plans future flights and research. The Office of Tracking and Data Acquisition provides a network for tracking flights and accumulating data.

http://www.nasa.gov/ Latest news from NASA, plus the most recent images from the *Hubble Space Telescope*. The site also contains answers to questions about NASA resources and the space programme, and a gallery of video and audio clips and still images.

http://www.lerc.nasa.gov/WWW/bpp/ Interesting discussion on the possibilities of interstellar travel. The pages document the work to develop new space travel technologies sponsored by NASA's Advanced Space Transportation Program.

http://history.nasa.gov/ Documents the history of NASA with a well-organized collection of articles, images, and technical drawings, and a chronology with mission transcripts. An A–Z index helps track down items of interest held by the History Office itself and across the NASA Internet.

http://neurolab.jsc.nasa.gov/trivia_pers.htm NASA site that answers all the questions you might have about the history of space exploration, including a section on 'Firsts'.

National Air and Space Museum (NASM) A museum belonging to the Smithsonian Institution, Washington, DC. It houses the world's largest collection of historic aircraft and spacecraft. It was established in 1946 as the National Air Museum (NAM) and was renamed in 1976 when it was relocated in a separate building. The museum has hundreds of artefacts on display, including the *Apollo 11* command module, a lunar rock sample that can be touched by visitors, the *Mercury Friendship 7* capsule, the *Apollo* lunar roving vehicle, spacesuits, the *Gemini 4* spacecraft, and the *Skylab 4* command module. An additional building, the Steven F Udvar-Hazy Center, named after a major donor, was officially opened on 10 December 2003 at the Washington Dulles International Airport. It has 70 611 sq m of space, and opened with dozens of space exhibits, including the *Enterprise* space shuttle, and 80 aircraft.

http://www.nasm.edu/ Browse the huge air and space collection of the Smithsonian Museum via this site's 'Collections and Research' pages. There is information about the various museum departments and their special exhibitions.

National Centre for Space Studies The English name for the French space organization *Centre National d'Etudes Spatiales.

National Institute for Space Research The English name for the *Instituto Naçional de Pesquisas Espaçiais (INPE), the Brazilian space agency.

National Oceanic and Atmospheric Administration (NOAA) A US government department that uses satellites to provide weather, water, and climate forecasts, and to monitor the environment. NOAA was established on 3 October 1970 under the US Department of Commerce. It currently operates two **Geostationary Operational Environmental Satellites (GOES)* and, in polar orbits, two advanced **Television Infrared Observation Satellites* (*Tiros*). *GOES* monitors such natural dangers as hurricanes and other severe storms, volcanic ash, and wildfires. *Tiros*, with its primary instrument of an *advanced very high-resolution radiometer, can track long-term weather conditions, including ozone levels in the atmosphere.

The National Weather Service, part of NOAA, is the main recipient of millions of bits of satellite data and images collected each day, and other customers include the United Kingdom Meteorological Office and members of the European Space Agency. Weather data from NOAA and other organizations are kept at the National Climatic Data Center in Asheville, North Carolina, the largest climate archive in the world. It is part of NOAA's National Environmental Satellite, Data, and Information Service. The National Environmental Satellite, Data and Information Service operates NOAA's National Data Centers.

http://www.noaa.gov/ Massive resource with news and information on all aspects of the National Oceanic and Atmospheric Administration's environmental monitoring activities. Earth-observation resources include the Geostationary Satellite Server with the latest visible and infrared images from the Geostationary Operational Environmental Satellite series, satellites used for short-range weather forecasting.

n

National Space Club (NSC) A US non-profit organization established in 1957 to stimulate the exchange of ideas and information concerned with rocketry and astronautics, and to promote recognition of US achievements in space. Its members are recruited from the space industry, educational institutions, the government, and the general public. The NSC presents several annual awards, including those to professionals who have made outstanding contributions to planetary exploration, and to those who have developed innovative uses of Earth-observation satellites. The NSC also maintains a Hall of Fame in recognition of individuals who have demonstrated exceptional efforts in the field of space exploration.

National Space Council A US government council that advised and assisted the US president on national space policy and strategy. It was created during the Kennedy administration, briefly revived during George Bush's term of office, but allowed to lapse during Bill Clinton's presidency.

National Space Development Agency (NASDA) The former name of Japan's national space agency which, on 1 October 2003, became the Japan Aerospace Exploration Agency (JAXA) by merging with the Institute of Space and Astronautical Science (ISAS) and the National Aerospace Laboratory of Japan (NAL). Established in 1969, it concentrated on the development and launch of communications and remote-sensing satellites, such as the **Advanced Earth-Observing Satellite*. It had also overseen the development of the *H rockets.

national space policy The official government policy for space exploration. The first official US national space policy was stated in the *National Aeronautics and Space Act 1958. The best known was the programme to put a person on the Moon that was announced by President John F Kennedy on 25 March 1961. The White House normally announces a new national space policy after it is reviewed by the National Science and Technology Council and the National Security Council. Russian national space policy announcements are often made as an edict from the president of the Russian Federation or as resolutions or laws adopted by the political assembly of the Russian Federation.

National Space Science Data Center (NSSDC) A centre that provides access to science data from NASA's space flight missions. This information, which can be seen online without cost, includes astrophysics, solar physics, space plasma, lunar, and planetary data. Established in 1966, NSSDC is part of the *Space Science Data Operations Office at NASA's *Goddard Space Flight Center.

National Space Society (NSS) An independent non-profit educational organization founded in 1974 by German rocket engineer Wernher *von Braun to increase public understanding and support of space exploration, and to support technical, economic, and political activities that increase human presence in space. The NSS has more than 22 000 members, and is based in the USA, with 75 chapters, including six in Australia, two in Canada, and one each in Mexico, Germany, Ireland, and the UK. The current executive director of the NSS is George Whitesides. In January 2004, NSS joined with the Satellite

Industry Association and the Washington Space Business Roundtable to create the National Space and Satellite Alliance (NSSA) to unify advocacy for issues concerning space policy.

The NSS publishes *Ad Astra*, a bimonthly magazine.

navigation The science and technology of finding the position, course, and distance travelled by a ship, plane, or other craft. Satellite navigation uses satellites that broadcast time and position signals.

The US *global positioning system (GPS) was introduced in 1992, featuring 24 *Navstar* satellites. The same year, 85 nations agreed to take part in trials of a new navigation system known as FANS, or Future Navigation System, using 24 Russian Glonass satellites and the 24 US GPS satellites.

Navstar A series of US satellites that provides navigation data to military units and thousands of civilian users worldwide. There were 38 satellites operating in 2003, each with a life expectancy of 7.5 years. The spacecraft orbit the Earth every 12 hours, providing customers with precise navigation details of location, velocity, and time. *Navstar*'s *global positioning system is part of the US Department of Defense and is operated by the 50th Space Wing at Falcon Air Force Base, Colorado. The system was a strategic part of the Allied victory during the Gulf War of 1991.

NEAP Acronym for *Near-Earth Asteroid Prospector.

Near-Earth Asteroid Prospector (NEAP) The first commercial attempt to build and fly a scientific mission into deep space. The US mission, to fly to a near-Earth asteroid, was planned by SpaceDev, the world's first commercial space exploration company, with a future launch date to be set. The 200-kg NEAP microsatellite will carry a mixture of scientific, engineering, and entertainment payloads.

To support the mission, SpaceDev contracted with NASA's *Jet Propulsion Laboratory to use its *deep-space network for tracking, communication, and other support. This is the first time in NASA's history that a commercial company has made a request to use its facilities.

SpaceDev's headquarters are in Poway, California, near San Diego.

Near-Earth Asteroid Rendezvous (*See* NEAR SHOEMAKER)

near-Earth space The area of space just above the Earth's atmosphere, where many artificial satellites orbit. In addition, the term 'near space' is sometimes used for the region that has received the most space probes, namely the Moon and the close planets of Mercury, Venus, and Mars.

http://www.nas.edu/ssb/elements.html Part of a National Research Council guide to space weather. Pages discuss the elements and the nature of near-Earth space. There are explanatory diagrams and photographs throughout the text and key words linked to a useful glossary.

near encounter phase (NE) The closest approach during a spacecraft's encounter with a target, especially a planet. It follows the *far encounter phase (FE) and is marked by intense high-priority, data-gathering operations for which the mission is intended. For example, NASA's *Voyager 2* probe began its NE with the planet Neptune on 24 August 1989 when it passed 4 827 km above the surface. This phase presents the best opportunity for the highest resolution possible. NE is followed by a **post encounter phase** (PE) when a spacecraft recedes from a planet, and PE is completed when the spacecraft's observations end.

Near-Infrared Mapping Spectrometer (NIMS) An instrument that combines spectroscopy and imaging for *remote-sensing devices. NIMS, which weighs 18 kg, uses mirrors and no lenses. It was launched on the **Galileo* spacecraft in 1989 to study the surfaces of Jupiter's moons and the atmosphere of the planet. NIMS mapped the surfaces of Europa, Ganymede, and Callisto to determine their mineral distributions, and monitored volcanic activity on Io. It measured infrared radiation on the near-infrared wavelength (0.7–5.2 microns) from Jupiter's atmosphere, providing data that contributed to an understanding of the atmosphere's motions and energy balances, and the nature of Jovian clouds.

near real time (NRT) The time that is somewhat delayed from real time (happening at the moment). 'Near real time' data is a designation of spacecraft data transmitted to ground stations a short time after its collection (normally a few hours). Other spacecraft data may be marked as 'real time', such as engineering data sent immediately from on-board computers. The absence of 'near real time' and 'real time' labels means that the scientific data has been stored on the spacecraft's tapes and transmitted at a later date.

n

Near Shoemaker A spacecraft launched on 17 February 1996 that carried out fly-bys of asteroids 253 *Mathilde on 17 June 1997 and asteroid 433 *Eros on 23 December 1998. It returned to orbit Eros on 14 February 2000. With an original orbit of 320×366 km above Eros, the orbit was lowered during the summer of 2000, to a near circular orbit of 35 km where the spacecraft spent several weeks. *NEAR Shoemaker* was the first spacecraft to soft land on an asteroid. The mission's science objectives were to map Eros's surface, and determine its size, shape, rotation rate, mass, density, and composition. Its instruments included an imager, infrared spectrometer, X-ray/gamma-ray spectrometer, magnetometer, and a lidar.

NEAR Shoemaker was built by the Johns Hopkins University Applied Physics Laboratory, Maryland, who also managed the project.

http://near.jhuapl.edu/ Images and scientific results from the *NEAR Shoemaker* mission to 433 Eros. Articles provide background on mission design and operations. The site has a good selection of asteroid and spacecraft animations and a comprehensive Frequently Asked Questions list.

nebula A cloud of gas and dust in space. Nebulae are the birthplaces of stars, but some nebulae are produced by gas thrown off from dying stars (*see* PLANETARY NEBULA; SUPERNOVA). Nebulae are classified depending on whether they emit, reflect, or absorb light.

An **emission nebula**, such as the Orion nebula, glows brightly because its gas is energized by stars that have formed within it. In a **reflection nebula**, starlight reflects off grains of dust in the nebula, such as surround the stars of the Pleiades cluster. A **dark nebula** is a dense cloud, composed of molecular hydrogen, which partially or completely absorbs light behind it. Examples include the Coalsack nebula in Crux and the Horsehead nebula in Orion.

Nelson, George Driver (1950–) US astronaut. His three space shuttle missions include the *Challenger* flight in April 1984 to deploy the **Long-Duration Exposure Facility*, when he undertook ten hours of extravehicular activity; and the *Discovery* mission in September 1988 to deploy the *Tracking and Data-Relay Satellite. Nelson was selected as an astronaut in 1978 and retired from NASA in 1989.

Neptune The eighth planet in average distance from the Sun. It is a giant gas (hydrogen, helium, methane) planet, with a mass 17.2 times that of Earth. It has the fastest winds in the Solar System.

mean distance from the Sun 4.5 billion km
equatorial diameter 49 530 km
rotation period 16 hours 7 minutes
year 164.8 Earth years
atmosphere methane in its atmosphere absorbs red light and gives the planet a blue colouring. Consists primarily of hydrogen (80%) with helium (19%) and methane (1–2%)

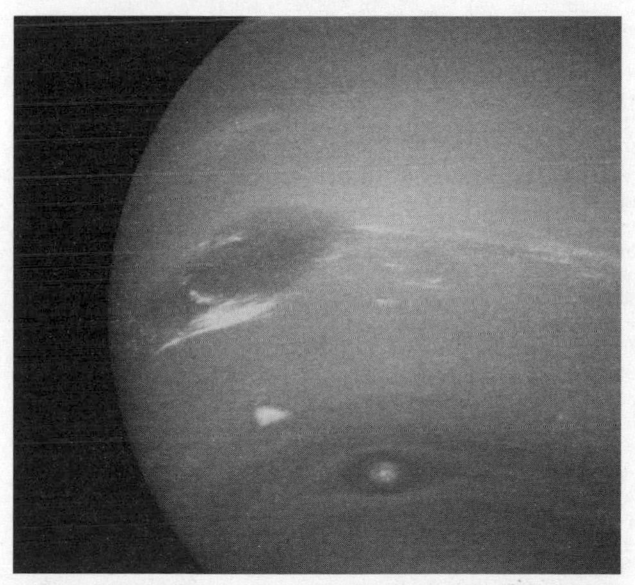

(Image © NASA)
Neptune reconstructed from two images taken by the *Voyager 2* probe in 1989. The Great Dark Spot is to the centre of the image, and the smaller dark spot DS2 (with its bright core) is to the bottom. Both features had disappeared by 1994. Between the two is the bright cloud feature known as the 'Scooter'.

surface hydrogen, helium, and methane. Its interior is believed to have a central rocky core covered by a layer of ice

satellites Neptune has over a dozen known moons, two of which (*Triton and Nereid) are visible from Earth. Six more were discovered by the *Voyager 2* probe in 1989, of which Proteus (diameter 415 km) is larger than Nereid (300 km)

rings there are five faint rings: Galle, Le Verrier, Lassell, Arago, and Adams (in order from Neptune). The Lassell ring is the widest at 4 000 km

Neptune was located in 1846 by German astronomers Johan Galle and Heinrich d'Arrest after calculations by English astronomer John Couch Adams and French mathematician Urbain Le Verrier had predicted its existence from disturbances in the movement of Uranus. *Voyager 2*, which passed Neptune in August 1989, revealed various cloud features, notably an Earth-sized oval storm cloud, the Great Dark Spot, similar to the Great Red Spot on Jupiter, but images taken by the **Hubble Space Telescope* in 1994 show that the Great Dark Spot has disappeared. A smaller dark spot, DS2, has also gone.

http://www.solarviews.com/solar/eng/neptune.htm Detailed description of the planet Neptune. The site includes a chronology of the exploration of the planet, along with statistics and information on its rings, moons, and satellites, supported by a good selection of images.

NERVA Acronym for *Nuclear Engine for Rocket Vehicle Application.

Nespoli, Paolo Angelo (1957–) Italian astronaut with the European Space Agency (ESA). He led the *EuroMir team that in 1995 developed a computer system for the Russian space station *Mir*. In 1996 he joined NASA's Johnson Space Center (JSC) to assist in the training of crews for the *International Space Station*. Nespoli became an ESA astronaut trainer in 1991. In 1998 he was selected as an astronaut by the Agenzia Spaziale Italiana (ASI; Italian Space Agency), joined the European Astronaut Corps, and began training at the JSC for a future space shuttle flight. He went on to join the Space Stations Operations Branch of the Astronaut Office.

Network Operations Control Center (NOCC) The operations hub for NASA's *deep-space network (DSN). It is located in Building 230 of the *Jet Propulsion Laboratory. The centre monitors the operations at the three *deep-space communications complexes at Goldstone, California, near Canberra, Australia, and near Madrid, Spain. It also oversees the performance of the network for flight-mission users, provides information for organizing and running the network, and takes part in network and mission testing.

neutral body position The body position naturally assumed by an astronaut floating in microgravity. The arms and legs are held out in front and the legs are raised and bent, as if falling forward from a sitting position.

Neutral Buoyancy Laboratory (NBL) An immense water tank to train astronauts in weightlessness. The tank used by NASA contains 23.5 million L and is 12.2 m deep. It is part of the Sonny Carter Training Facility at the Johnson Space Center, Texas. Trainee astronauts practise extravehicular activities

(EVAs) around mock-ups of a shuttle payload bay and airlock. Astronauts wearing EVA spacesuits practised assembly of the *International Space Station* using mock-ups sunk in the tank. The facility is also used to test spacecraft designs and EVA equipment. Russian cosmonauts train in a neutral buoyancy tank at Star City in Russia.

neutron star A very small, 'superdense' star composed mostly of neutrons. They are thought to form when massive stars explode as *supernovae, during which the protons and electrons of the star's atoms merge, owing to intense gravitational collapse, to make neutrons. A neutron star has a mass two to three times that of the Sun, compressed into a globe only 20 km in diameter.

If its mass is any greater, its gravity will be so strong that it will shrink even further to become a *black hole. Being so small, neutron stars can spin very quickly. The rapidly flashing radio stars called *pulsars are believed to be neutron stars. The flashing is caused by a rotating beam of radio energy similar in behaviour to a lighthouse beam.

New General Catalogue A catalogue of star clusters and nebulae compiled by the Danish astronomer John Louis Emil Dreyer and published in 1888. Its main aim was to revise, correct, and expand upon the *General Catalogue* compiled by English astronomer John Herschel, which appeared in 1864.

New Horizons A NASA mission to Pluto and the Kuiper Belt of comets beyond, planned for launch in 2006. It will swing past Jupiter for a gravity boost in 2007, and reach Pluto and its moon, Charon, in 2015. Then the spacecraft will head into the Kuiper Belt to study one or more of the icy mini-worlds in that region at the edge of the Solar System. The probe will map the surface appearance of these bodies, measure their surface compositions and temperatures, and study Pluto's thin atmosphere.

Newman, James Hansen (1956–) US astronaut. His four space shuttle flights include the *Discovery* mission in September 1993 that deployed the *Advanced Communications Technology Satellite; and the first *International Space Station* assembly mission on *Endeavour* in December 1998, during which he undertook three space walks. He went on to become the director of Human Space Flight Programs—Russia.

Newman joined NASA's Johnson Space Center in 1985 to train crews in space shuttle propulsion, guidance, and control. He was selected as an astronaut in 1990.

New Millennium Program A NASA initiative established in 1995 to develop and test new technologies and instruments for space exploration with strict budgets and schedules. The first of the series was *Deep Space 1*, launched in October 1998, which tested a xenon ion drive on a mission past asteroid Braille and comet Borrelly. *Deep Space 2*, also known as the *Mars Microprobe Project*, consisted of two surface penetrators carried aboard the ill-fated *Mars Polar Lander* craft. Other spacecraft in the New Millennium Program are members of the Earth Observing and Space Technology series.

n

Nicollier, Claude (1944–) Swiss astronaut with the European Space Agency (ESA). His four space shuttle flights include two missions to service the **Hubble Space Telescope*, on *Endeavour* in December 1993 and on *Discovery* in December 1999. He has logged more than 1000 hours in space with one space walk of eight hours. Nicollier joined the ESA as a research-scientist in 1976. He was selected a member of the first group of European astronauts in 1978 for training at NASA's Johnson Space Center in 1980.

Nikolayev, Andrian Grigoriyevich (1929–) Soviet cosmonaut. He flew aboard *Vostok 3* in August 1962, as part of the first simultaneous flight of two spacecraft, with *Vostok 4*. With Vitali *Sevastyanov on *Soyuz 9* in June 1970, he set a record (at the time) for the longest space flight of 424 hours and 59 minutes. Nikolayev was selected as a cosmonaut in 1960 and left the programme in 1982.

Nimbus The generic name of seven meteorological satellites launched by NASA from 1964 to 1978 to test advanced systems and collect Earth's atmospheric data. The craft were each launched in near-polar, *Sun-synchronous orbits, and carried such instruments as infrared radiometers and microwave spectrometers. *Nimbus 1* was launched in August 1964 and *Nimbus 7* in October 1978. The latter's experiments included using the *coastal zone colour scanner (CZCS) to map chlorophyll in coastal waters and ocean currents.

NOAA Abbreviation for *National Oceanic and Atmospheric Administration.

NOCC Acronym for *Network Operations Control Center.

Node 2 A component of the *International Space Station* (*ISS*) that attaches to one end of the US Destiny laboratory module. It acts as a connecting hub to three other science experiment facilities: the Kibo Japanese Experiment Module, the European Columbus Laboratory, and the Centrifuge Accommodation Module. It also provides attachment ports for Multi-Purpose Logistics Modules (MPLMs), the Japanese H II Transfer Vehicle, and the Pressurized Mating Adapter to which space shuttles dock. Node 2 is an aluminium cylinder 6.7 m long and 4.5 m in diameter. It was built in Europe in exchange for the launch of the European Columbus Laboratory to the *ISS*. Node 1, known as *Unity, was launched in December 1998 and is attached to the opposite end of the Destiny module. The arrival of Node 2 will complete the assembly of the US-owned parts of the *ISS*.

Noguchi, Soichi (1965–) Japanese astronaut with the *National Space Development Agency (NASDA). He was selected as an astronaut by NASDA in 1996 and began training at NASA's Johnson Space Center, handling technical duties at NASA's Astronaut Office while awaiting assignment to a space shuttle flight. He flew aboard *Discovery* in July–August 2005 on the first space shuttle mission following the *Columbia* disaster of 2003.

Nomex cloth A type of nylon used for the *Mercury* and *Gemini* *spacesuits. Created by the DuPont company in 1963, it has a high temperature resistance. The *Apollo 7* astronauts, the first crew to orbit the Moon, wore suits inside the spacecraft with an inner layer of Nomex cloth. The material was also used with cotton for the inflight clothing of space shuttle crew members.

Noriega, Carlos Ismael (1959–) US astronaut. He flew on the *Atlantis* space shuttle mission in May 1997 that docked with the Russian space station *Mir*; and on the *Endeavour* mission in November 2000 to assemble the *International Space Station*, performing three space walks totalling 19 hours. Noriega was assigned to the US Space Command in 1990 and was selected as an astronaut in 1994. His assigned flight was postponed following the *Columbia* space shuttle disaster in 2003.

North American Aviation A US company that became the principal contractor for the * *Apollo* project and produced its command module (CM) and service module (SM). It also built the second stage for the Saturn V launch vehicle (*see* SATURN ROCKET) and the docking module for the 1975 **Apollo–Soyuz* Test Project.

The company was established in 1928 as North American Aviation and formed Rocketdyne as a separate division in 1955. North American merged with Rockwell Standard Corporation in 1967 to become North American Rockwell and in 1973 became Rockwell International. In 1996 its space and defence divisions merged with the *Boeing Company.

Its other projects included the X-10 supersonic aircraft in the 1950s and the X-15 rocket plane in the 1960s.

northern lights Common name for the *aurora borealis.

nosecone The conical part at the front of a rocket or other space vehicle, such as the space shuttle. It can serve as a nose *fairing to protect the payload. A missile's nosecone covers its warhead and is protected by a *heat shield for re-entry to the Earth's atmosphere.

Nova rocket An early NASA idea for a super rocket, powerful enough to fly astronauts directly to the Moon and return them to the Earth. The plan for a vehicle capable of a *direct ascent mode was supported by German rocket engineer Wernher *von Braun and informally known as the 'Dream Rocket'. The Nova, if it had materialized, would have been the ultimate *Saturn rocket, 183 m long with a first stage made up of 8 to 12 *F-1 engines providing more than 8.6 million kg of thrust. The project was dropped when US president John F Kennedy set a tight deadline for landing on the Moon, and NASA turned to a *lunar-orbit rendezvous mode.

http://www.astronautix.com/lvfam/nova.htm Compares designs proposed during the 1960s for a heavy booster to supersede the Saturn V launcher. The site includes historical notes, technical information, and simple drawings of the rockets.

***Nozomi* (Japanese: 'hope')** A Japanese space probe to study the atmosphere and ionosphere of the planet Mars, launched in July 1998. *Nozomi* was to go into orbit around Mars in January 2004 and gather data for a period of two years, but its thrusters failed to put it into orbit and on 4 December 2003 the spacecraft flew by the planet at a distance of 1 000 km into orbit around the Sun.

nozzle An opening at the end of a rocket engine through which the gases escape to produce thrust. Nozzles, which are shaped like bells, gimbal (swivel) to guide a rocket or spacecraft. Computers on board the space shuttle give commands to the engines' nozzles to change their angles.

N rocket Japan's first large rocket. The N-1, the country's first liquid-fuel N rocket, was launched on 9 September 1975 measuring 32.6 m in length, with a circumference of 2.4 m, and weighing 81 tonnes. It had first and third stages developed by imported US technology, and the second stage was Japanese made. In 1977, the N rocket carried Japan's first geostationary satellite into orbit. The first N-2 rocket was launched on 11 February 1981 and is capable of carrying a 35-kg satellite into orbit.

NSC Abbreviation for *National Space Club.

NSS Abbreviation for *National Space Society.

NSSDC Abbreviation for *National Space Science Data Center.

Nuclear Engine for Rocket Vehicle Application (NERVA) A joint space programme involving NASA and the Atomic Energy Commission (AEC) 1959–72 to develop nuclear propulsion for a rocket. NASA and the AEC oversaw the building of 20 nuclear-reactor rocket engines that were tested at the US government's Nevada Test Site. The programme was slowed in the 1960s by public reactions against nuclear power, resulting in its cancellation in 1972. NERVA was an outgrowth of the Project Rover nuclear rocket programme begun in 1955 by the AEC and the US Air Force.

OAO Abbreviation for *Orbiting Astronomical Observatory.

Oasis A greenhouse aboard the Soviet space station **Salyut*. It was set up by Viktor *Patsayev, one of the first three crew members to reside there. Scientists were uncertain that plants would grow naturally in space, and Patsayev was delighted to see seeds sprouting during his *Soyuz 11* mission in 1971.

OB Abbreviation for *observatory phase.

OBS Acronym for *operational bioinstrumentation system.

observatory phase (OB) During a spacecraft's *fly-by operation, the moment when a celestial target can be better resolved by the spacecraft's on-board instruments than by those on Earth. This phase usually occurs a few months before the fly-by. Encounter command-sequences on board begin to activate observations of the target, and ground stations become operational in support. The observatory phase follows the interplanetary *cruise phase, and occurs before the *far encounter phase (FE).

occultation A temporary obscuring of a star or spacecraft as it passes behind a celestial object, or its atmosphere or rings. Occultations are used to provide information about changes in an orbit, and the structure of objects in space, such as radio sources. When a spacecraft is obscured, it can rotate so that its radio signal is refracted towards the Earth. The point of refraction is known as 'virtual Earth'.

ocean colour and temperature scanner (OCTS) An optical radiometer that measured the ocean colours and sea surface temperatures from its position aboard the Japanese **Advanced Earth-Observing Satellite*. OCTS provided data on global ocean colours from its launch in August 1996 until it lost power in June 1997. Its information increased the understanding of marine ecosystems, the marine carbon cycle, and the carbon exchange between the atmosphere and ocean. OCTS was a successor to the *coastal zone colour scanner.

Ochoa, Ellen (1958–) US astronaut. Her four space shuttle flights include the *Discovery* mission in May 1999 for the first docking with the **International Space Station* (*ISS*), and the April 2002 *Atlantis* flight to the *ISS*, where she operated the robotic arm. She has logged more than 978 hours in space. She went on to become the deputy director of the Flight Crew Operations Directorate at the Johnson Space Center. Ochoa was chief of the Intelligent Systems Technology Branch at NASA's *Ames Research Center when selected as an astronaut in 1991.

Ockels, Wubbo Johannes (1946–) Dutch astronaut with the European Space Agency (ESA). He was one of three European payload specialists aboard the first German **Spacelab* mission on the *Challenger* space shuttle in October 1985. He was selected as an astronaut in 1978 and retired from ESA in 1985.

O'Connor, Bryan Daniel (1946–) US astronaut. He was pilot of the *Atlantis* space shuttle in November 1985, the first mission to deploy four satellites. In June 1991, he commanded the *Columbia* space shuttle on the first mission dedicated to life science studies. O'Connor was selected as an astronaut in 1980. He became director of NASA's Space Shuttle Program in 1994, resigning in 1996.

Ofeq (Hebrew: 'horizon') A series of Israeli satellites. *Ofeq 1* was launched by the Israeli Space Agency (ISA) and Israel Aircraft Industries (IAI) on 19 September 1988, making Israel the eighth country to launch its own satellite. The vehicle conducted experiments in transmissions from space and on the Earth's magnetic field. *Ofeq 2* was launched on 3 April 1990, *Ofeq 3* on 5 April 1995, *Ofeq 4* on 22 January 1998, and *Ofeq 5* on 28 May 2002. The later satellites engaged in military surveillance.

offgassing The release of a possibly harmful gas within a crewed spacecraft. NASA requires that all materials used in habitable areas be tested for offgassing, flammability, and odour. This includes materials in experiments, stowed equipment, and parts of the spacecraft itself. Tests are carried out by the Toxicology Laboratory at the Johnson Space Center and at other NASA facilities. Examples range from battery chargers to urine kits. Gas is sometimes vented out into space; gas in the wet-trash stowage compartment is removed by connecting the container to a vent in the *waste management system to send it overboard.

Office National d'Etudes et de Recherches Aérospatiales (ONERA; National Aerospace Research Establishment) A French centre for aeronautics and space research, established in 1946. It has taken part in all major French and European space and aeronautical programmes, including the *Ariane rocket, the Concorde supersonic aircraft, and the **Mars Express* and **Rosetta* space probes. Under the French Ministry of Defence, ONERA serves as a scientific, technical, and public establishment with industrial and commercial responsibilities. It conducts research involving the design of aircraft and spacecraft, including aerodynamics, structural strength, electronics, robotics, and information processing. It operates several laboratories and a range of research and industrial wind tunnels. ONERA cooperates on research with the Centre National de la Recherche Scientifique (CNRS; National Scientific Research Centre) and several universities.

http://www.onera.fr/english.html News of the wide-ranging research activities of the French National Aerospace Research Establishment.

Office of Outer Space Affairs (UNOOSA) A United Nations (UN) office that is the secretariat for the UN's *Committee on the Peaceful Uses of Outer Space (UNCOPUOS). UNOOSA coordinates international discussions on the political and legal implications of the exploration of outer space, such as the

1968, 1982, and 1999 UN Conferences on the Exploration and Peaceful Uses of Outer Space (named UNISPACE). UNOOSA also maintains a Register of Objects Launched into Outer Space and an index to the Status of United Nations Treaties Governing Activities in Outer Space. It has a Programme on Space Applications (PSA) and produces reports and publications on space science and technology applications, international space law, and international cooperation in space.

Office of Space Science and Applications (OSSA) A NASA office responsible for research and development activities in Earth resources, meteorology, life sciences, and communications. It also has the general goal of increasing knowledge of the universe through the use of space missions and ground-based observations. Australia has a similarly named office that runs the country's Earth-Observation Centre and is part of Australia's Commonwealth Scientific and Industrial Research Organization (CSIRO).

Olivas, John Daniel (1966–) US astronaut. He joined NASA's *Jet Propulsion Laboratory as a senior research engineer in 1996 and subsequently became a programme manager. He was selected as an astronaut in 1998, and assigned to the Robotics Branch of NASA's Astronaut Office until chosen for a space shuttle flight.

Olympus A European Space Agency communications satellite launched in 1989, the only such craft known to have been disabled by a meteor. *Olympus* was launched in 1989 and was hit while passing through the Perseid meteor shower in 1993. As a result, the craft lost its directional control and a subsequent electrical failure ended its transmissions. *Olympus* had previously lost a solar panel in 1991, and later that year ground stations lost control of the satellite for two months.

Originally named *L-Sat*, *Olympus* was developed by European companies including British Aerospace, who was the prime contractor. The satellite was designed for television broadcasts, telephone routing, video conferences, and other business communications.

OMS Acronym for *orbital manoeuvring system.

OMS burn NASA term for the ignition of a space shuttle's *orbital manoeuvring system engines.

ONERA Acronym for *Office National d'Etudes et de Recherches Aérospatiales.

one-way light time (OWLT) The elapsed time taken for light or a radio signal to go either way between Earth and a spacecraft or other body in space. It is measured in milliseconds. OWLT changes as a spacecraft changes its distance from the Earth. OWLT added to *transmission time is equal to *spacecraft event time.

Onizuka, Ellison Shoji (1946–86) US astronaut. He was a mission specialist aboard the January 1985 flight of the space shuttle *Discovery*, the first Department of Defense mission. On his second flight, Onizuka was killed with other crew members aboard *Challenger* when the craft exploded soon after its launch in January 1986. He was selected as an astronaut in 1978.

Onufrienko, Yuri Ivanovich (1961–) Russian cosmonaut. He was commander of the resident crew aboard the Russian space station *Mir* from February to September 1996, during which time he made six space walks and conducted many research experiments. He was commander of the fourth resident crew of the **International Space Station*, staying from 5 December 2001 to 19 June 2002. He has logged 389 days in space and taken eight space walks. Onufrienko was selected as a cosmonaut in 1989.

Oort cloud A spherical cloud of comets beyond Pluto, extending out to about 100 000 astronomical units (approximately one light year) from the Sun. The gravitational effect of passing stars and the rest of our Galaxy disturbs comets from the cloud so that they fall in towards the Sun on highly elongated orbits, becoming visible from Earth. As many as 10 trillion comets may reside in the Oort cloud, named after Dutch astronomer Jan Oort who postulated its existence in 1950.

operational bioinstrumentation system (OBS) An electrocardiograph worn by either of two designated crew members on the space shuttle during ascent and re-entry of the vehicle. The system can also be used during orbit in special situations. Three OBS electrodes are placed on the skin with electrode paste and double-sided adhesive tape from an electrode application kit. An amplified analogue signal, monitoring the astronaut's heart, is transmitted to the ground in real time or stored on tape.

Operations and Checkout Building (O and C Building) A NASA unit at the *Kennedy Space Center that is the site for the assembly and integration of payloads before their placement into a space shuttle at the *Orbiter Processing Facility. Located east of the centre's headquarters building, the five-storey O and C Building has a total area of 55 800 sq m, including a *clean room that runs the entire length and height at the rear. Two cranes weighing 24.8 tonnes are used to move the payloads.

operations phase (or mission operations and data analysis phase) A NASA designation for the flight of a spacecraft and the task of obtaining science data from its mission. The phase lasts from the spacecraft's launch to its re-entry.

OPF Abbreviation for *Orbiter Processing Facility.

Opportunity One of NASA's twin *Mars Exploration rovers, launched on 8 July 2003. It landed on 25 January 2004 in Meridiani Planum, an area once thought to have been covered in water.

optical navigation image (opnav image) An image used to determine a space probe's trajectory. Normal on-board imaging instruments are used to observe a target planet or other destination body against a known background of stars. The observation commands are uplinked in advance. In contrast, opnav images are downlinked in *telemetry to be assessed by the navigation team. The data helps them update information about the vehicle's trajectory.

optical solar reflector (OSR) A quartz mirror attached to a spacecraft to keep it from overheating. The OSR reflects sunlight and provides shade. It was used on the **Magellan* space probe.

optional programme A programme of activities of the European Space Agency that allows each member nation to decide if it wishes to participate and what financial contribution it will make. Examples include the *Ariane rocket and **Spacelab*. The alternative to the optional programme is the *mandatory programme.

orbit The path of one body in space around another, such as the orbit of the Earth around the Sun or of the Moon around the Earth. When the two bodies are similar in mass, as in a *binary star, both bodies move around their common centre of mass. The movement of objects in orbit follows *Kepler's laws, which apply to artificial satellites as well as to natural bodies.

As stated by the laws, the orbit of one body around another is an ellipse. The ellipse can be highly elongated, as are comet orbits around the Sun, or it may be almost circular, as are those of some planets. The closest point of a planet's orbit to the Sun is called **perihelion**; the most distant point is **aphelion**. (For a body orbiting the Earth, the closest and furthest points of the orbit are called **perigee** and **apogee**.)

orbital debris Alternative name for *space debris, the defunct spacecraft (or pieces of spacecraft) that remain in orbit after the end of a mission.

orbital manoeuvring system (OMS) The two rockets on a space shuttle that lift the craft into orbit, change its attitude while orbiting, and decelerate the spacecraft for re-entry into the upper edge of the atmosphere. The rockets are located on either side of the aft fuselage near the main engines. They have their own hydrazine fuel, oxidized by nitrogen tetroxide. After launching, when the main engines have burned out and the external tank has dropped away, the shuttle is lifted into an elliptical orbit by a 2.5-minute first firing, the OMS-1 burn. As the vehicle reaches the *apogee of this orbit, a shorter second ignition, the OMS-2 burn, places it into a circular orbit.

Orbital Sciences Corporation A US corporation that had built and launched more than 90 satellites by 2004. Its Hyper-X rocket launched NASA's X-43A hypersonic research aircraft that broke the aeronautical speed record on 27 March 2004. It also produces small launch vehicles, such as the air-launched *Pegasus rocket and the ground-launched Taurus and Minotaur rockets. It has completed more than 150 missions for government and commercial customers. Its science instruments includes the *total ozone mapping spectrometer (TOMS). Orbital is also developing the X-34 reusable rocketplane.

The company's satellites are manufactured at its facility in Dulles, Virginia, which has 11 625 sq m of space, including several *clean rooms and a high bay with test equipment.

Orbital Transport und Raketen AktienGesellschaft Full name of the *OTRAG rocket.

orbital velocity The velocity required to place a spacecraft in Earth orbit, which is 28 000 kph. This escape velocity is required to overcome the Earth's gravitational force.

orbiter A spacecraft that orbits the Sun, a planet, or a moon. The best-known orbiters are those that have travelled to a distant planet to provide an intense study of its atmosphere and surface. These include **Ulysses* around the Sun, **Cassini* around Saturn, and **Magellan* around Venus.

An orbiter must be able to survive communication cut-offs when the object it orbits moves to a position between the craft and the Earth. It is also designed to overcome breaks in electricity production from its solar panels when the planet or moon blocks off the Sun.

orbiter NASA's official designation for the *space shuttle vehicle. Although 'space shuttle' is the term used by the press and public for the orbiting spacecraft, NASA only applies that name to the orbiter and its rockets together.

Orbiter Processing Facility (OPF) Any of three buildings at the Kennedy Space Center, Florida, used for maintenance of the space shuttle, which is towed there immediately after landing. NASA employees remove the remaining fuel and then inspect, test, and refurbish the shuttle for its next mission. The OPF-1 and OPF-2 buildings have high bay areas of 2 700 sq m, and are connected by a low bay with an area of 2 130 sq m. OPF-3, formerly called the Orbiter Maintenance and Refurbishment Facility (OMRF), has an area of 4 645 sq m.

Orbiting Astronomical Observatory (OAO) A series of four NASA orbiting observatories including **Copernicus* (*OAO-3*), the fourth and last of the craft to be launched. *OAO-2* was launched in December 1968 as the heaviest satellite at that time, weighing 1 890 kg. It carried 11 telescopes, including 4 ultraviolet and 4 photoelectric. During its mission, *OAO-2* discovered a supernova (May 1972) and carried out imaging of young stars. *OAO-3* was launched in August 1972, with the largest telescope at that time, and studied ultraviolet radiation from stars and interstellar gas and dust. The other two missions were not successful: *OAO-1* was launched in April 1966 but quickly suffered a power failure; *OAO-B* (the third in the series) failed during its launch in November 1970.

Orbiting Solar Observatory (OSO) Any of eight nearly identical solar observatories launched by NASA from 1962 to 1975. OSOs are designed mainly to point several ultraviolet and X-ray telescopes at the Sun from a platform mounted on a cylindrical wheel. *OSO 1*, active from its launch in March 1962 to August 1963, was a 200-kg spacecraft deployed to measure solar electromagnetic radiation in the ultraviolet, X-ray, and gamma-ray wavelengths, and to investigate dust particles in space. The University of Minnesota gamma-ray experiment aboard measured the intensity and direction of low-energy gamma rays in space.

OSO 5 had the longest lifespan, from January 1969 to July 1975. It studied solar X-ray bursts and measured the diffuse cosmic background. *OSO 8*, launched in June 1975 and active until October 1978, had four additional instruments to study other celestial X-ray sources.

orbit trim manoeuvre (OTM) A slight change in a spacecraft's orbit around a planet, such as an increase in altitude to avoid orbit *decay. The required change is calculated in a similar way as a *trajectory correction manoeuvre and must be small to preserve fuel.

'Original Seven' The first group of astronauts selected by NASA in 1959. The term began to be used after other groups were chosen. The seven astronauts were, from the US Air Force, Gordon *Cooper Jr, Gus *Grissom, and Deke *Slayton; from the US Navy, Scott *Carpenter, Walter *Schirra Jr, and Alan *Shepard Jr; and from the US Marines, John *Glenn Jr.

They were chosen from 32 finalist candidates selected to take physical exams. The original criteria required applicants to be less than 40 years old, less than 5 ft 11 in tall, in excellent physical condition, holding a bachelor's degree or equivalent, a graduate of a test pilot school and a qualified jet pilot, and to have logged at least 1 500 hours of flying time.

http://history.nasa.gov/40thmerc7/intro.htm Absorbing biographies of the 'Original Seven' astronauts. The site includes an image gallery, a chronology of events, links, and downloadable copies of contemporary documents.

Origins Program A NASA endeavour to answer the two most fundamental questions in astronomy and cosmology: Where did we come from, and are we alone? The Origins Program aims to explain how galaxies formed in the early Universe, to study the formation and evolution of stars and planetary systems, to understand how life begins and evolves, and to determine whether habitable or life-bearing planets exist around other stars.

Current space missions that come under the Origins Program are the *Hubble Space Telescope* (*HST*), the *Far-Ultraviolet Spectroscopic Explorer* (*FUSE*), and the *Spitzer Space Telescope*. Forthcoming launches are *Kepler*, a planet detector, in 2008, and SIM Planet Quest in 2011. Next in line are the *James Webb Space Telescope*, a replacement for *HST*, in 2012, and the Terrestrial Planet Finder in 2015. Beyond that are plans for the *Single Aperture Far-Infrared Observatory*, a follow-on to *Spitzer*, and the Large UV/Optical Telescope. After 2020 will come more ambitious telescopes such as Life Finder, to search for chemical signatures of life in the atmospheres of extrasolar planets, and Planet Imager, an array of space telescopes to achieve the imaging power of a telescope 360 km wide.

http://origins.jpl.nasa.gov/ Beginner's-level introductions to cosmology, star and planet formation, and astrobiology, with a full, updated listing of all missions that come under the umbrella of the Origins Program.

O-ring One of paired rubber seals (primary and back-up) on the joints of a solid rocket booster for the space shuttle. The O-rings, each with a diameter of 71 mm, prevent hot gases from escaping through the joints of the rocket's four segments. An investigation revealed that the **Challenger* space shuttle disaster in 1986 was caused by a failure of both rings at the 'aft field' joint between the bottom segment and the one above on the right booster. The failure was thought to be due to a faulty design of the O-rings that proved unable to withstand the combination of previous use and the launch vehicle's dynamic bending during launch.

orthostatic hypotension A medical ailment suffered by some astronauts after a space mission. Blood pools in the legs and results in a feeling of dizziness on return to Earth. Affected astronauts often need help in walking, and a brief loss of consciousness is possible. Orthostatic hypotension is caused by low blood pressure due to an upright posture in *microgravity during a long flight.

OSO Abbreviation for *Orbiting Solar Observatory.

OSR Abbreviation for *optical solar reflector.

Osumi Japan's first national satellite, launched in February 1970 by a L-4S rocket. It made Japan the fourth nation to launch a satellite using its own technology, after the USSR, USA, and France. Named after Japan's Osumi peninsula, the satellite was launched under the direction of Japan's *Institute of Space and Aeronautical Science.

Oswald, Stephen Scott (1951–) US astronaut. He piloted two flights aboard the space shuttle *Discovery* in January 1992 and April 1993. In March 1995, he commanded the *Endeavour* flight on the second mission of the *Astro Observatory, during which he established a space shuttle duration record of 17 days. He was then assigned to NASA headquarters in Washington, DC, as deputy associate administrator for Space Operations. He returned to the Astronaut Office in 1998 and retired from NASA in 2001.

Oswald was selected as an astronaut in 1985. He served as deputy associate administrator for Space Operations at NASA headquarters in Washington, DC, 1995–8, retiring from NASA in 2000.

OTM Abbreviation for *orbit trim manoeuvre.

OTRAG rocket (Orbital Transport und Raketen AktienGesellschaft) A low-cost rocket produced by the OTRAG company of Stuttgart, Germany, and test-launched from 1977 to 1983. It used four rocket modules, the engines of which could be individually controlled. The cheap propellant was kerosene with an oxidizer of nitric acid. Windscreen-wiper motors operated the fuel valves. Supposedly intended to launch satellites, the OTRAG rocket aroused suspicions about its military use.

The company established a rocket range in Zaire (now Democratic Republic of Congo) in 1975 but this agreement was cancelled after objections from the USSR. A new rocket range was founded in Libya in 1979, but US and German protests ended tests there. OTRAG was discontinued in 1987.

Outer Space Treaty An international agreement adopted by the United Nations in 1963 and coming into force on 10 October 1967. Proposed by US president Dwight Eisenhower, it was based in part on the Antarctic Treaty of 1961, and sought to promote the peaceful use of outer space, stating that the exploration and use of outer space should be carried out for the benefit and in the interests of all countries, irrespective of their degree of economic or scientific development.

Overmyer, Robert Franklyn (1936–1996) US astronaut. He was pilot of *Columbia* in November 1982, the first fully operational flight of a space shuttle, on a five-day mission. In April 1985, he commanded the *Challenger* shuttle carrying the third **Spacelab* mission. Overmyer was assigned to the **Manned Orbital Laboratory* programme in 1966, and selected as an astronaut when the programme was cancelled in 1969. He retired from NASA in 1986.

overshoot boundary The upper edge of a spacecraft's re-entry corridor. If the vehicle enters above this level, the low atmospheric density will not slow it down enough, and the spacecraft will skip up into space again.

OWLT Abbreviation for *one-way light time.

PAC Acronym for *Processing and Archiving Centre.

packetizing A method of transmitting data from a spacecraft. Packetizing is a type of *multiplexing, the sending of different data at the same time. During a *downlink, a flow of data known as a packet comes from one instrument followed by a packet from others, in no particular order. The packets, with identifications of their measurements, are put tightly into a structure of frames for the downlink.

Padalka, Gennadi Ivanovich (1958–) Russian cosmonaut. He was commander of the resident crew of the Russian space station *Mir* from August 1998 to February 1999. He commanded the ninth resident crew of the **International Space Station*, arriving in the *Soyuz* spacecraft in April 2004. He was selected as a cosmonaut in 1980.

***Palapa* (Indonesian: 'fruits of labour')** A series of eight Indonesian communication satellites. The first was launched in 1976 by the government-owned company Perumtel. The system was subsequently operated by Satelindo, a private Indonesian company established in 1993. *Palapa B2*, the fourth satellite, was launched by the space shuttle *Discovery* in February 1984 but failed to reach its proper orbit. The satellite was recovered by *Discovery* in November 1984 in the first satellite retrieval and repair, then relaunched in April 1990. The first of the third-generation *Palapa C* satellites was launched in 1996.

pallet A U-shaped platform on which instruments and equipment are mounted. The pallets are directly exposed to space and covered with aluminium. They are controlled by astronauts and hold large instruments for experiments or systems that require sweeping views, such as telescopes, sensors, and antennae. The pallets were developed for **Spacelab*, with up to five flown during a single mission. If an *igloo (pressurized service container) is attached, a pallet can supply electrical power, cool equipment, and provide connections for acquiring data from experiments.

PAM Acronym for *payload assist module.

Pan-American Space Organization (PASO) An organization developed by a pan-American group of scientists and engineers. The group envisaged the unification of space activities in the Americas and the peaceful uses of space exploration. Their efforts resulted in the Space Conferences of the Americas: Prospects in Cooperation, sponsored by the United Nations. Conferences were held in Costa Rica in March 1990, in Chile in April 1993, and in Uruguay in November 1996. The aim of these conferences is to encourage cooperation in the areas of science and technology for peaceful uses of space among the pan-American countries for the betterment of humanity.

pantry food The food aboard a space shuttle supplementary to the regular menu food. Pantry food is a two-day contingency supply that also includes snacks and drinks for the duration of the mission. It provides 2,100 calories daily for each crew member. The supply includes fresh, rehydratable, irradiated, and thermo-stabilized food.

PAR Acronym for *photosynthetically active radiation.

parachute Any canopied device used to slow down descent from a high altitude. NASA's **Mercury* project craft used one 19-m parachute and a back-up to land capsules in the Atlantic Ocean, while the **Apollo* project employed three, each 27 m in diameter, for landings in the Pacific Ocean. Parachutes are also used for space atmospheric probes, such as **Galileo*'s probe on Jupiter in 1995.

Space shuttle astronauts were originally given parachute packs on the back of their launch and entry suits, and these packs also contained a locator beacon for rescue.

parallax A change in the apparent position of an object against its background when viewed from two different positions. In astronomy, nearby stars show a shift owing to parallax when viewed from different positions on the Earth's orbit around the Sun. A star's parallax is used to deduce its distance from the Earth.

Nearer bodies such as the Moon, Sun, and planets also show a parallax caused by the motion of the Earth.

parallel staging A launch system that uses boosters on the sides of a rocket rather than in stacked stages. The boosters are jettisoned outwards after their fuel is burned. The space shuttle launch is an example of the use of parallel staging.

Parazynski, Scott Edward (1961–) US astronaut and physician. His four space shuttle flights include the *Endeavour* mission in April 2001 to deliver the robotic arm *Canadarm 2 to the **International Space Station*. During the mission he undertook took two space walks to unfold the device. His flight in 2004 was postponed because of the *Columbia* space shuttle disaster the previous year. He has logged 1 019 hours in space, with three space walks totalling 20 hours. While studying at medical school in the 1980s, Parazynski conducted research at NASA's *Ames Research Center on human fluid shifts during space flight. He was selected as an astronaut in 1992.

Parker, Robert Alan Ridley (1936–) US astronaut. His two space shuttle flights include the first **Spacelab* mission on board *Columbia* in November 1983. Parker was selected as a scientist-astronaut in 1967, and held several posts with NASA before being appointed director of the NASA Management Office at the Jet Propulsion Laboratory in 1997.

Parkes Observatory A radio observatory near the town of Parkes, New South Wales, Australia, site of a radio telescope of 64-m aperture opened in 1961 and owned by the Commonwealth Scientific and Industrial Research Organization (CSIRO). As well as radio astronomy, the Parkes dish has been used to track space missions on behalf of NASA, including the *Mariners* and

Apollos, *Voyager 2*, *Galileo*, and most recently various NASA Mars probes. NASA has funded upgrades to the telescope and its receivers.

www.parkes.atnf.csiro.au/ Guide to the work of Australia's main radio telescope, plus the true story of its role in receiving TV pictures of the *Apollo 11* Moon walk as dramatized in the feature film *The Dish*.

parking orbit The orbit used by a spacecraft as a waiting position between two phases, for example before it begins a new trajectory. A space shuttle might go into a parking orbit, circling the globe in a *low Earth orbit (LEO) until it deploys a payload to a higher orbit. *Apollo 11* reached an Earth parking orbit in 11 minutes and completed 1.5 orbits before its Saturn thrusters sent it towards the Moon with the first crew to land there.

parking spot An informal NASA term for the final point reached by a satellite going into *geosynchronous orbit or *geostationary orbit.

parsec (symbol: pc) A unit used for distances to stars and galaxies. One parsec is equal to 3.2616 *light years, 2.063×10^5 *astronomical units, and 3.857×10^{13} km.

A parsec is the distance at which a star would have a *parallax (apparent shift in position) of one second of arc when viewed from two points the same distance apart as the Earth's distance from the Sun; or the distance at which one astronomical unit subtends an angle of one second of arc.

PASO Acronym for *Pan-American Space Organization.

Patsayev, Viktor (1933–1971) Soviet cosmonaut. He was a member of the first crew to visit the first space station, *Salyut 1*. However, he and the rest of the crew were killed during the return flight. Patsayev was launched as flight engineer in June 1971 in *Soyuz 11* with commander Georgi Dobrovolsky and flight engineer Vladislav Volkov. The crew resided at the space station for 23 days, achieving a duration record. They were killed just before re-entry when a faulty valve emptied the *Soyuz 11* capsule of air. Spacesuits were not worn because of the capsule's space limitations.

Patsayev was selected as a civilian engineer cosmonaut in 1968. He was first appointed as a back-up research engineer for *Soyuz 10*.

Payette, Julie (1963–) Canadian astronaut. She flew on the *Discovery* mission in May 1999 for the first manual docking with the **International Space Station*, operating the *Canadarm during the flight. She is the chief astronaut for the Canadian Space Agency and went on to become a CapCom (spacecraft communicator) at the Mission Control Center in Houston. Payette was selected as an astronaut for the Canadian Space Agency in 1992, and began training at NASA's Johnson Space Center in 1996.

payload An item on a spacecraft that is not part of the vehicle's structure or its operating systems and subsystems. Payloads include satellites to be deployed, as well as materials and instruments for science experiments.

payload assist module (PAM; formerly spinning solid upper stage) A rocket attached to a satellite deployed into a geostationary orbit from the space shuttle's cargo bay. The PAM establishes the satellite's orbit.

payload bay Alternative name for the *cargo bay of a spacecraft.

payload developer A person, group, or organization that develops a satellite payload. At NASA, the primary payload developer trains the payload specialist on that mission. For *International Space Station* (*ISS*) payloads, the US payload developer receives training from the Payload Systems Group at the *ISS* Payload Operations Integration Center (POIC) in the *Marshall Space Flight Center.

payload fairing (or payload shroud) A protective covering put over a payload on the top of a rocket for the launch. The covering is designed to be jettisoned when it has risen above the atmosphere. An aluminium-lithium payload fairing that weighed 518 kg was used for the April 1999 launch of the IKONOS 1 satellite, but it did not separate and the mission failed.

payload module (PLM) A module for the instruments and support equipment on the European Space Agency's *Polar Platform satellite. The module's payload carrier (PLC) is the main structural support for ten externally mounted instruments that have a mass of 2 200 kg. The PLM's **U**-shaped payload equipment bay (PEB) houses internally mounted equipment and provides such services as electrical distribution and data management. Other parts of the PLM include a solar array, star sensors, and an antenna.

payload operations control centre (POCC) Any of several centres used by owners of payloads or experiments carried into space by NASA's space shuttle or interplanetary vehicles. Operated in conjunction with flight-control rooms, the POCCs allow the owners to monitor and control their equipment. Payloads that are deployed, retrieved, or serviced by the shuttle are monitored by a POCC at the *Goddard Space Flight Center. The POCC for **Spacelab* missions was at the *Marshall Space Flight Center. Payloads on board interplanetary and planetary missions are monitored from the POCC at the *Jet Propulsion Laboratory. Other POCCs are maintained at locations chosen by private owners and foreign governments.

payload specialist A crew member aboard a space shuttle who handles specialized *payloads, experiments, and equipment. Payload specialists are not NASA astronauts, although they are subject to NASA physical examinations and training from the *payload developer concerning their special missions. Up to four payload specialists, who can include scientists, engineers, teachers, and physicians, may be assigned to a flight.

PDI Abbreviation for *powered-descent initiation.

PDS Abbreviation for *planetary data system.

PEAP Acronym for *personal egress air pack.

Pegasus rocket The world's first privately-developed spacecraft-launch vehicle. It carried out its first mission in 1990, and had conducted a total of 35 by 2004. The three-stage, delta-wing rocket is first carried to about 12 200 m by aeroplane, and then released with its payload to free-fall for five seconds before its motor is fired. The satellites, which each weigh up to

450 kg, are placed into a low orbit. Recent satellite launches include NASA's **Reuven Ramaty High-Energy Solar Spectroscopic Imager* (*RHESSI*) on 5 February 2002 and the *Canadian Space Agency's Science Satellite* (*SCISAT-1*) on 12 August 2003.

The launcher was created and built by the US *Orbital Sciences Corporation, and it is Orbital's Stargazer L-1011 aircraft that carries Pegasus. The rocket is the first winged vehicle to reach eight times the speed of sound. Orbital's Hyper-X launch vehicle was flight-proven on a Pegasus rocket before it launched NASA's X-43A hypersonic research aircraft which set the aeronautical speed record on 27 March 2004.

http://www.dfrc.nasa.gov/gallery/photo/Pegasus/index.html Captioned photographs of the Pegasus air-launched space booster rocket.

penetrator spacecraft A space probe that penetrates the surface of a planet, moon, or other body in space, such as a comet. The craft must be able to survive a great impact and still transmit data relating to the subsurface being explored. In 1999, NASA lost contact with its ambitious *Mars Polar Lander* which carried two penetrators that were to be fired in to Mars's soil, but its twin Mars Exploration rovers that landed in 2003 had rock abrasion tools and used their wheels to scrape the surface. ESA's **Rosetta* spacecraft, launched on 2 March 2004, will land on a comet and drill into its surface. NASA's *Deep Impact* mission, fired an impactor into the nucleus of comet Tempel 1 in July 2005, excavating a crater, and Japan's **Lunar-A* is scheduled, to drop two penetrators into the Moon.

penumbra The region of partial shade between the totally dark part (umbra) of a shadow and the fully illuminated region outside. It occurs when a source of light is only partially obscured by a shadow-casting object. The darkness of a penumbra varies gradually from total darkness at one edge to full brightness at the other. In astronomy, a penumbra is a region of the Earth from which only a partial *eclipse of the Sun can be seen.

periapsis The point of an orbit that is closest to the planet or other celestial body being orbited. Spacecraft often carry out mapping of planetary surfaces during the time of periapsis passage. The spacecraft's periapsis altitude can be increased by firing its thrusters when *apoapsis is reached, and the craft's altitude can be reduced by using the thrusters to slow the vehicle, or by *aerobraking. Selective firing of the thrusters at periapsis can also increase or decrease the vehicle's altitude at apoapsis.

perigee The point at which an object, travelling in an elliptical orbit around the Earth, is at its closest to the Earth. The point at which it is furthest from the Earth is the *apogee.

perihelion The point at which an object, travelling in an elliptical orbit around the Sun, is at its closest to the Sun. The point at which it is furthest from the Sun is the *aphelion.

personal egress air pack (PEAP) An emergency air-supply pack connected to the helmets of space shuttle astronauts while the vehicle is on the launch pad and during lift-off. It is intended for emergencies on the ground. Following the 1986 *Challenger* disaster, NASA recovered four PEAPs from the bottom of the Atlantic Ocean and found that three of them had been activated manually by the crew members.

personal satellite assistant (PSA) A robot that monitors environmental conditions aboard a spacecraft. It was developed by NASA's *Ames Research Center for installation on board the space shuttle and the *International Space Station*. The PSA is approximately the size of a cricket ball and has sensors to monitor a spacecraft's pressure, temperature, oxygen, carbon dioxide, and other gases. It also has navigation sensors and its own propulsion to operate autonomously throughout the spacecraft, as well as a camera for video conferencing.

Peterson, Donald Herod (1933–) US astronaut. He flew on the *Challenger* space shuttle mission in April 1983 that deployed the first *Tracking and Data-Relay Satellite. Peterson was selected as an astronaut in 1969, and retired from NASA in 1984 to work as a consultant in crewed aerospace operations.

phase In a space mission, any of its predefined periods. An example is the launch and early orbit phase. A fly-by has an observatory phase, far encounter phase, near encounter phase, and post encounter phase.

phase The apparent shape of the Moon or a planet when all or part of its illuminated hemisphere is facing the Earth.

As the Moon orbits the Earth its appearance from Earth changes as different amounts of its surface are illuminated by the Sun. During one orbit of the Earth (29.5 days—a lunar month) the Moon undergoes a full cycle of phases from new, to first quarter, to full, to last quarter.

The Moon does not reflect sunlight onto the Earth when it is between the Earth and the Sun, and the Moon is not seen. This is the new Moon phase. As the Moon orbits the Earth, part of the Moon reflects sunlight onto the Earth and a crescent is seen, starting from the right-hand side of the Moon. This is the waxing crescent phase. The half Moon phase occurs when half of the Moon's disc is illuminated as seen from Earth. As the Moon continues its orbit, gradually more of the Moon's surface becomes illuminated. At the waxing gibbous phase, three-quarters of the Moon's disc is visible. The full Moon phase occurs when the Earth is between the Moon and the Sun, and the Moon's disc is fully illuminated.

After the full Moon, gradually less of the Moon's surface is illuminated as it continues its orbit. It passes through the waning gibbous phase where three-quarters of its disc is visible, to the third quarter where half is seen, followed by the waning crescent, and finally the new Moon phase again.

The planets whose orbits lie within that of the Earth can also undergo a full cycle of phases, as can an asteroid passing inside the Earth's orbit.

Mars can appear gibbous at quadrature (when it is at right angles to the Sun). The gibbous appearance of Jupiter is barely noticeable.

http://tycho.usno.navy.mil/vphase.html Image display of the current phase of the Moon, updated every four hours. You can also find the Moon phase for any date and time between AD 1800 and AD 2199.

Phillips, John Lynch (1951–) US astronaut. He flew on the *Endeavour* space shuttle mission in April 2001 that delivered the *Canadarm 2 to the **International Space Station*. His flight scheduled for 2005 faced delay following the *Columbia* space shuttle disaster in 2003. In 1989, Phillips became the principal investigator for the Solar Wind Plasma Experiment aboard the **Ulysses* space probe, based at the Los Alamos National Laboratory. He was selected as an astronaut in 1996.

Phobos Either of two of Soviet spacecraft launched towards *Mars in July 1988 to land on Phobos, one of the planet's two moons. Communication was lost with *Phobos 1* before it reached Mars. *Phobos 2* went into orbit around Mars in January 1989 but contact was lost on 27 March when it had approached to within 800 km/497 mi of Phobos.

Phoenix A NASA probe planned to land in the northern polar region of Mars between 65° and 75° latitude. The lander will deploy a robotic arm and dig trenches into the layers of water ice present there. *Phoenix* will analyse soil samples collected by the robotic arm in search of the chemicals of life. *Phoenix*'s stereo camera will build up a three-dimensional picture of its surroundings while meteorology instruments scan the Martian atmosphere up to 20 km in altitude, obtaining data about the formation, duration, and movement of clouds, fog, and dust. It will also measure temperatures and atmospheric pressure. *Phoenix* is scheduled for launch in 2007.

photocell (or photoelectric cell) A device for measuring or detecting light or other electromagnetic radiation, since its electrical state is altered by the effect of light. In a **photoemissive** cell, the radiation causes electrons to be emitted and a current to flow (photoelectric effect); a **photovoltaic** cell causes an electromotive force to be generated in the presence of light across the boundary of two substances. A **photoconductive** cell, which contains a semiconductor, increases its conductivity when exposed to electromagnetic radiation.

Photocells are used for photographers' exposure meters, burglar and fire alarms, automatic doors, and in solar energy arrays.

photodiode A semiconductor *p–n* junction diode used to detect light or measure its intensity. The photodiode is encapsulated in a transparent plastic case that allows light to fall onto the junction. When this occurs, the reverse-bias resistance (high resistance in the opposite direction to normal current flow) drops and allows a larger reverse-biased current to flow through the device. The increase in current can then be related to the amount of light falling on the junction.

Photodiodes that can detect small changes in light level are used in alarm systems, camera exposure controls, and optical communication links.

photoelectric cell Alternative name for *photocell.

photosynthetically active radiation (PAR) A range of wavelengths of the Sun's spectrum that plants use for growth. Daily PAR readings can be ordered from NASA. As part of its Earth Science Enterprise, NASA launched the *Terra satellite with the moderate resolution imaging spectroradiometer (MODIS) in 1999, which measures the available radiation in the photosynthetically active wavelength that leaves absorb.

pilot (P or PLT) In a space shuttle, a crew member who assists the commander to operate and control the vehicle. A pilot-astronaut may also serve as commander of a mission and is able to use the *Canadarm to help deploy and retrieve satellites. The first 15 shuttle pilots were selected in 1978. Pilot-astronaut candidates must be between 162.5 and 193 cm tall and be able to pass NASA's Class 1 space medical, having at least 20/50 vision uncorrected (20/20 corrected). Candidates must have a minimum of a Bachelor of Science degree (in engineering, biological science, physical science, or mathematics) and have flown at least 1 000 hours as an air force jet pilot, with aircraft test-pilot experience desirable.

pilot chute A small auxiliary canopy or parachute, the ejection and utilization of which assists in the stable deployment of the *drogue parachute and main parachute.

Pioneer Any of a series of US Solar System space probes launched 1958–78. The probes *Pioneer 1–3*, all launched in 1958, were intended Moon probes, but the launch of *Pioneer 1A* and *Pioneer 2* failed, and *Pioneer 1B* and *Pioneer 3* failed to reach their target, although they did fly deep enough into space to measure the *Van Allen radiation belts. *Pioneer 4* began to orbit the Sun after passing the Moon. *Pioneer 5* (1960) was the first of a series to study the solar wind between the planets. *Pioneer 6* (1965), *Pioneer 7* (1966), *Pioneer 8* (1967), and *Pioneer 9* (1968) went into orbit around the Sun and monitored solar activity. *Pioneer 10* (March 1972) was the first probe to reach Jupiter (December 1973) and to leave the Solar System (1983). *Pioneer 11* (April 1973) passed Jupiter (December 1974) and was the first probe to reach Saturn (September 1979), before also leaving the Solar System. NASA ceased to operate *Pioneer 10* in April 1997, but re-established contact in March 2002, when the probe had reached a distance of 11.9 billion km from the Earth. *Pioneer 11* ceased to function in 1995.

Pioneer 10 and *11* carry plaques containing messages from Earth in case they are found by other civilizations among the stars. *Pioneer Venus* probes were launched May and August 1978. One orbited Venus, and the other dropped three probes onto the surface. The orbiter finally burned up in the atmosphere of Venus in 1992.

Pioneer Venus NASA's mission of two spacecraft to the planet Venus in 1978, an orbiter launched in May and a multiprobe launched in August. The first spacecraft went into orbit around Venus on 4 December, carrying 12 instruments that included a gamma-ray burst detector and a surface radar-mapper. The orbiter's primary objectives were to investigate the solar wind in Venus's environment, study the planet's upper atmosphere and ionosphere, and map the topography of the surface. Designed to operate for one Venusian year (243 days), the orbiter and most of its instruments lasted until 1992 when it fell into the atmosphere. The second spacecraft encountered Venus on 9 December and separated into five pieces: the transporter, an atmospheric

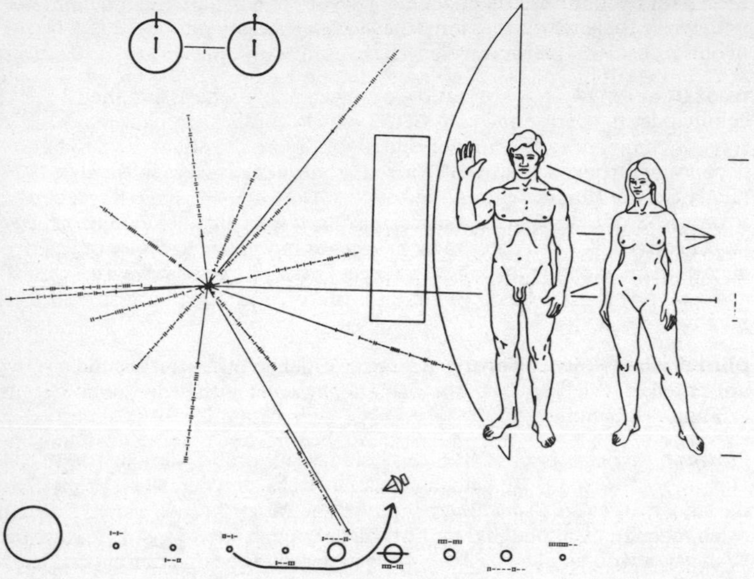

(Image © NASA)

The design etched into the gold-anodized plate carried by *Pioneer 10* and *Pioneer 11*, and which might—in millions of years—be found by other intelligent spacefaring beings. The symbols are designed to identify the Solar System, show the position of Earth, the time the spacecraft was launched, and the type of creatures who built it. The man is shown making a gesture of goodwill.

entry probe, and three identical small probes. All entered the atmosphere and sent back data for an hour as they descended.

http://nssdc.gsfc.nasa.gov/planetary/pioneer_venus.html
Straightforward guide to the *Pioneer Venus* mission. As well as the descriptions of the spacecraft instruments and data, there are artists' views of the spacecraft. The page also provides links to many important Venus exploration resources.

pitch The upwards or downwards motion, relative to the body of the craft, of the nose of a spacecraft or rocket.

pitchover The instant after launch when a spacecraft or rocket turns from its vertical ascent to curve into a trajectory towards orbit. A space shuttle and its launch vehicle roll into a high arc over the Atlantic Ocean within a minute of launch from the Kennedy Space Center.

Planck A European Space Agency mission scheduled for launch in 2007 to examine details of the *cosmic background radiation that was the first electromagnetic radiation to fill the Universe after the Big Bang. Planck carries a telescope of 1.5 m diameter with detectors that will map the cosmic microwave background with far greater resolution and sensitivity than before. It will be positioned at the L2 *Lagrangian point of the Earth's orbit, 1.5 million km from Earth in a direction opposite that of the Sun. Planck's results should help evaluate important cosmological values such as the Hubble constant as well as elucidating the nature of the dark matter that dominates the present Universe.

The spacecraft is named after the German theoretical physicist and Nobel prize winner Max Planck.

http://sci.esa.int/science-e/www/area/index.cfm?fareaid=17 European Space Agency site with information on all aspects of the Planck mission, including pages examining the fundamental questions Planck should help answer.

planet (Greek: 'wanderer') A large celestial body in orbit around a star, composed of rock, metal, or gas. There are nine planets in the *Solar System orbiting the *Sun: Mercury, Venus, Earth, Mars, Jupiter, Saturn, Neptune, Uranus, and Pluto. The inner four, called the **terrestrial planets**, are small and rocky, and have few natural *satellites. The outer planets, with the exception of Pluto, are called the **major planets**, and have denser atmospheres consisting mainly of hydrogen and helium gases, and many natural satellites. The largest planet in the Solar System is Jupiter (about 780 million km from the Sun) with a diameter of 140 000 km, which contains a mass greater than all the other planets combined. The smallest (and furthest from the Sun at about 5 900 million km) is Pluto with a diameter of 2 300 km.

Planets of other stars are now being discovered, and are known as *extrasolar planets.

Space probes to planets include *Mariner 2* (launched in 1962 to Venus), *Pioneer 10* (launched in 1972 to Jupiter), *Mariner 10* (launched in 1973 to Venus and Mercury), *Viking 1* and *Viking 2* (launched in 1975 to Mars), *Voyager 1* and *Voyager 2* (launched in 1977) to Jupiter and Saturn, *Galileo* (launched in 1989 to Jupiter), and *Cassini* (launched in 1997 to Saturn), and numerous missions to Mars in the 1990s.

http://nssdc.gsfc.nasa.gov/planetary/ NASA archive of planetary data and images. The site includes fact sheets and press releases, information on forthcoming missions, and a useful chronology of lunar and planetary exploration. There is also information about the impact of the comet Shoemaker-Levy 9 with Jupiter.

planetary data system (PDS) A NASA system, sponsored by its Office of Space Science, that maintains archives of images from its lunar and planetary missions, astronomical observations, and laboratory measurements. PDS also distributes the images to scientists and the public to ensure their long-term use and to stimulate advanced research. The system operates Regional Planetary Imaging Data Facilities (RPIF) at more than a dozen sites around the USA and in other nations. Each has a complete library of images that can be viewed by appointment, selected, and ordered.

planetary mapper A space probe that makes intense observations of a planet, especially one that obtains data from the surface. Examples include the **Mars Global Surveyor*, which used processes such as altimetry and spectroscopy to map the planet Mars, and the **Magellan* probe that carried out detailed mapping of 99% of the surface of Venus using altimetry, radiometry, synthetic aperture radar (SAR), and other instruments. A planetary mapper is placed into a high-inclination orbit, such as a polar one, because the planet below rotates to reveal its entire surface.

planetary nebula A shell of gas thrown off by a star at the end of its life. Planetary nebulae have nothing to do with planets. They were named by German-born English astronomer William Herschel, who thought their rounded shape resembled the disc of a planet. After a star such as the Sun has expanded to become a *red giant, its outer layers are ejected into space to form a planetary nebula, leaving the core as a *white dwarf at the centre.

Planetary Society An international organization that encourages the exploration of the Solar System and the search for extraterrestrial life. The non-profit society has more than 100 000 members from over 140 countries, making it the world's largest space-interest organization. It sponsors space-related projects that are funded by its members. These have included four Optical SETI telescope projects searching for extraterrestrial life, the Mars Microphone aboard the **Mars Polar Lander*, which was the first privately-funded instrument aboard a NASA mission, and the *Red Rover project for students.

The Planetary Society was founded in 1980 by the astronomer Carl *Sagan; Bruce Murray, professor of Planetary Science at the *California Institute of Technology and president of the society in 2001; and Louis Friedman, executive director of the society in 2001, who was formerly with NASA's *Jet Propulsion Laboratory.

http://planetary.org/ Home page of the Planetary Society, which was founded in 1980 to encourage the exploration of our Solar System and the search for extraterrestrial life. The site gives the latest news stories on space exploration and encourages visitors to get involved in one of the projects by searching for extraterrestrial life from home.

p

planetary system explorer A space probe that uses remote-sensing and direct-sensing instruments to explore a planetary system, observe a planet, its atmosphere, magnetosphere, satellites, and any rings. Such a mission usually lasts for several years. Examples include the **Galileo* probe that explored the planet Jupiter and its system, including rings and satellites, from 1995, and **Cassini*, scheduled to reach the planet Saturn in 2004. A planetary system explorer is placed into a low-inclination orbit, such as an equatorial one, because this provides coverage of the planet and its *magnetosphere and a good observation of orbiting moons in the equatorial plane.

planetesimal A body of rock in space, smaller than a planet. Scientists study planetesimal images from telescopes in Earth and in space, adhering to the modern theory that planetesimals were the building blocks of planets.

The Sun and the planets are thought to have formed from a rotating dust cloud generated by a *supernova explosion. On condensation, this cloud formed a central sun and a rotating disc, the material of which separated into rings of dust grains that began to stick together. Larger and larger clumps formed in each ring and eventually collected into bodies the size of present-day asteroids, or planetesimals. After numerous collisions, these bodies eventually formed the nucleus of the various planets of our Solar System.

plasma The ionized gas produced at extremely high temperatures, as in the Sun and other stars. It contains positive and negative charges in equal numbers. It is a good electrical conductor. In thermonuclear reactions the plasma produced is confined through the use of magnetic fields.

http://www.plasmas.org/ Impressive guide to plasmas. The site has lots of good articles on plasma basics, plasma technology, and the applications of plasma science. The pages contain lots of photographs and there is a gallery of colourful and interesting plasma pictures.

plasma detector A spacecraft instrument used to measure plasmas in interplanetary space and within planetary *magnetospheres. Plasmas are a hot state of matter made up of positive ions and electrons. Detectors measure the composition, temperature, density, velocity, and distribution of plasmas. The instruments also collect information on the *solar wind and its interaction with planetary systems. A plasma detector was carried on board the **Cassini* space probe.

plasma wave instrument An instrument that measures the changes with time (waves) of the electric and magnetic fields within the ionized *plasma gases produced by the Sun and other stars. There are of two types of plasma waves: electrostatic oscillations similar to sound waves, and electromagnetic waves.

NASA's plasma wave subsystem (PWS), for example, is designed to identify plasma waves by measuring the properties of varying electric fields over a certain frequency range.

Plesetsk Cosmodrome A rocket-launching site 170 km south of Arkhangel, Russia. From 1966 the USSR, and later Russia, launched (mainly military) artificial satellites from here. Many are now from other countries, such as Nigeria's *NigeriaSat 1*, launched on 27 September 2003, and Japan's SERVIS-1, launched on 30 October 2003. The Cosmodrome is controlled by the *Voenno-Kosmicheski Sily (VKS; Military Space Forces).

Plexiglas display screen A transparent plastic screen on a space shuttle that displays the spacecraft's speed and altitude. It is positioned in the window in front of the pilot who is required to focus on the runway when landing.

PLM Abbreviation for *payload module.

PLSS Abbreviation for *primary life-support system.

Pluto The smallest and, usually, outermost planet of the Solar System. The existence of Pluto was predicted by calculation by US astronomer Percival

Lowell and the planet was located by US astronomer Clyde Tombaugh in 1930. Its highly elliptical orbit occasionally takes it within the orbit of Neptune, as in 1979–99. Pluto has a mass about 0.002 of that of Earth.

mean distance from the Sun 5.9 billion km

equatorial diameter 2 390 km

rotation period 6.39 Earth days

year 247.9 Earth years

atmosphere thin atmosphere with small amounts of methane gas

surface low density, composed of rock and ice, primarily frozen methane; there is an ice cap at Pluto's north pole

satellites one moon, Charon, discovered in 1978 by US astronomer James Christy. It is about 1 200 km in diameter, half the size of Pluto, making it the largest moon in relation to its parent planet in the Solar System. It orbits about 20 000 km from the planet's centre every 6.39 days—the same time that Pluto takes to spin on its axis. Charon is composed mainly of ice. Some astronomers have suggested that Pluto was a former moon of Neptune that escaped, but it is more likely that it was an independent body that was captured. Pluto is the only planet that has not been visited by a space probe, although a visit is planned with the **New Horizons* mission in 2015

http://www.solarviews.com/solar/eng/pluto.htm Detailed description of the planet Pluto and a selection of images.

POCC Acronym for *payload operations control centre.

pogo The oscillations and longitudinal vibrations of a rocket during firing, due to engine vibration and the movement of fuel. NASA's *Gemini* programme experienced delays caused by pogo problems with the Titan II launch vehicle. The term was first used by engineers who likened the vibrations to those of a pogo stick.

Pogue, William Reid (1930–) US astronaut. He was the command module pilot of the third and final mission to **Skylab*. The mission lasted a record 84 days, during which the **Apollo Telescope Mount* logged 338 hours operation time. Pogue was selected as an astronaut in 1966, and left NASA in 1975.

Polansky, Mark Lewis (1956–) US astronaut. He was pilot of the *Atlantis* space shuttle mission in February 2001 that delivered and connected the Destiny module to the **International Space Station*. Since 2002, he has been chief of NASA's CapCom (spacecraft communicator) Branch. His flight in 2004 was postponed because of the *Columbia* space shuttle disaster the previous year. Polansky joined NASA as an aerospace engineer and research pilot in 1992, and was selected as an astronaut in 1996.

polar orbit A spacecraft orbit that has a 90° inclination to the equator, passing over both poles in a north–south direction. An Earth polar orbit allows a spacecraft to pass over the entire surface of the planet as the Earth rotates beneath the orbital path. These *low Earth orbits are especially used for mapping or surveillance missions. A spacecraft going into a polar orbit requires more fuel and power than one that uses the *free ride of the Earth's rotation.

Space probes can also be placed in polar orbits, such as NASA's **Lunar Prospector* around the Moon and the **Magellan* spacecraft in a near polar orbit around the planet Venus.

ESA's **Ulysses* probe is in polar orbit around the Sun.

Poleshchuk, Alexander Fedorovich (1953–) Russian cosmonaut. He completed a 179-day residency on the Russian space station *Mir* in January–March 1993, during which time he performed two space walks totalling nearly ten hours. Poleshchuk was a test engineer for the *Energiya Rocket and Space Complex from 1977, and was selected as a cosmonaut in 1989.

Poliakov, Valeri Vladimirovich (1942–) Soviet cosmonaut and physician. He was launched aboard *Soyuz* TM-18 in January 1994, to join the Russian space station *Mir*. He stayed 478 days, the longest single mission flight time. Having made an earlier space flight to *Mir* in 1988, Poliakov accumulated 678 days of space experience. He was selected as a cosmonaut in 1972.

Popovich, Pavel Romanovich (1930–) Soviet cosmonaut. He flew aboard *Vostok 4* in August 1962, as part of the first joint space flight, with *Vostok* 3. The mission ended a day early because Popovich reported seeing thunderstorms over the Gulf of Mexico, forgetting that 'thunderstorms' was a code word for returning early because of motion sickness. He also flew aboard the July 1974 *Soyuz 14* mission that docked with the *Salyut* space station. Popovich was selected as a cosmonaut in 1960, and left the programme in 1982.

powered-descent initiation (PDI) The moment that thrusters are fired to assist a spacecraft's descent on to the surface of a planet or moon.

The best known powered descents were made by the *Apollo* astronauts to the lunar surface, using the descent propulsion system of the lunar module (LM). The full power of the thrusters were used during the last few minutes before touchdown.

prebreathing A procedure undertaken by an astronaut before beginning an *extravehicular activity (EVA). Pure oxygen is breathed for three hours from the *life-support system to purge nitrogen from the bloodstream. This prevents the onset of the *bends (decompression sickness) when the air pressure is reduced during the EVA.

precessing orbit Alternative name for *walking orbit.

precision ranging A method used to find a distant spacecraft's position in the sky by angular position. Two *deep-space network stations that are widely separated, such as those located in the USA and Australia, conduct ranging measurements so that a spacecraft can then be found by triangulation. Precision ranging can pinpoint both its location and distance (accurate to within metres at a distance as far as the planet Jupiter).

Precourt, Charles Joseph, Jr (1955–) US astronaut. By 2001, he had logged more than 932 hours in space on four space shuttle flights, including three docking missions with the Russian space station *Mir* (*Atlantis*, June 1995 and May 1997; *Discovery*, June 1998). He is currently deputy manager of the *International Space Station* Program at the Johnson Space Center. Precourt was selected as an astronaut in 1990, and later became chief of the Astronaut Corps.

predicts The sets of data used by NASA's *deep-space network stations to predict a spacecraft's locations in the sky and the radio frequencies needed to track it.

preflare The action of levelling the glideslope of a space shuttle before landing. The preflare is initiated 32 seconds before touchdown at an altitude of 526 m, and it becomes a complete preflare 17 seconds before touchdown at 41 m. After this, the commander begins the landing flare by lifting the shuttle's nose.

preflight systems check A late check made before the launching of a spacecraft. Before a space shuttle is launched, at *T minus 1 hour and 30 minutes, crew members (strapped into their seats) begin their preflight systems check. This includes checking gauges and other instruments, such as those measuring cabin pressure and temperature, and communication links with launch control at the Kennedy Space Center and mission control at the Johnson Space Center.

President's Scientific Advisory Committee (PSAC) A committee that existed from 1957 until 1973 whose space panel advised the US president on space exploration advances and priorities. It was created in the 1950s by President Dwight *Eisenhower and also provided intelligence information on Soviet space capabilities. In 1967, PSAC (pronounced 'pea-sack') anticipated the space shuttle by recommending that: 'Studies should be made of more economical ferrying systems, presumably involving partial or total recovery and use.'

pressure bladder A layer of a spacesuit that retains the pressure for extravehicular activities. It is made of urethene-coated nylon and is protected by a layer made of dacron.

The early *Gemini* spacesuit had a rubber pressure bladder that was manufactured by the David Clark Company.

primary contact Any NASA official or other astronaut who is the main contact for crew members on a space flight. During the final hours of isolation before a launch, the crew may only communicate with their families and the primary contact.

primary jet The main thruster on a space shuttle that is used for quick rotations and straight-line *translation manoeuvres. The shuttle has 38 primary jets, and 6 small thrusters called 'vernier jets'.

primary life-support system (PLSS) A self-contained spacesuit system that provides an astronaut with oxygen and removes carbon dioxide. It is worn on the back, and the oxygen goes into the *hard upper torso and then the back of the helmet, so the visor does not fog up. As it passes down the astronaut's face, the air removes carbon dioxide and humidity, continuing down the body before a fan pulls it back into the PLSS where a contaminant control cartridge filters out the carbon dioxide. The oxygen then passes through a carbon dioxide sensor before returning to the astronaut. A secondary oxygen pack (SOP) is attached at the bottom of the PLSS to activate if the first system fails.

The PLLS was developed in 1976 by the Hamilton Standard Division of United Technologies. PLLS originally was the abbreviation for **portable life-support system** (used by *Apollo* astronauts) until a secondary oxygen system was added.

principal investigator A scientist or engineer responsible for developing a concept for a space science mission, or leading a research investigation or grant. Under the new NASA Explorer and Discovery Program the principal investigator carries the overall responsibility for the success and completion of the science objectives of a mission or research project.

Processing and Archiving Centre (PAC) One of five centres maintained by member states of the European Space Agency to process and archive data from spacecraft instruments. The British centre (UK-PAC) at Farnborough, Surrey, for example, has responsibility for the *advanced along-track scanning radiometer (AATSR). Other centres are located in France (F-PAC), Germany (D-PAC), Italy (I-PAC), and Sweden (S-PAC).

prograde orbit (or direct-motion orbit) In an orbiting satellite, an orbit in which the satellite travels in the same direction as the planet's rotation. It is the opposite of a *retrograde orbit.

A planet has a prograde orbit or rotation if the sense of rotation is the same as the general sense of rotation of the Solar System. On the *celestial sphere, it refers to motion from west to east against the background of stars.

Progress A robot supply-vehicle developed by the USSR and used from 2001 to transport provisions to the **International Space Station* (*ISS*). The craft is 8 m long and 2.2 m wide, and can fly a payload of 2300 kg. Its *thrusters can also be fired to change the attitude of the *ISS*. The first Progress was launched in January 1978 to the Russian space station *Salyut 6* and Progress craft later made regular trips to the Russian space station **Mir*.

projectile A particle that travels with both horizontal and vertical motion in the Earth's gravitational field. If the frictional forces of air resistance are ignored, the two components of its motion can be analysed separately: its vertical motion will be accelerated due to its weight in the gravitational field; its horizontal motion may be assumed to be at constant velocity. In a uniform gravitational field and in the absence of frictional forces the path of a projectile is a parabola.

Project RAND (contraction of Project Research and Development) A US military project established in 1946 to develop the German V-2 rocket into an advanced military rocket. The McDonnell Douglas company in Santa

Monica, California, was home to the project, which produced the document entitled *Preliminary Design of an Experimental World-Circling Spaceship*. The project was expanded in 1948 to become the RAND Corporation.

promethazine A medication that NASA makes available to astronauts to treat *space sickness due to motion. The drug alleviates the symptoms in most crew members but can cause side-effects, such as a slower reaction time and a negative impact on moods.

prominence A bright cloud of gas projecting from the Sun into space 100 000 km or more.

Quiescent prominences last for months, and are held in place by magnetic fields in the Sun's corona.

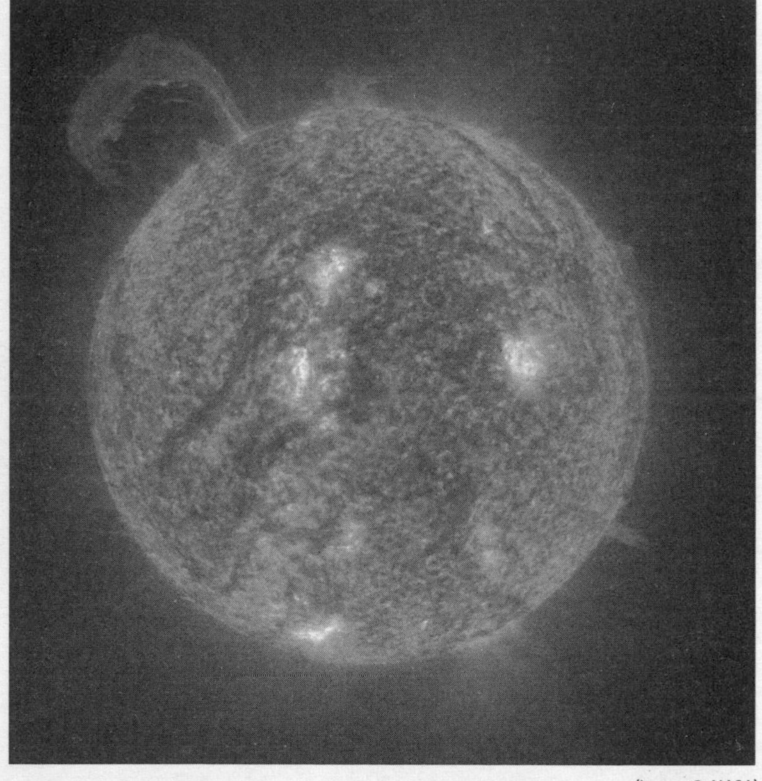

(Image © NASA)
Extreme-Ultraviolet Imaging Telescope (EIT) Image of a solar prominence, taken on 14 September 1999. Prominences are huge clouds of relatively cool dense plasma suspended in the Sun's corona. The lightest areas seen on the Sun's surface are the hottest and the darkest are the coolest.

Surge prominences shoot gas into space at speeds of 1 000 kps.

Loop prominences are gases falling back to the Sun's surface after a solar *flare.

propellant A substance burned in a rocket for *propulsion. With **bipropellant**, two propellants are used: oxidizer and fuel are stored in separate tanks and pumped independently into the combustion chamber. Liquid oxygen (oxidizer) and liquid hydrogen (fuel) are common propellants, used, for example, in the space shuttle main engines. The explosive charge that propels a projectile from a gun is also called a propellant.

propulsion (or reaction propulsion) A forward movement of a rocket or spacecraft. This is due to the pressure of the exhaust produced by the burning propellant.

Prospero Britain's first national satellite, launched on 28 October 1971 at Woomera, Australia, by a *Black Arrow rocket. Built by the British Aircraft Corporation, the spin-stabilized *Prospero* was launched to test basic systems for future satellites. It also performed a single scientific experiment to detect micrometeoroids.

Proton rocket A Soviet space rocket introduced in 1965, used to launch heavy satellites, space probes, and the *Salyut* and *Mir* space stations.

Proton is a three-stage rocket that can lift 22 tonnes into low Earth orbit; a fourth stage is added for higher orbits.

Six Proton rockets launched Soviet missions to Mars from 1971 to 1973, and a Proton-K (with a Pizza Hut logo on its side) launched the Zvezda module to the **International Space Station* in July 2000. The more powerful Proton-M was first launched in April 2001. The Proton series has never been used to launch humans into space.

(Image © Research Machines plc)

The Soviet Proton 1 rocket was originally designed to launch atomic bombs, though the weapons-carrying version was never built.

protostar An early formation of a star that has recently condensed out of an interstellar cloud and which is not yet hot enough for hydrogen burning to start. Protostars derive their energy from gravitational contraction.

PSA Abbreviation for *personal satellite assistant.

PSAC Abbreviation for *President's Scientific Advisory Committee.

pulsar A celestial source that emits pulses of energy at regular intervals, ranging from a few seconds to a few thousandths of a second. Pulsars are thought to be rapidly rotating *neutron stars, which flash at radio and other wavelengths as they spin. They were discovered in 1967 by Jocelyn Bell Burnell and Antony Hewish at the Mullard Radio Astronomy Observatory, Cambridge, England. Over 1 000 pulsars are now known.

Pulsars slow down as they get older, and eventually the flashes fade. Of the 500 known radio pulsars, 20 are millisecond pulsars (flashing 1 000 times a second). Such pulsars are thought to be more than a billion years old. Two pulsars, one (estimated to be 1 000 years old) in the Crab nebula and one (estimated to be 11 000 years old) in the constellation Vela, give out flashes of visible light.

Pulsars were first identified as compact radio sources regularly emitting very rapid, intense pulses of radiation. The discovery of the first, which had a period of 1.33728 seconds, was regarded with incredulity by Bell's colleagues. It was only in 1968, after three more had been located, that the Cambridge team of which she was a member announced their existence. Within a year of their discovery pulsars were identified with rapidly rotating neutron stars formed by the gravitational collapse of stars in supernovae explosions.

pyrometer Any instrument used for measuring high temperatures by means of the thermal radiation emitted by a hot object. In a **radiation pyrometer** the emitted radiation is detected by a sensor such as a thermocouple. In an **optical pyrometer** the colour of an electrically heated filament is matched visually to that of the emitted radiation. Pyrometers are especially useful for measuring the temperature of distant, moving, or inaccessible objects.

pyrotechnic device (or pyro) A mechanical device on a spaceship that is fired by a high current initiated by an on-board pyrotechnic switching unit (PSU). A pyro is used for such actions as separating a spacecraft and its rocket, releasing instrument covers, deploying a *boom (swing arm) or parachutes, and jettisoning the *aeroshell. More than 100 pyros may be used by a spacecraft during its ascent. The devices include *'explosive' bolts, pin pullers, and cable cutters.

A pyro is a simple, lightweight, and reliable device, but can only be used once and is dangerous when handled. New technologies are being developed to replace them.

quarantine (French: *quarantaine* '40 days') Any period for which people, animals, plants, or vessels may be detained in isolation to prevent the spread of contagious disease. After returning from the Moon, the *Apollo 11* astronauts were quarantined for three weeks in the *Lunar Receiving Laboratory at the *Johnson Space Center. White mice were placed with them to provide an early sign of any disease, but the Moon proved to be a sterile environment.

quasar (contraction of quasi-stellar object; or QSO) One of the most distant extragalactic objects known, discovered in 1963. Quasars appear starlike, but each emits more energy than 100 giant galaxies. They are thought to be at the centre of galaxies, their brilliance emanating from the stars and gas falling towards an immense *black hole at their nucleus. The **Hubble Space Telescope* revealed in 1994 that quasars exist in a remarkable variety of galaxies.

Quasar light shows a large *redshift, indicating that the quasars are very distant. The furthest are over 10 billion light years away. A few quasars emit radio waves (*see* RADIO ASTRONOMY), which is how they were first identified.

QuikSCAT A NASA satellite mission, the *Quick Scatterometer*, launched in June 1999 to measure wind speed and direction near the Earth's ocean surfaces. QuikSCAT studies interactions between the air and sea, movements and changes in the Arctic and Antarctic ice packs, and changes in rainforest vegetation. The data collected also improves weather forecasts near coastlines, as well as storm warnings and monitoring. In March 2002, it detected shifts in the Pacific winds that indicated the formation of the disruptive El Niño weather pattern.

Measurements, taken over land and ice, are made by one instrument, the SeaWinds *Scatterometer, a specialized microwave radar that can make measurements in all weather and cloud conditions.

RA Abbreviation for *radar altimeter.

rad A unit of absorbed radiation dose, now replaced in the SI system by the gray (one rad equals 0.01 gray). It is defined as the dose when one kilogram of matter absorbs 0.01 joule of radiation energy.

radar altimeter (RA) An instrument that transmits a microwave pulse and measures the return echo. Earth-observation satellites use them to gather information on the characteristics of the Earth's surface. Measurements of the echo's delay, power, and shape will indicate land features, wind speeds, sea elevation, and sea ice. This is used to monitor global climate changes. The first spaceborne radar altimeter flew on the **Skylab* space station in 1973. The European Space Agency installed the instrument on ERS-1 and ERS-2, and put its RA-2 instrument, which has improved measurement performance, on the **Envisat*.

radar astronomy The bouncing of radio waves off objects in the Solar System, with reception and analysis of the 'echoes'. Radar contact with the Moon was first made in 1945 and with Venus in 1961. The travel time for radio reflections allows the distances of objects to be determined accurately. Analysis of the reflected beam reveals the rotation period and allows the object's surface to be mapped topographically. The rotation periods of Venus and Mercury were first determined by radar. Radar maps of Venus were obtained first by Earth-based radar and subsequently by orbiting space probes.

RADARSAT An advanced Earth-observation satellite system implemented and operated by the Canadian Space Agency (CSA) to monitor environmental changes and the Earth's resources. Its *RADARSAT-1* satellite was launched by NASA in November 1995 to become the first fully operational civilian radar satellite system having large-scale production. *RADARSAT-2* is scheduled for launch in late 2005 as the world's most advanced commercial *synthetic aperture radar (SAR) satellite, having multipolarization to identify surface features to aid mapping and surveillance.

The system is funded by the private and governmental sectors, but *RADARSAT-2* will eventually be privatized to be managed and operated by its builder, MacDonald, Dettwiler, and Associates of Richmond, British Columbia, Canada.

radial velocity The velocity of an object, such as a star or galaxy, along the line of sight, moving towards or away from an observer. The amount of *Doppler shift (apparent change in wavelength) of the light reveals the object's velocity. If the object is approaching, the Doppler effect causes a *blueshift in its light. That is, the wavelengths of light coming from the object appear to be shorter, tending toward the blue end of the *spectrum. If the object is receding, there is a *redshift, meaning the wavelengths appear to be longer, toward the red end of the spectrum.

radiant heat The energy that is radiated by all warm or hot bodies. It belongs to the infrared part of the *electromagnetic spectrum and causes heating when absorbed. Radiant heat is invisible and should not be confused with the red glow associated with very hot objects, which belongs to the visible part of the spectrum.

Infrared radiation can travel through a vacuum and it is in this form that the radiant heat of the Sun travels through space. It is the trapping of this radiation by carbon dioxide and water vapour in the atmosphere that gives rise to the greenhouse effect.

radiation The emission of radiant energy as particles or waves—for example, heat, light, alpha particles, and beta particles. An astronaut's exposure to radiation must be monitored constantly with dosimeters. Dangerous amounts of radiation have apparently not been encountered yet on space missions. Crewed spacecraft around the Earth usually orbit below the *Van Allen radiation belts to avoid radiation, but intergalactic cosmic rays and solar flares are of concern.

http://www.nsbri.org/Radiation/ General information about radiation in space and its effects on humans. There are articles that discuss radiation research aboard the *International Space Station* and how NASA mitigates the risk to astronauts.

radiation sickness A sickness resulting from exposure to radiation, including X-rays, gamma rays, neutrons, and other nuclear radiation, as from weapons and fallout. Such radiation ionizes atoms in the body and causes nausea, vomiting, diarrhoea, and other symptoms. The body cells themselves may be damaged even by very small doses, causing leukaemia and other cancers.

radio astronomy The study of radio waves emitted naturally by objects in space, by means of a *radio telescope. Radio emission comes from hot gases (**thermal radiation**); electrons spiralling in magnetic fields (**synchrotron radiation**); and specific wavelengths (**lines**) emitted by atoms and molecules in space, such as the 21-cm line emitted by hydrogen gas.

Radio astronomy began in 1932 when US astronomer Karl Jansky detected radio waves from the centre of our Galaxy, but the subject did not develop until after World War II. Radio astronomy has greatly improved our understanding of the evolution of stars, the structure of galaxies, and the origin of the universe. Astronomers have mapped the spiral structure of the Milky Way from the radio waves given out by interstellar gas, and they have detected many individual radio sources within our Galaxy and beyond.

Among radio sources in our Galaxy are the remains of *supernova explosions, such as the Crab nebula and *pulsars. Short-wavelength radio waves have been detected from complex molecules in dense clouds of gas where stars are forming. Searches have been undertaken for signals from other civilizations in the Galaxy, so far without success.

Strong sources of radio waves beyond our Galaxy include radio galaxies and *quasars. Their existence far off in the universe demonstrates how the universe has evolved with time. Radio astronomers have also detected weak cosmic background radiation, thought to be from the *Big Bang explosion that marked the birth of the universe.

radio blackout A loss of radio contact with a spacecraft. This can occur with instrument problems or the apparent destruction of a spacecraft, such as the radio blackout experienced when NASA's Mars Climate Orbiter was lost in 1998. Spacecraft also routinely have a radio blackout when passing behind a celestial object, such as a planet, or when they re-enter the atmosphere and the radio waves cannot penetrate the hot ionized gases that create an electromagnetic atmosphere around the vehicle. The space shuttle experiences a radio blackout of about 12 minutes during re-entry, regaining communications with the ground after it slows to about 12 870 kph.

radio galaxy A galaxy that is a strong source of electromagnetic waves of radio wavelengths. All galaxies, including our own, emit some radio waves, but radio galaxies are up to a million times more powerful.

In many cases the strongest radio emission comes not from the visible galaxy but from two clouds, invisible through an optical telescope, that can extend for millions of light years either side of the galaxy. This double structure at radio wavelengths is also shown by some *quasars, suggesting a close relationship between the two types of object. In both cases, the source of energy is thought to be a massive black hole at the centre. Some radio galaxies are thought to result from two galaxies in collision or recently merged.

radioisotope thermoelectric generator (RTG) A spacecraft generator using the radioactive element plutonium to provide electricity. This is especially useful when a space probe travels far from the Sun where solar arrays cannot be used for power. RTGs are on the *Galileo* and *Cassini* spacecraft. An RTG contains several kilograms of an isotopic mixture of plutonium in a ceramic fuel pellet. Radioactive decay produces heat to generate electricity, and waste heat is radiated into space. RTGs are designed to survive a launch accident without releasing plutonium.

Each launch must undergo a stringent safety analysis by the US Department of Energy and be approved by the US president.

radio relay An activity of an artificial satellite that *uplinks and processes data from another satellite or surface vehicle, such as a lander or rover. The **Mars Global Surveyor* is designed for radio relay using its Mars relay antenna, and NASA's proposed Mars Network of satellites will increase communications for future probes. These spacecraft are able to have extended missions, since radio relay requires little propellant and attitude control. A satellite on its own scientific mission, however, must reduce or interrupt some of its own data gathering activities to serve as a relay.

radio science system (RSS) A system used to receive science data that has been acquired by ground measurements of a spacecraft's radio signal. Investigators can use remote controls to command equipment in NASA's *deep-space network to record data of this nature.

Radio science experiments involve the effects that bodies and phenomena in space have on a spacecraft's radio signal. Such effects can be caused by the Sun, a planet, a moon, planetary rings, or gravitational fields.

radio telescope An instrument for detecting radio waves from the universe in *radio astronomy. Radio telescopes usually consist of a metal bowl that collects and focuses radio waves the way a concave mirror collects and focuses light waves. Radio telescopes are much larger than optical telescopes, because the wavelengths they are detecting are much longer than the wavelength of light. The largest single dish, 305 m in diameter, is at Arecibo Observatory, Puerto Rico.

A large dish such as that at Jodrell Bank, Cheshire, England, can see the radio sky less clearly than a small optical telescope sees the visible sky.

Interferometry is a technique in which the output from two dishes is combined to give better resolution of detail than with a single dish.

Very long baseline interferometry (VBLI) uses radio telescopes spread across the world to resolve minute details of radio sources.

In **aperture synthesis**, several dishes are linked together to simulate the performance of a very large single dish. This technique was pioneered by English radio astronomer Martin Ryle at the Mullard Radio Astronomy Observatory, Cambridge, England, site of a radio telescope consisting of eight dishes in a 5-km line. The Very Large Array in New Mexico consists of 27 dishes arranged in a **Y**-shape, which simulates the performance of a single dish 27 km in diameter. Other radio telescopes are shaped like long troughs, and some consist of simple rod-shaped aerials.

radio wave An electromagnetic wave possessing a long wavelength (ranging from about 10^{-3} to 10^{4} m) and a low frequency (from about 10^{5} to 10^{11} Hz) that travels at the speed of light. Included in the radio wave part of the spectrum are: microwaves, used for both communications and for cooking; ultra high- and very high-frequency waves, used for television and FM (frequency modulation) radio communications; and short, medium, and long waves, used for AM (amplitude modulation) radio communications. Radio waves that are used for communications have all been modulated to carry information. Certain astronomical objects emit radio waves, which may be detected and studied using *radio telescopes.

There is a layer in the atmosphere, called the ionosphere, where gas molecules are separated from their electrons by radiation from the Sun. When radio waves reach the ionosphere, they produce movements or oscillations of the electrons. As the electrons oscillate they produce electromagnetic waves that are identical to the radio waves with which the electrons were stimulated. The radio waves are reflected back and can be detected by a receiver. It is important to note that the reflected radio waves come from the oscillating electrons in the ionosphere.

Raffaello The second of three *Multi-Purpose Logistics Modules (MPLMs) used to transport supplies to the *International Space Station* (*ISS*). *Raffaello* had its first flight on STS-100 in April 2001. The MPLMs were built for NASA by the Italian Space Agency. The other two are called **Leonardo* and **Donatello*.

(Image © NASA)
Logistics module *Raffaello* mated with the Unity module on the *International Space Station*. STS-100 pilot Jeffery Ashby used the shuttle *Endeavour*'s robot arm to orientate *Raffaello* in the payload bay prior to docking.

Raketoplan (Russian: 'rocket plane') Early Soviet version of a space shuttle that was flight tested in 1961 and 1963 but cancelled in 1965. It had delta wings and a reusable launch vehicle. *Raketoplan*, designed to carry a crew of two on 24-hour missions, was conceived as a military spacecraft that could gather intelligence and also intercept, investigate, and destroy US satellites. The concept was later expanded to include scientific missions and even flights to the Moon. The advanced prototype, *Raketoplan Chelomei*, was flown in March 1963 and travelled 1 900 km, reaching a maximum altitude of 400 km.

RAL Acronym for *Rutherford Appleton Laboratory.

RAND corporation The name from 1948 of the US military project *Project RAND.

Ranger A series of nine US space probes, only the final three of which were successful. All were intended to photograph the surface of the Moon before crashing onto it. *Ranger 7* produced 4 000 lunar photographs, and *Ranger 8* yielded more than 7 000.

RCS Abbreviation for *reaction control system.

reach aid (or swizzle stick) A device used by space shuttle astronauts to touch control switches that would otherwise be out of reach. The aid is an adjustable bar that can be extended by pressing a tab. It is used to pull toggle switches or pull or push circuit breakers, and is sometimes required when crew members are strapped in their seats.

reaction control system (RCS) An array of small *thrusters on a spacecraft, used for *attitude control, altitude, and speed. A space shuttle has 44 thrusters on its nose and tail. The pilot moves hand controllers that instruct computers to fire the thrusters in specific combinations in order to control the shuttle by pitch, roll, and yaw movements.

Out of the 44 thrusters, 6 are quiet vernier thrusters (2 forward and 4 aft) that move the shuttle slowly; they are normally the only ones required. The remaining 38 are loud primary thrusters (14 forward and 24 aft) that give a strong, rapid movement.

reaction propulsion A scientific term for propulsion that causes a craft to move by equal and opposite reaction to the discharge of exhaust fumes. All types of rocket propulsion use this type of thrust.

Readdy, William Francis (1952–) US astronaut. His three space shuttle flights include the *Atlantis* mission in September 1996 that docked with the Russian space station *Mir*. During the docking astronauts were exchanged there for the first time. He has logged 672 hours in space and went on to become the associate administrator for space flights at NASA's headquarters in Washington, DC. Readdy joined NASA in 1986 as an aerospace engineer and instructor pilot, and was selected as an astronaut in 1987.

real-time commanding The sending of commands to spacecraft in the present time. This may be necessary to add an activity to a mission, or to correct mistakes discovered in the on-board command sequences. Real-time commanding, however, is a high-risk operation because of the possibility that an incorrect command file may be uplinked or that typographical errors may occur. In contrast, the planned command sequences already stored on the vehicle have been carefully devised and tested.

recovery The retrieval of a spacecraft and, if crewed, its astronauts. *Apollo* missions splashed down in the Pacific Ocean recovery zone, and the astronauts were taken from the sea by US Navy ships. Recoveries of cosmonauts and their spacecraft have occurred on land. Orbiting satellites are also said to be recovered when taken aboard a space shuttle.

red dwarf Any star that is cool, faint, and small (about one-tenth the mass and diameter of the Sun). Red dwarfs burn slowly, and have estimated lifetimes of 100 billion years. They may be the most abundant type of star, but are difficult to see because they are so faint. Two of the closest stars to the Sun, Proxima Centauri and Barnard's Star, are red dwarfs.

red fuming nitric acid (RFNA) A chemical used as an oxidizer in rocket engines. RFNA, or Inhibited RFNA, is an oxidizer for hypergolic engines, which use propellants such as unsymmetrical dimethyl hydrazine (UDMH). When RFNA oxidiser and UDMH propellants combine, they ignite spontaneously, requiring no external energy source. RFNA is often used in remotely-controlled propulsion systems on orbiting satellites and in *Titan rockets.

red giant Any large bright star with a cool surface. It is thought to represent a late stage in the evolution of a star like the Sun, as it runs out of hydrogen fuel at its centre and begins to burn heavier elements, such as helium, carbon, and silicon. Because of more complex nuclear reactions that then occur in the red giant's interior, it eventually becomes gravitationally unstable and begins to collapse and heat up. The result is either explosion of the star as a *supernova, leaving behind a *neutron star, or loss of mass by more gradual means to produce a *white dwarf.

Red giants have diameters between 10 and 100 times that of the Sun. They are very bright because they are so large, although their surface temperature is lower than that of the Sun, about 2 000–3 000 K (1 700–2 700 °C). *See also* RED SUPERGIANT.

Red Rover (or Red Rover Goes to Mars) The first project to involve the public in a planetary exploration mission. In 2000, the *Planetary Society and the LEGO Company conducted a worldwide essay competition for students (9–15 years old) with competitors elected as student scientists and student navigators. The winners used data from the **Mars Global Surveyor* (*MGS*) and **Viking* missions to choose a possible landing site on the planet Mars. They then went to Malin Space Science Systems in San Diego, California, to create commands for the *MGS* Mars Orbiter Camera to take images of the chosen site.

redshift The lengthening of the wavelengths of light from an object as a result of the object's motion away from us. It is an example of the *Doppler effect. The redshift in light from galaxies is evidence that the Universe is expanding.

Lengthening of wavelengths causes the light to move or shift towards the red end of the *spectrum, hence the name. The amount of redshift can be measured by the displacement of lines in an object's spectrum. By measuring the amount of redshift in light from stars and galaxies, astronomers can tell how quickly these objects are moving away from us. A strong gravitational field can also produce a redshift in light; this is termed **gravitational redshift**.

Redstone rocket A short-range US military missile, modified for use as a space launcher. Redstone rockets launched the first two flights of the *Mercury* project. A modified Redstone, Juno 1, launched the first US satellite, *Explorer 1*, in 1958.

The rocket was developed at Redstone Arsenal in Huntsville, Alabama, by a team headed by German rocket engineer Wernher *von Braun.

redundancy The condition of having back-up equipment and instruments. Main units might fail during the lifetime of a mission, and the most crucial devices have redundant back-ups. They can be programmed to function automatically in such situations, or can be activated by *real-time commanding. Redundancy is used for such equipment as antennae, gyroscopes, and transmitters and receivers. The lifetime of a number of spacecraft has been extended by redundancy, including the *Voyager*, *Galileo*, and *Pioneer* probes.

re-entry The return of a spacecraft into the Earth's atmosphere. The narrow re-entry corridor is the safe path a spacecraft must follow, since too steep a re-entry, exceeding the *undershoot boundary, would burn up the vehicle, whereas a too-shallow angle, above the *overshoot boundary, would send it back into space again. An *ablation shield is used to protect a spacecraft from destruction by the intense re-entry friction heat.

Regional Planetary Imaging Data Facility (RPIDF) Any of NASA's 18 facilities that make up an international system of planetary image libraries. Established in 1977, they keep photographic, digital, and cartographic data, and mission documentation. The facilities are open to scientists, educators, students, and the public.

The ten US facilities are located at Arizona State University, Phoenix, Arizona; Brown University, Providence, Rhode Island; Cornell University, Ithaca, New York; the *Jet Propulsion Laboratory, Pasadena, California; the Lunar and Planetary Institute, Houston, Texas; the *National Air and Space Museum, Washington, DC; the US Geological Survey, Flagstaff, Arizona; the University of Arizona, Tucson, Arizona; the University of Hawaii at Manoa, Hawaii; and Washington University at St Louis, Missouri. Other facilities are located at the Instituto di Astrofisica Spaziale, Rome, Italy; the Deutsches Zentrum für Luft- und Raumfahrt, Berlin, Germany; the *Institute of Space and Astronomical Sciences, Kanagawa, Japan; the Israeli Regional Planetary Image Facility at Beer-Sheva; University College London, England; the University of New Brunswick, Fredericton, Canada; the University of Oulu, Finland; and the Université de Paris-Sud, Paris, France.

rehydration station An electronic dispensing system to add water to dried foods and beverages aboard a space shuttle. An astronaut places a food or beverage container into the station, a water dispenser needle penetrates it, and the astronaut rotates a switch to select the quantity of water. Once rehydrated, the food and drink can be heated, if desired.

Reightler, Kenneth Stanley, Jr (1951–) US astronaut. He flew with the *Discovery* space shuttle mission in September 1991 that deployed the **Upper-Atmosphere Research Satellite*. In February 1984, he flew aboard *Discovery* on the first joint US–Russian space shuttle mission. Reightler was selected as an astronaut in 1987.

Reilly, James Francis, II (1954–) US astronaut. He flew with the *Endeavour* space shuttle mission in January 1998 that docked with the Russian space station *Mir* for the fifth and last exchange of an astronaut. In July 2001, he flew aboard *Atlantis* to install a new airlock module on the **International Space Station*. He has logged over 517 hours in space, with three space walks totalling over 16 hours. He later moved to the Astronaut Office to lead shuttle training, and was assigned to a flight in 2003 that was postponed because of the *Columbia* space shuttle disaster. Reilly was selected as an astronaut in 1994.

relativity The theory of the relative rather than absolute character of mass, time, and space, and their interdependence, as developed by German-born US physicist Albert Einstein in two phases:

special theory of relativity (1905)
Starting with the premisses that (1) the laws of nature are the same for all observers in unaccelerated motion and (2) the speed of light is independent of the motion of its source, Einstein arrived at some rather unexpected consequences. Intuitively familiar concepts, like mass, length, and time, had to be modified. For example, an object moving rapidly past the observer will appear to be both shorter and more massive than when it is at rest (that is, at rest relative to the observer), and a clock moving rapidly past the observer will appear to be running slower than when it is at rest. These predictions of relativity theory seem to be foreign to everyday experience merely because the changes are quite negligible at speeds less than about 1 500 kps and only become appreciable at speeds approaching the speed of light.

general theory of relativity (1915)
The geometrical properties of space-time were to be conceived as modified locally by the presence of a body with mass. A planet's orbit around the Sun (as observed in three-dimensional space) arises from its natural trajectory in modified space-time. Einstein's general theory accounts for a peculiarity in the behaviour of the motion of the perihelion of the orbit of the planet Mercury that cannot be explained in Newton's theory. The new theory also said that light rays should bend when they pass by a massive object. The predicted bending of starlight was observed during the eclipse of the Sun in 1919. A third corroboration is found in the shift towards the red in the spectra of the Sun and, in particular, of stars of great density—white dwarfs such as the companion of Sirius.

Einstein showed that, for consistency with the above premisses (1) and (2), the principles of dynamics as established by Newton needed modification; the most celebrated new result was the equation $E = mc^2$, which expresses an equivalence between mass (m) and energy (E), c being the speed of light in a vacuum. In 'relativistic mechanics', conservation of mass is replaced by the new concept of conservation of 'mass-energy'. General relativity is central to modern *astrophysics and *cosmology; it predicts, for example, the possibility of *black holes. General relativity theory was inspired by the simple idea that it is impossible in a small region to distinguish between acceleration and gravitation effects (as in a lift one feels heavier when the lift accelerates upwards), but the mathematical development of the idea is formidable. Such is not the case for the special theory, which a non-expert can follow up to $E = mc^2$ and beyond.

A famous relativity problem is the 'twin paradox'. If one of two twins, *O′*, sets off at high speed in a spaceship, travels for some years, and then turns around and returns at high speed to rejoin the other twin, *O*, who stayed at home, he will have aged less than *O*, because his clocks have been running slower. If his speed is $v = 0.995c$ in each direction, he will have aged only one year for every 10 years that *O* has aged. This conclusion has been hotly disputed, because it appears to contradict the symmetry that ought to exist between *O* and *O′*. But now there *is* no such symmetry: *O* has remained in an unaccelerated state, whereas *O′* has suffered three very large changes in velocity, and both will in fact agree that *O* has aged more than *O′*. The observed slowing down of the decay of mesons when moving conclusively shows the existence of this effect, called time dilation.

http://www.damtp.cam.ac.uk/user/gr/public/index.html Discussion of quantum gravity, strings, black holes, inflation, and cosmology, on Cambridge's relativity pages. The site also has information on COSMOS, the national cosmology supercomputer, and links to English physicist Stephen Hawking's pages.

Remek, Vladimir (1948–) Czechoslovakian cosmonaut. He was the first Czechoslovakian to become a cosmonaut, and the first person in space from a country other than Russia or the USA. Selected in 1976, he had one flight, in March 1978, as a research cosmonaut aboard *Soyuz 28* as it formed a working unit in space with *Soyuz 27* and *Salyut 6*. Remek left the cosmonaut programme in 1978.

remote-control arm Alternative name for the *Canadarm payload deployment system.

remote manipulator system (RMS) Alternative name for the payload deployment system more commonly known as the *Canadarm.

remote sensing The gathering and recording of information from a distance. Remote sensing usually refers to photographing the Earth's surface with orbiting satellites, but space probes have sent back photographs and data about planets as distant as Neptune.

Satellites such as *Landsat* have surveyed all of the Earth's surface from orbit.

http://rsd.gsfc.nasa.gov/rsd/RemoteSensing.html Directory of remote-sensing resources on the Internet. They include weather and climate observations, views of the Earth from space, and imagery from specific instruments, such as *Landsat*, the Geostationary Operational Environmental Satellite, and the advanced very high-resolution radiometer imaging system.

remote terminal unit (RTU) A spacecraft device that controls and monitors functions for relatively simple equipment. The *Envisat* scheduled for launch by the European Space Agency in January 2002 has three RTUs in the payload equipment bay. They are connected to the payload management computer.

rendezvous A planned meeting of two spacecraft in orbit or of a spacecraft and celestial target, such as a comet. A rendezvous in Earth orbit usually precedes the *docking of spacecraft or the retrieval of a satellite.

Rene 41 A nickel-steel alloy used in the construction of spacecraft. It is a high-temperature material that can withstand *re-entry temperatures. Rene 41 has been used for many vehicle parts, such as the conical walls of the *Mercury* spacecraft capsule and the internal load-carrying structure of the **Dyna-Soar*. Tests on spacecraft integral tank and fuselage structures found that Rene 41 panels have a life cycle of 500 ascents and re-entries.

rescue ball An early NASA idea developed to rescue astronauts in the event of a space shuttle being disabled. The rescue ball was a strong, deflated bag with a small viewing port. The marooned astronaut curled up inside, zipped up the bag, and activated the life-support system that inflated the bag and supplied air for 20 minutes. Astronauts arriving on a rescue shuttle would recover the stranded crew members by taking space walks to manoeuvre the balls into the second vehicle.

Resnik, Judith Arlene (1949–86) US astronaut. She flew on the *Discovery* space shuttle mission in August–September 1984 during which three satellites were deployed. She was killed along with six fellow crew members on the *Challenger* space shuttle when it exploded soon after its launch on 28 January 1986. Resnik was a NASA support engineer working on the *sounding rocket and *telemetry system programmes before her selection as an astronaut in 1978.

retrograde Describing the orbit or rotation of a *planet or *satellite if the sense of rotation is opposite to the general sense of rotation of the Solar System. On the *celestial sphere, it refers to motion from east to west against the background of stars.

retrorocket A small auxiliary rocket used to slow down a spacecraft. Either it has its exhaust nozzle pointed in the direction of the flight or the spacecraft turns around to point the retrorocket in the direction of travel. A **retrofire**, or **retroburn**, can be used to return an orbiting spacecraft to Earth, to make a soft landing (as by a lunar module), or to decelerate a space probe into a planetary orbit.

return vehicle A spacecraft used by crew members to return to Earth, usually from a space station. For instance, the return vehicle for the first resident crew of the **International Space Station* was the space shuttle *Discovery*, after its arrival at the station with the second crew. The term 'return vehicle' is also used for the X-38 *crew return vehicle (CRV).

reusable launch vehicle (RLV) A launch vehicle that can be recovered, reconditioned, and reused. The space shuttle's solid rocket boosters burn out and fall away two minutes after blast-off, and parachutes open to slow the boosters' descent into the Atlantic Ocean. They are recovered about 280 km off the coast by waiting tugboats and returned to the Kennedy Space Center and prepared for future flights.

***Reuven Ramaty High-Energy Solar Spectroscopic Imager* (*RHESSI*)** A small NASA explorer satellite to investigate the basic physics of particle acceleration and the explosive energy released in solar flares. It was launched on 5 February 2002 with Reuven Ramaty's name added to the mission; he was a founding member of the *HESSI* team and died in 2001. Its objectives include discovering the location, frequency, and evolution of impulsive energy in the Sun's corona; studying the acceleration of electrons, protons, and heavier ions in flares; and investigating the heating of plasma to tens of millions of degrees. The sixth Small Explorer (SMEX) mission, it is expected to have a life of up to three years.

Hard X-ray (high-energy X-rays) and gamma-ray imaging will be used during the mission. Besides the Sun, *RHESSI* will also study the Crab nebula and other areas of the sky.

http://hesperia.gsfc.nasa.gov/hessi/ Full details of the workings, scientific goals, and status of the High-Energy Solar Spectroscopic Imager (HESSI). The site's articles are clearly laid out and easy to read. There are animations, an image gallery, and even a downloadable paper model of the HESSI spacecraft.

revolution One complete journey of a spacecraft around the Earth. Athough an orbit is measured by an imaginary point in space, a revolution is measured from an imaginary line on the Earth's surface (longitude 80° west for NASA's launch site at the Kennedy Space Center in Florida). Thus, a satellite travelling west to east, the direction the Earth turns, will take longer to complete a revolution than one heading east to west.

Richards, Paul William (1964–) US astronaut. He flew with the *Discovery* space shuttle mission in March 2001 that took the second resident crew to the **International Space Station*. During the flight, he performed an extravehicular activity of more than six hours. He left NASA in 2002 to pursue private interests. Richards joined NASA's *Goddard Space Flight Center as an engineer in 1987. He was selected as an astronaut in 1996.

Richards, Richard Noel (1946–) US astronaut. His four space shuttle flights include commanding *Discovery* (October 1990) to launch the **Ulysses* spacecraft; and commanding *Columbia* (June 1992) on a two-week flight conducting microgravity experiments. He logged more than 33 days in space. Richards was selected as an astronaut in 1980, and resigned in 1995 to manage missions for NASA's Space Shuttle Program Office.

Ride, Sally (Kristen) (1951–) US astronaut and astrophysicist. In 1983 she became the first US woman in space, working aboard the space shuttle *Challenger* as a mission specialist. She was also mission specialist on two further shuttle flights in 1984 and 1985. She served on a presidential commission investigating the *Challenger* accident in 1986.

Born in Los Angeles, California, Ride was selected for NASA's astronaut programme out of 1000 female candidates in 1978, the year she was awarded a PhD in physics from Stanford University, California. She was special assistant to the administrator of NASA in 1987. She left NASA in the same year and in 1989 became director of the California Space Institute and professor of physics at the University of California, San Diego.

From 1999 to 2000, Ride was president of the Web-site company, space.com. She has written four children's books about space.

riding the stack (or riding the fire) Astronauts' expression for being launched in a spacecraft.

rigidized NASA jargon for the final tight connection made when spacecraft dock.

rigid sleep station A solid tier system used by astronauts for sleeping. On a space shuttle, the sleep station is installed on the starboard side of the mid-deck. Each tier includes a sleeping bag, personal stowage containers, a light, and a ventilation inlet and outlet.

The station is built in two versions: the plastic three-tier system for three astronauts weighs about 93 kg, while the metal four-tier version for four astronauts weighs 78 kg. The lower tier of both versions can be removed for access to the *cabin debris trapdoor.

RLV Abbreviation for *reusable launch vehicle.

RMS Abbreviation for remote manipulator system, an alternative name for the *Canadarm.

Robertson, Patricia (1963–2001) US astronaut and physician. She was a crew-support astronaut for the second resident crew of the **International Space Station*. Robertson joined NASA's Johnson Space Center in 1997 to provide health care for astronauts and their families. She was selected as an astronaut in 1998.

Robinson, Stephen Kern (1955–) US astronaut. He flew on the *Discovery* space shuttle mission in August 1997, and was the payload commander aboard the *Discovery* flight in October 1998. He was flight engineer on *Discovery* in July–August 2005, the first space shuttle mission since the *Columbia* disaster of 2003. Robinson joined NASA's *Ames Research Center in 1979 as a research scientist, became chief of the Experimental Flow Physics Branch at the *Langley Research Center in 1990, and returned to Langley in 1994 as leader of Aerodynamics and Acoustics for NASA's General Aviation Technology programme.

robot arm Alternative name for the *Canadarm payload deployment system.

robotic spacecraft An uncrewed spacecraft. Although *landers and *rovers are most often identified with robots, all uncrewed spacecraft receiving commands from on-board computers and ground stations are robotic. Such vehicles are more suited than astronauts for lengthy travel to other planets and for repetitive measurements. Human beings can best handle tasks involving analytical decisions and quick adjustments, such as retrieving or repairing satellites.

Since the *Challenger* space shuttle disaster in 1986, NASA has launched as many robotic spacecraft as possible on expendable rockets rather than from space shuttles.

http://telerobotics.jpl.nasa.gov/ Presentation of the activities of the JPL robotics group. The site has information about missions and the ongoing development of robotic craft, as well as photographs and information about the 'Mars Yard'—a simulated Mars surface designed to test rovers like Sojourner.

robotic vertigo A NASA term describing a spacecraft computer that becomes disoriented and makes the wrong decisions. When *Voyager 2* was launched, its main computer switched to back-up sensors in response to an emergency that did not exist, and later began to switch thrusters to cope with another non-existent emergency. Its executive program eventually reprogrammed the faulty sensors.

rocat (contraction of rocket catapult) A device designed to eject astronauts from the *Gemini* spacecraft in the event of the explosion of its Titan II rocket on the launch pad. The rocat was never used, but was designed to eject a crew member and his or her seat about 92 m upwards and 305 m to the side of the craft. A parachute would then open, stabilized by a small balloon-parachute called a ballute.

rocket A projectile driven by the reaction of gases produced by a fast-burning fuel. Unlike jet engines, which are also reaction engines, rockets carry their own oxygen supply to burn their fuel and do not require any surrounding atmosphere. For warfare, rocket heads carry an explosive device.

Rockets have been valued as fireworks since the middle ages, but their intensive development as a means of propulsion to high altitudes, carrying payloads, started only in the interwar years with the state-supported work in Germany (primarily by German-born US rocket engineer Wernher *von Braun) and the work of US rocket pioneer Robert Hutchings Goddard. Being the only form of propulsion available that can function in a vacuum, rockets are essential to exploration in outer space. *Multistage rockets have to be used, consisting of a number of rockets joined together.

Two main kinds of rocket are used: one burns liquid propellants, the other solid propellants. The fireworks rocket uses gunpowder as a solid propellant. The *space shuttle's solid rocket boosters use a mixture of powdered aluminium in a synthetic rubber binder. Most rockets, however, have liquid propellants, which are more powerful and easier to control. Liquid hydrogen and kerosene are common fuels, while liquid oxygen is the most common oxygen provider, or oxidizer. One of the biggest rockets ever built, the Saturn V Moon rocket (*see* SATURN ROCKET), was a three-stage design, standing 111 m high. It weighed more than 2 700 tonnes on the launch pad, developed a take-off thrust of some 3.4 million kg, and could place almost 140 tonnes into low Earth orbit. In the early 1990s, the most powerful rocket system was the Soviet *Energiya, capable of placing 190 metric tonnes into low Earth orbit. The US space shuttle can only carry up to 29 metric tonnes of equipment into orbit.

http://www.rocketryonline.com/ Searchable site with reference materials, organizations, clubs, upcoming launches, products, and books devoted to rocketry.

Rodnik (Russian: 'spring') A hot and cold water system used by cosmonauts on the *Salyut* and *Mir* space stations.

roentgen (or röntgen; symbol R) A unit of radiation exposure, used for X-rays and gamma rays. It is defined in terms of the number of ions produced in one cubic centimetre of air by the radiation. Exposure to 1 000 roentgens gives rise to an absorbed dose of about 870 rads (8.7 grays), which is a dose equivalent of 870 rems (8.7 sieverts).

The annual dose equivalent from natural sources in the UK is 1 100 microsieverts.

Rogers Dry Lake A dry lake-bed in California's Mojave Desert that is the landing area for space shuttles. It is part of *Edwards Air Force Base.

Rohini A series of Indian satellites. The 35-kg *Rohini 1B* experimental satellite was launched on 18 July 1980 from Sriharikota at Bengal Bay and re-entered the Earth's atmosphere on 20 May 1981. A previous satellite, *Rohini 1A*, had failed on 10 August 1979. India subsequently launched *Rohini 2* on 31 May 1981, which stayed in orbit for eight days. *Rohini 3* was launched on 17 April 1983 and orbited the Earth for seven years.

roll The movement of any air- or spacecraft about the axis running down the length of the craft.

roll reversal The manoeuvres made by a space shuttle to slow it down during re-entry to the Earth's atmosphere. This **S**-shaped flight involves rolling the vehicle up to 90° for a left-turn bank, and then reversing to roll on the other side in a similar right-turn bank.

Rominger, Kent Vernon (1956–) US astronaut. He logged 1 611 hours in space on his five space shuttle flights. He was crew commander of the *Discovery* mission in May 1999 that delivered supplies to the **International Space Station* (*ISS*); and of the *Endeavour* mission in April–May 2001 that installed *Canadarm 2 on the *ISS*. He went on to become chief of the Astronaut Office at the Johnson Space Center. Rominger was selected as an astronaut in 1992.

***Röntgen Satellite* (*ROSAT*)** A joint US–German–UK satellite launched in 1990 to study cosmic sources of X-rays and extreme-ultraviolet (EUV) wavelengths, named after German physicist Wilhelm Röntgen, the discoverer of X-rays. After more than 9000 observations of objects including comets, quasars, black holes, clusters of galaxies, and supernovae, the satellite was finally switched off on 12 February 1999.

ROSAT was designed to produce the first all-sky fully imaging surveys in the X-ray and extreme ultraviolet parts of the spectrum. The satellite used a German X-ray telescope developed under the leadership of the Max Planck Institute, and the Wide-Field Camera (extreme-ultraviolet telescope) constructed by a UK team led by the University of Leicester and funded by the UK's Particle Physics and Astronomy Research Council (PPARC). The USA provided an additional X-ray camera and the launch vehicle, and the main spacecraft and mission operations were funded by Germany.

http://astroe.gsfc.nasa.gov/docs/rosat/rosgof.html Profiles the nine-year *Röntgen Satellite* (*ROSAT*) mission and describes the spacecraft and its science payload. There is extensive technical documentation, including the *ROSAT* users' handbook, access to data from the mission, and a very good image gallery.

Roosa, Stuart Allen (1933–94) US astronaut. He was the command module pilot of *Apollo 14* (January–February 1971), circling the Moon for 35 hours while Alan *Shepard and Edgar *Mitchell explored the lunar surface. During the flight to the Moon, Roosa overcame a problem with docking the command module and lunar module, and was successful on the sixth attempt. He was selected as an astronaut in 1966, retiring from NASA in 1976.

ROSAT Contraction of **Röntgen Satellite*.

Rosaviakosmos (RK) A Russian space agency established in 1992 after the break-up of the former USSR in 1991. The facilities of the former USSR civilian space programme were transferred to the RK. Yuri *Koptev headed the organization from its founding until 2004. It has a staff of about 300, with much work contracted to the *Energiya Rocket and Space Complex and other companies. RK was in charge of the space station *Mir*, and is a major partner in the **International Space Station*. It has played a more major role since NASA's space shuttles were grounded in 2003 following the *Columbia* space shuttle disaster. On 21 October 2003, it signed an agreement with the European Space Agency (ESA) to launch two uncrewed Foton capsules for scientific experiments in 2005 and 2006.

Rosetta A European Space Agency probe, launched on 2 March 2004, which will encounter comet Churyumov–Gerasimenko in 2014 and go into orbit around its elliptical nucleus, which is estimated to measure 5 × 3 km. *Rosetta* will encounter the comet at a distance of around 600 million km from the Sun, four times the distance of the Earth from the Sun. After mapping the nucleus *Rosetta* will release a lander, named *Philae*, in November 2014 to make the first landing on a comet's nucleus. *Rosetta* and *Philae* will observe the comet's activity as it approaches perihelion in August 2015.

On the way to the comet, *Rosetta* will gain speed by making three fly-bys of the Earth and one of Mars. It will also pass two asteroids, one called Steins in September 2008 and another called Lutetia in July 2010.

Rosetta takes its name from the Rosetta Stone, an engraved stone that provided the key to deciphering Egyptian hieroglyphs.

Ross, Jerry Lynn (1948–) US astronaut. He logged over 1 393 hours in space on seven space shuttle flights, which is a record. These included the December 1998 *Endeavour* flight that was the first assembly mission to the **International Space Station* (*ISS*), during which Ross performed three space walks totalling more than 21 hours. In April 2002, he flew on *Atlantis* to the *ISS* doing assembly work that required two space walks. He had logged more than 58 hours on space walks. He went on to become the chief of the Vehicle Integration Test Office at the Johnson Space Center. Ross joined to NASA's Johnson Space Center in 1979, and was chosen as an astronaut in 1980.

***Rossi X-Ray Timing Explorer* (*RXTE*)** A NASA spacecraft launched in December 1995 to explore the variability of X-ray sources. It carries the proportional counter array (PCA) to cover the lower part of the energy range, and the high-energy X-ray timing experiment (HEXTE) for the upper energy range. It also carries the all-sky monitor (ASM) that scans about 80% of the sky on every orbit. *RXTE* continues to operate, although it had an original lifetime goal of five years. Its many discoveries include the closest black hole to the Earth, 1 600 light years away, detected in February 2000, and the first transient magnetar, a rare class of extremely magnetic neutron star, discovered in January 2004.

RXTE was launched from the *Kennedy Space Center by a Delta II rocket that placed it into a *low Earth orbit of 580 km from the Earth. The mission is managed by the *Goddard Space Flight Center.

RXTE was named the X-Ray Timing Explorer when launched, and the Rossi name was added in 1996 to honour Professor Bruno Rossi of the Massachusetts Institute of Technology who, with his colleagues, discovered the first non-solar source of X-rays in 1963.

round-trip light time (RTLT) The elapsed time taken by a signal travelling from the Earth to a spacecraft or other celestial body, then immediately transmitted or reflected back to the original transmission point. This is about equal to twice the *one-way light time, but it varies because the motions of the Earth and space objects create different travelling times each way. RTLT to the Sun is about 17 minutes, and it was about 23 hours to the *Voyager 1* probe in October 2001.

rover A spacecraft that moves across the surface of a planet or moon. Its tasks can include investigating and collecting soil samples, and transmitting pictures of the landscape. Rovers are programmed to move and survive on an alien planet, but they can be remotely steered from Earth stations. NASA's *Sojourner was a successful 'microrover' carried to the planet Mars in 1997 by the *Mars Pathfinder mission. More dramatic Mars data was collected in 2004 by the twin rovers Spirit and Opportunity launched separately in June and July 2003 on the *Mars Exploration Rover mission.

RPIDF Abbreviation for *Regional Planetary Imaging Data Facility.

RSS Abbreviation for *radio science system.

RTG Abbreviation for *radioisotope thermoelectric generator.

RTLT Abbreviation for *round-trip light time.

RTU Abbreviation for *remote terminal unit.

rudder On a space shuttle, the movable section at the rear of the large vertical tail. The rudder is split to move outwards on both sides. When a returning shuttle re-enters the atmosphere, the rudder controls yaw and breaks the vehicle's speed. Immediately prior to touchdown, the rudder is opened into a wide '**V**' to act as a brake.

The rudder is controlled by flight-control software until the shuttle's wings are level before touchdown. Then two pairs of rudder pedals, one each for the pilot and commander, are used.

Runco, Mario, Jr (1952–) US astronaut. His three space shuttle flights include the January 1993 mission of *Endeavour* to deploy a *Tracking and Data-Relay Satellite, and another *Endeavour* flight in May 1996 during which he deployed two satellites. He has logged over 551 hours in space, including four and a half hours on the space walk. Runco went on to become an earth and planetary scientist in the Human Exploration Science Office at the Johnson Space Center. He also worked on the NASA programme to develop optical-quality windows for the **International Space Station* and was assigned as a CapCom (spacecraft communicator) in the *mission control team. Runco was selected as an astronaut in 1987.

Rutherford Appleton Laboratory (RAL) A UK laboratory engaged in space and other science activities, especially in support of academic research. The laboratory's facilities and services include its Space Science and Technology Department and the Natural Environment Research Council (NERC) Earth-Observation Data Centre (NEODC). Another of its facilities is the RAL Ground Station, a satellite control centre that supports orbiting missions, including those of NASA. RAL is part of the Council for the Central Laboratory of the Research Councils (CLRC), an independent body of the UK Office of Science and Technology. The CLRC awarded the laboratory £3 million on 23 March 2004 to install an Astra laser, providing the UK with a unique dual-beam facility that can create temperatures as high as those on the Sun's surface as well as magnetic fields like those on the polar regions of neutron stars. The facility will be completed in three years.

RAL was formed when the Rutherford High-Energy Laboratory (established in 1957 under the National Institute for Research in Nuclear Science) merged with the Atlas Laboratory in 1975 and the Appleton Laboratory in 1979. It is located near Didcot, Oxfordshire, England.

RXTE Abbreviation for **Rossi X-Ray Timing Explorer*.

Ryumin, Valery Victorovitch (1939–) Russian cosmonaut. His three space flights include two long missions aboard the *Salyut* 6 space station: 175 days in 1979 and 185 days in 1980. In June 1998, he was on the *Discovery* flight that docked with *Mir*. He was flight director for the *Salyut* 7 and *Mir* space stations 1981–9, and in 1992 became director of the Russian portion of the NASA–*Mir* programme. Ryumin was employed at the *Energiya Rocket and Space Complex when selected as an cosmonaut in 1973. He has remained there through his career, holding several positions, including deputy general designer for testing.

SAF Acronym for *Spacecraft Assembly Facility.

safing The activation of the fault-protection system on a spacecraft. Intricate safing routines are also called 'contingency modes'. The system is operated by an on-board computer that is able to shut down or reconfigure the spacecraft's components to prevent damage during an emergency. The safing system may also automatically search to re-establish broken communications. Switching to safing may temporarily interrupt scientific observations on board the spacecraft.

SAFIR Acronym for **Single Aperture Far-Infrared Observatory.*

SAG Acronym for *Science Advisory Group.

SAGE Acronym for *stratospheric aerosol and gas experiment.

***Sakigake* (Japanese: 'pioneer')** Japan's first deep-space probe, designed to measure the *solar wind and magnetic field during a fly-by of Halley's Comet. The 138-kg spacecraft was launched on 7 January 1985 as *MS-T5*, and was subsequently renamed. On 11 March 1986, *Sakigake* made its closest approach to the comet at a distance of 6.99 million km. The probe made several Earth fly-bys until 1995, before depletion of its fuel supply.

Sakigake was identical to Japan's second deep-space probe, **Suisei*, except for the scientific instruments it carried on board.

***Salyut* (Russian: 'salute')** A series of seven space stations launched by the USSR 1971–86. *Salyut* was cylindrical in shape, 15 m long, and weighed 19 tonnes. It housed two or three cosmonauts at a time, for missions lasting up to eight months.

Salyut 1 was launched on 19 April 1971. It was occupied for 23 days in June 1971 by a crew of three, who died during their return to Earth when their **Soyuz* ferry craft depressurized. In 1973 *Salyut 2* broke up in orbit before occupation. The first fully successful *Salyut* mission was a 14-day visit to *Salyut 3* in July 1974. In 1984–5 a team of three cosmonauts endured a record 237-day flight in *Salyut 7*. In 1986 the *Salyut* series was superseded by **Mir*, an improved design capable of being enlarged by additional modules sent up from Earth.

Crews observed Earth and the sky, and carried out processing of materials in weightlessness. The last in the series, *Salyut 7*, crashed to Earth in February 1991, scattering debris in Argentina.

http://www.astronautix.com/project/salyut.htm History, description, and specifications of the *Salyut* space station modules. There are lots of excellent photographs of *Salyut 6* and its interior, as well as pictures and technical drawings.

Samara Space Centre (or TsSKB-Progress) A Russian space complex created in 1996 by a Russian presidential decree. It combined the TsSKB Central Specialized Design Bureau and the Progress production plant, each of which had primary roles in the Soviet and Russian space programmes for more than 40 years. In 1959, the OKB-1 design bureau (now the *Energiya Rocket and Space Complex) was established at Samara (known as Kuibyshev between 1935 and 1991). Also located in Samara is the European–Russian organization *Starsem, established in 1966.

SAMPEX Acronym for **Solar, Anomalous, and Magnetospheric Particle Explorer*.

SAR Acronym for *synthetic aperture radar.

SAS Acronym for *Small Astronomy Satellite.

SAS-1 Alternative name for **Uhuru*.

satellite Any small body that orbits a larger one.

Natural satellites that orbit planets are called **moons**. The first **artificial satellite**, *Sputnik 1*, was launched into orbit around the Earth by the USSR in 1957. Artificial satellites can transmit data from one place on Earth to another, or from space to Earth. *Satellite applications include science, communications, weather forecasting, and military use.

An **active satellite** is one that transmits signals that return with information. An example was **Seasat*, which used radar and radio-sensing instruments to monitor the Earth's oceans. A **passive satellite** does not send out signals, but only records information, such as images, about the target object.

At any time, there are several thousand artificial satellites orbiting the Earth, including active satellites, satellites that have ended their working lives, and discarded sections of rockets. The brightest artificial satellites can be seen by the naked eye. Artificial satellites eventually re-enter the Earth's atmosphere. Usually they burn up by friction, but sometimes debris falls to the Earth's surface, as with **Skylab* and *Salyut 7*.

http://liftoff.msfc.nasa.gov/RealTime/JTrack/Spacecraft.html Real-time tracking system that displays on a world map the current position and orbit information for the *International Space Station*, *Hubble Space Telescope*, the space shuttle and other major satellites. Pressing the shift button on your keyboard while clicking on a craft will take you to a page with information about that craft.

satellite applications The uses to which artificial satellites are put. These include:

scientific experiments and observation
Many astronomical observations are best taken above the disturbing effect of the atmosphere. Satellite observations have been carried out by the *Infrared Astronomy Satellite* (1983) which made a complete infrared survey of the skies, and **Solar Max* (1980), which observed solar flares. The **Hipparcos* satellite, launched in 1989, measured the positions of many stars. The **Röntgen Satellite*, launched in 1990, examined ultraviolet and X-ray radiation. In 1992, the **Cosmic Background Explorer* satellite detected details of the Big Bang that mark the first stage in the formation of galaxies. Medical experiments have been carried out aboard crewed satellites, such as the Russian space station *Mir* and the US *Skylab*.

reconnaissance, land resource, and mapping applications
Apart from military use and routine mapmaking, the US *Landsat, the French *Satellite Pour l'Observation de la Terre*, and equivalent Russian satellites have provided much useful information about water sources and drainage, vegetation, land use, geological structures, oil and mineral locations, and snow and ice.

weather monitoring
The US *National Oceanic and Atmospheric Administration series of satellites, and others launched by the European Space Agency, Japan, and India, provide continuous worldwide observation of the atmosphere.

navigation
The US Global Positioning System uses 24 **Navstar* satellites that enable users (including walkers and motorists) to find their position to within 100 m. The US military can make full use of the system, obtaining accuracy to within 1.5 m. The *Transit system, launched in the 1960s, with 12 satellites in orbit, locates users to within 100 m.

communications
A complete worldwide communications network is now provided by satellites such as the US-run *Intelsat system.

***Satellite Pour l'Observation de la Terre* (*SPOT*) (French: 'Satellite for Earth Observation')** A series of five French remote-sensing satellites designed by the *Centre National d'Etudes Spatiales and operated by the commercial company Spot Image, which claims the largest commercial market for Earth geographical images from satellites. The satellites' mission includes monitoring crops for the European Union and measuring ozone and aerosols at the polar regions. The data is also used for other environmental studies, forestry, surveillance, cartography, and oil and gas exploration.

The satellites are in a near polar orbit at an altitude of 832 km, and their real-time data is sent to 18 ground receiving stations around the world. *SPOT-1* was launched in 1986, *SPOT-2* in 1990, *SPOT-3* in 1993, *SPOT-4* in 1998, and *SPOT-5* in April 2002. *SPOT-4* has two high-resolution imagers (visible-light and infrared). An image from each sensor covers 60 sq km.

satellite retrieval and redeployment The act of a spacecraft that rendezvous with an orbiting satellite to capture it for maintenance and repairs before returning it to orbit. This function is performed by the space shuttle; the first retrieval and deployment was done by *Discovery* launched on 12 April 1985. Satellites are often captured by astronauts during an extravehicular activity and manoeuvred into the shuttle's cargo bay. They can be worked on there and redeployed or transported back to Earth and later relaunched.

Saturn The sixth planet from the Sun, and the second-largest in the Solar System, encircled by bright and easily visible equatorial rings. Viewed through a telescope it is ochre. Its polar diameter is 12 000 km smaller than its equatorial diameter, a result of its fast rotation and low density, the lowest of any planet. Its mass is 95 times that of the Earth and its magnetic field 1 000 times stronger.

mean distance from the Sun 1.427 billion km

equatorial diameter 120 500 km

rotational period 10 hours 14 minutes at equator, 10 hours 40 minutes at higher latitudes

year 29.45 Earth years

atmosphere visible surface consists of swirling clouds, probably made of frozen ammonia at a temperature of −170 °C, although the markings in the clouds are not as prominent as Jupiter's. The **Voyager* probes, visiting in 1980 and 1981, found winds reaching 1 800 kph

surface Saturn is believed to have a small core of rock and iron, encased in ice and topped by a deep layer of liquid hydrogen

satellites Over 30 known moons. The largest moon, *Titan, has a dense atmosphere

rings The rings visible from Earth begin about 14 000 km from the planet's cloudtops and extend out to about 76 000 km. Made of small chunks of ice and rock (averaging 1 m), they are 275 000 km rim to rim, but only 100 m thick. The *Voyager* probes showed that the rings actually consist of thousands of closely spaced ringlets, looking like the grooves in a gramophone record

From Earth, Saturn's rings appear to be divided into three main sections. Ring A, the outermost, is separated from ring B, the brightest, by the Cassini division, named after its discoverer Italian astronomer Giovanni Cassini (1625–1712), which is 3 000 km wide; the inner, transparent ring C is also called the Crepe Ring. Each ringlet of the rings is made of a swarm of icy particles like snowballs, a few centimetres to a few metres in diameter. Outside the A ring is the narrow and faint F ring, which the *Voyagers* showed to be twisted or braided. The rings of Saturn could be the remains of a shattered moon, or they may always have existed in their present form.

(Image © NASA)

Cassini captured this image of Saturn from a distance of 5 million km on 13 July 2004. The line separating day from night indicates the tilt of its axis and that it is summer in its southern hemisphere.

The *Cassini* space probe, developed jointly by NASA and the European Space Agency, was launched in October 1997, and went into orbit around Saturn in July 2004.

http://www.solarviews.com/solar/eng/saturn.htm How many rings does Saturn have? How many satellites? Find out the answers to these questions and more at this site, which also features a video of a storm in the planet's atmosphere and information on the international *Cassini* mission to Saturn and Titan.

(Image © NASA)

The rings of Saturn taken by *Cassini* at a distance of 6.4 million km from the planet.

Saturn rocket A family of large US rockets, developed by German-born US rocket engineer Wernher von Braun for the *Apollo* project. The two-stage Saturn IB was used for launching *Apollo* spacecraft into orbit around the Earth. The three-stage Saturn V sent *Apollo* spacecraft to the Moon, and launched the *Skylab* space station. The take-off thrust of a Saturn V was 3.4 million kg. After *Apollo* and *Skylab*, the Saturn rockets were retired in favour of the *space shuttle.

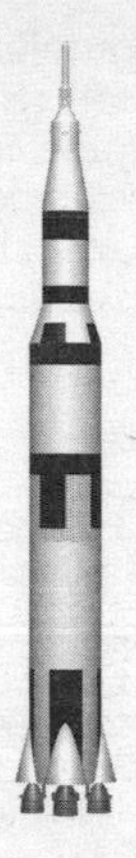

(Image © Research Machine)

When it was built the Saturn V was the largest operational launch vehicle ever produced. With its *Apollo* spacecraft payload it was over 111 m high.

http://www.apollosaturn.com/ *Apollo–Gemini* project resource. There are contemporary press releases, technical reports and diagrams, and photographs. The site features a page with some truly amazing *Apollo* facts and figures.

Savitskaya, Svetlana Yevgeniyena (1948–) Soviet cosmonaut, the second woman in space. She made two *Soyuz* missions in 1982 and 1984. On her second mission, she became the first woman to make a space walk, which lasted 3 hours 35 minutes. Savitskaya was selected as a cosmonaut in 1980. In 1989 she became a member of the Soviet parliament, and in 1993 she retired as a cosmonaut.

scanning image absorption spectrometer for atmospheric cartography (SCIAMACHY) An imaging spectrometer developed by the European Space Agency (ESA) to measure trace gases in the Earth's troposphere (lowest region of the atmosphere) and the stratosphere above it. These measurements, using three different viewing angles, can be used to investigate events that influence atmospheric chemistry, such as pollution, burning, dust storms, and volcanic eruptions. SCIAMACHY operates on **Envisat* which was launched on 1 March 2002.

scan platform A spacecraft appendage to which an *imaging instrument is attached. It can be commanded to point in different directions, independent of the spacecraft's attitude, as on the *Voyager* and *Galileo* probes. Scan platforms are not included on most spacecraft today, having been gradually phased out because of their large mass and the chance of mechanical problems. Instead, spacecraft rotate to point imagers mounted directly on the spacecraft.

scatterometer The *Doppler radar sensor in an artificial satellite, used to measure the scattering effect, or reflection, produced as the radar scans the Earth's surface.

Ocean scatterometers study the interaction between seas and the atmosphere, ocean circulation, and other phenomena that affect weather patterns and global climate. Scatterometers can also identify other environmental effects, such as the reduction of sea ice in the polar regions and the effects of deforestation of the Earth's rainforests.

A scatterometer was installed aboard Japan's **Advanced Earth-Observing Satellite*, launched in 1996.

SCET Abbreviation for *spacecraft event time.

Schirra, Walter (Marty) (1923–) US astronaut. He was one of the *'Original Seven' astronauts chosen by NASA in 1959, and the only one to participate in all three US pioneering missions into space, flying on *Sigma 7* of the **Mercury* project (1962), and commanding *Gemini 6* (1965) and *Apollo 7* (1968). Schirra resigned from NASA and the navy in 1969.

Schlegel, Hans (1951–) German astronaut with the European Space Agency (ESA). He was payload specialist aboard the space shuttle *Columbia* in April and May 1993 as part of the German-sponsored **Spacelab* D-2 mission. Schlegel began training as an astronaut in 1988 at the Deutsches Zentrum für Luft- und Raumfahrt (DLR; German Aerospace Centre). He became an ESA astronaut in 1998 and in the same year began training at NASA's Johnson Space Center as a mission specialist. He has since worked on mechanical systems, crew equipment, and robotic issues in the Astronaut Office at the Johnson Space Center.

Schmitt, Harrison Hagan (1935–) US scientist-astronaut. He is the only scientist to have walked on the Moon, and he and Eugene *Cernan were the last astronauts on the Moon, flying in *Apollo 17* in December 1972. They remained on the lunar surface for three days and took three excursions in the lunar roving vehicle. The 22 hours and 6 minutes they spent in *extravehicular activities on the lunar surface was a record, as was their retrieval of 115 kg of lunar soil and rocks. Schmitt discovered orange glass spherules in the lunar soil.

Schmitt was selected for NASA's first group of scientist-astronauts in 1965. After his one flight, Schmitt left the programme in 1975 and was elected a senator in 1976. He has stayed active in lunar research.

Schwarzschild radius In *astrophysics, the radius of the event horizon surrounding a *black hole within which light cannot escape its gravitational pull.

For a black hole of mass m, the Schwarzschild radius Rs is given by $Rs = 2gm/c^2$, where g is the gravitational constant and c is the speed of light. The Schwarzschild radius for a black hole of solar mass is about 3 km. It is named after Karl Schwarzschild, the German astronomer who deduced the possibility of black holes from *Einstein's general theory of relativity in 1916.

Schweickart, Russell Louis (1935–) US astronaut. He was the lunar module pilot aboard the *Apollo 9* flight, a ten-day Earth orbit in March 1969 to test *Apollo*'s lunar-landing hardware. During the mission, he also undertook a

46-minute space walk to test the new *Apollo* spacesuit. Schweickart was selected as an astronaut in 1963 and left NASA in 1979.

SCIAMACHY Abbreviation for *scanning image absorption spectrometer for atmospheric cartography.

Science Advisory Group (SAG) A team of scientists organized by the European Space Agency to oversee a spacecraft mission or scientific instrument. Each group is headed by principal investigators. The development of the *scanning image absorption spectrometer for atmospheric cartography (SCIAMACHY), for example, was overseen by a group (SSAG) led by three scientists from Germany, the Netherlands, and Belgium.

science data The data that is collected by a spacecraft's scientific instruments and transmitted to ground stations.

Science Working Group (SWG) A science group formed for the preliminary phase of each NASA space mission. After developing science goals and requirements for the mission, the group prepares an overall scientific conception of the mission. The SWG members' work is the basis for NASA's *announcement of opportunity, calling for proposals for science experiments to be carried on the mission.

***Scientific Satellite Atmospheric Chemistry Experiment* (*SCISAT*)** A Canadian space agency satellite that measures global ozone processes. It was launched on 12 August 2003 from NASA's Vandenberg air force base in California. It is in a low polar orbit of 650 km and travels around the Earth 15 times a day. *SCISAT* helps scientists improve their understanding of the depletion of the ozone layer, with special emphasis on changes occurring over Canada and in the Arctic.

SCISAT's primary instrument is a Fourier transform spectrometer and another is named MAESTRO (Measurements of Aerosol Extinction in the Stratosphere and Troposphere Retrieved by Occultation). The mission is to last two years.

scientist-astronaut A NASA category of astronaut with an academic background in science or medicine. A scientist-astronaut is required to have a doctorate in science, engineering, or medicine. As with all astronauts, they must demonstrate excellent physical and psychological capabilities. All receive training as jet pilots in preparation for some of the features of space travel. The first six scientist-astronauts were chosen in 1965.

scintillation counter An instrument for measuring very low levels of radiation. The radiation strikes a scintillator (a device that emits a unit of light when a charged elementary particle collides with it), whose light output is 'amplified' by a photomultiplier; the current pulses of its output are in turn counted or added by a scaler to give a numerical reading.

SCISAT Abbreviation for *Scientific Satellite Atmospheric Chemistry Experiment*, a Canadian satellite that measures global ozone conditions.

SCLK Contraction of *spacecraft clock.

Scobee, Francis Richard (1939–86) US astronaut. He was pilot of the *Challenger* space shuttle flight in April 1984 that deployed the *Long-Duration Exposure Facility. He was killed with his six fellow crew members in January 1986 aboard the *Challenger* flight that exploded soon after its launch. Scobee was selected as an astronaut in 1978.

Scott, David Randolph (1932–) US astronaut. He made three space flights including the *Apollo 15* mission to the Moon launched in July 1971, which he commanded. He and James *Irwin landed on the lunar surface and were the first to use the *lunar roving vehicle. Scott was selected as an astronaut in 1963. He was director of NASA's *Dryden Flight Research Center 1975–7, and retired from NASA in 1977.

Scott, Winston Elliott (1950–) US astronaut. His two space shuttle flights include the November 1997 *Columbia* mission in which he performed two space walks to capture a *Spartan 201 satellite and to test space station assembly techniques. Scott was selected as an astronaut in 1992, and retired from NASA in 1999.

Scout Acronym for *Solid Controlled Orbital Utility Test rocket.

scrub An informal term for postponing a spacecraft launch. This could be due to problems with the spacecraft or adverse weather conditions.

SCS Abbreviation for *stabilization control service.

SCTF Abbreviation for *Sonny Carter Training Facility.

search for extraterrestrial intelligence Full name of *SETI.

Searfoss, Richard Alan (1956–) US astronaut. He piloted the *Columbia* flight launched on 18 November 1993, piloted the *Atlantis* space shuttle mission to dock with the Russian space station *Mir* in March 1996, and commanded the last **Spacelab* mission launched in April 1998 on board the *Columbia* space shuttle. Searfoss was selected as an astronaut in 1990. He retired from the US Air Force and as an astronaut in 1998, was later a research test pilot at NASA's Dryden Flight Research Center, then left in February 2003 to pursue private business interests.

Seasat A NASA satellite, the first to use *synthetic aperture radar (SAR) for *remote sensing of the Earth's oceans. It was launched in June 1978 with a scheduled mission of up to three years. However, its electric power system failed in October after sending only about 42 hours of data.

Seasat was to have measured the sea surface winds and temperatures, wave heights, internal waves, and water in the atmosphere, and investigate sea ice and ocean topography. Despite its failure, the mission successfully demonstrated the use of microwave sensors to monitor the seas.

SeaWIFS The Sea-viewing Wide Field-of-view Sensor instrument, an ocean colour scanner launched in August 1997 aboard the *SeaStar* satellite (the satellite was later renamed *Orbview*). The SeaWIFS instrument is comparable and complementary to the *Medium Resolution Imaging Spectrometer

(MERIS) on *ESA's *Envisat*, and monitors subtle changes in ocean colour to identify pollution and concentrations of marine phytoplankton. The spacecraft and instrument are owned by *Orbital Sciences Corporation which markets the data for commercial use.

secondary oxygen pack A small pack containing a 30-minute emergency oxygen supply for extravehicular activity (space walks outside a spacecraft). The pack also contains a valve and oxygen regulator, and is mounted at the base of the *primary life-support system (backpack) worn by the astronaut.

Seddon, (Margaret) Rhea (1947–) US astronaut and physician. Her three space shuttle flights include the life-science research mission aboard *Columbia* in October 1993 that was considered to be the most successful **Spacelab* flight made. Seddon was selected as an astronaut in 1978 and retired from NASA in 1997.

See, Elliott McKay, Jr (1927–66) US astronaut. He was pilot of the back-up crew for the *Gemini 5* mission in 1965. He was chosen as command pilot for *Gemini 9* in 1966, but was killed in February (three months before the spacecraft's launch date) in the crash of a *T-38 jet trainer. See was selected as an astronaut in 1962.

Sega, Ronald Michael (1952–) US astronaut. His three space shuttle flights include the first joint US–Russian space shuttle mission, in February 1994, aboard *Discovery*. He also flew on board *Atlantis* for the third docking with the Russian space station *Mir*. Sega was selected as an astronaut in 1990 and left NASA in 1996.

SEI Abbreviation for *Space Exploration Initiative.

***Selenological and Engineering Explorer* (*SELENE*)** Japan's first large lunar probe. With a planned launch date of 2006, the mission will collect scientific data to help ascertain the origin and evolution of the Moon. It will also aid the development of technology for future lunar exploration. The probe will be launched by an H-IIA rocket (*see* H ROCKET).

http://www.isas.ac.jp/e/enterp/missions/selene/index.html Simple site profiling the planned *SELENE* lunar mission. The mission's scientific goals, design, and instrumentation are sketched out, accompanied by illustrations and diagrams.

separation During a launch, the moment when a rocket burns all of its fuel and falls away from the remaining launch vehicle or spacecraft.

sequence of events (SOE) An itemized list of planned events for a spacecraft and of the ground commands needed to execute these. It is generated by a **spacecraft event file** (SEF) and by schedules for each tracking station of the *deep-space network.

At NASA, the *ace managing a mission needs the SOE list to compare predicted events with those that actually happen. In addition, the list of commands is used to *uplink commands and times to the spacecraft's computer, and to produce an individual SOE for each tracking station.

service module (SM) A module of a spacecraft that provides the *command module with electricity, air, and water. It is also the location of the thrusters used for manoeuvring. It is usually identified with the *Apollo* missions, which had cylindrical service modules, 7.5 m long and 3.9 m in diameter. The service module was launched sitting above the lunar module and under the command module, to which it remained connected in space to form the *command and service modules (CSM). The service module had the only engine power available to return the *Apollo* astronauts to Earth.

service propulsion system (SPS) A single, large rocket engine of the *command and service module (CSM) used for the *Apollo* space missions that began in the 1960s. It took the form of a cone projecting from the rear of the *service module. The SPS was burned for 6 minutes to slow the spacecraft as it neared the Moon, and for 2.5 minutes to power the CSM out of lunar orbit for the return journey. There was no back-up, so any malfunction of the SPS would have stranded the astronauts in permanent lunar orbit.

service tower Alternative name for *umbilical tower.

SETI (acronym for Search for ExtraTerrestrial Intelligence) A programme originally launched by *NASA in 1992, using powerful *radio telescopes to search the skies for extraterrestrial signals. NASA cancelled the SETI project in 1993, but other privately funded SETI projects continue.

http://www.seti.org/ SETI (Search for ExtraTerrestrial Intelligence) Institute research into the question 'Are we alone in the universe?' The site has information on current and past projects, including their connection with the films *Independence Day* and *Contact*. There is also the opportunity to devote some of your own computer's time to 'crunching' the enormous amounts of data the Institute collects and so doing a little to help the search.

http://www.setileague.org/ Home page of a SETI organization, of, by, and for the amateur radioastronomer, whose aim is to try to find extraterrestrial life. The premiss of doing so is explained right at the beginning of the page. The page contains links to lots of relevant material, as well as to a screen saver that enables the user to join the search for ET!

Sevastyanov, Vitali I (1935–) Soviet cosmonaut. He flew with Andrian *Nikolayev aboard *Soyuz 9* in June 1970, setting a record at that time for the longest space flight, nearly 18 days. He later flew on the *Soyuz 18* mission in May–July 1975, docking with *Salyut 4* to conduct joint experiments. Sevastyanov was selected as a cosmonaut in 1967, became a member of the Soviet parliament in 1989, and left the cosmonaut programme in 1993.

Seyfert galaxy A galaxy in which a small, bright centre is caused by hot gas moving at high speed around a massive central object, possibly a *black hole. Almost all Seyferts are spiral galaxies. They seem to be closely related to *quasars, but are about 100 times fainter. They are named after their discoverer, US astronomer Carl Seyfert.

SFOF Abbreviation for *Space Flight Operations Facility.

Sharipov, Salizhan Shakirovich (1964–) Russian cosmonaut. He was a mission specialist on NASA's space shuttle *Endeavour* flight in January 1998, the eighth shuttle docking with the Russian space station *Mir*. The mission delivered scientific equipment and carried out the last exchange of an astronaut. He has been assigned to the tenth resident crew on the *International Space Station*, with the launch set for October 2004. Sharipov was selected as a cosmonaut in 1990.

Sharman, Helen (1963–) English astronaut and the first Briton to fly in space, chosen from 13 000 applicants for a 1991 joint UK–Russian space flight. A research chemist, she was launched on 18 May 1991 in *Soyuz* TM-12 and spent six days with Russian cosmonauts conducting scientific experiments aboard the **Mir* space station.

Born in Sheffield, England, she was awarded a Bachelor of Science (BSc) degree in chemistry by Sheffield University. Whilst working as a research technologist for Mars Confectionery, she heard a radio advertisement: 'Astronaut wanted—no experience necessary'. After her training and flight, Sharman returned to scientific work, presenting numerous science programmes on radio and television. She was awarded an OBE in 1992.

Shaw, Brewster Hopkinson, Jr (1945–) US astronaut and head of the Space Shuttle Operations Branch at NASA 1989–95. His three space shuttle flights included piloting the *Columbia* mission in November 1983 with the first **Spacelab* and first international shuttle crew. Shaw was selected as an astronaut in 1978, and also became the space shuttle deputy programme manager in 1992. He left NASA and the US Air Force in 1996 to join Rockwell. That year Rockwell was acquired by Boeing for whom Shaw served in several positions. In 2003 he became chief operating officer of United Space Alliance (USA), the prime contractor for the Space Shuttle Program.

Shen Zhou China's spacecraft that launched the country's first astronaut on 15 October 2003. The first uncrewed *Shen Zhou 1* was launched on a test flight in November 1999 that lasted for less than a day. *Shen Zhou 2* was launched in January 2001 by a *Long March rocket from the *Shuang Cheng Tzu launch centre. It carried out experiments involving space materials, astronomy, and life sciences. Several animals were also on board. The spacecraft's re-entry module landed after 6 days and 108 orbits, leaving an orbital module to remain in space for several months. *Shen Zhou 5* was the upgraded vehicle for the first crewed flight.

Shen Zhou's design is based on Russia's **Soyuz*, with modifications.

http://www.spacedaily.com/china.html News of China's space programme. The site regularly reports on the status of the crewed *Shen Zhou* spacecraft as well as China's remote-sensing and other aerospace projects.

Shepard, Alan Bartlett (1923–98) US astronaut, the first American in space and the fifth person to walk on the Moon. He piloted the sub-orbital *Mercury-Redstone 3* mission on board the **Freedom 7* capsule on 5 May 1961, and commanded the *Apollo 14* lunar landing mission in 1971.

Shepard was born in East Derry, New Hampshire, and graduated in 1944 from the US Naval Academy in Annapolis. He served as a fighter pilot, test pilot, and aircraft readiness officer of the Atlantic fleet before being selected to become a *NASA astronaut in 1959. He resigned in 1974.

http://www.achievement.org/autodoc/page/she0pro-1 Description of the life and career of the first American in space, Alan Shepard. The site contains not only a profile and biographical information, but also holds a lengthy interview with Shepard from 1991 accompanied by a large number of photographs, video sequences, and audio clips.

Shepherd, William McMichael (1949–) US astronaut. In October 2000, he was launched from the *Baikonur Cosmodrome to become commander of the first resident crew of the **International Space Station*. He returned in March 2001 aboard the space shuttle *Discovery*. Shepherd had previously flown on three space shuttle missions as a *mission specialist, including the *Discovery* flight in October 1990 that deployed the *Ulysses* probe. He logged more than 159 hours in space. He was selected as an astronaut in 1984 and has since left NASA.

Shoemaker-Levy 9 A comet that crashed into Jupiter in July 1994. The fragments crashed at 60 kps over the period 16–22 July 1994. The impacts occurred on the far side of Jupiter, but the impact sites came into view of Earth about 25 minutes later. Analysis of the impacts shows that most of the pieces were solid bodies about 1 km in diameter, but that at least three of them were clusters of smaller objects.

When first sighted on 24 March 1993 by US astronomers Carolyn and Eugene Shoemaker and David Levy, it was found to consist of at least 21 fragments in an unstable orbit around Jupiter. It is believed to have been captured by Jupiter in about 1930, and fragmented by tidal forces on passing within 21 000 km of the planet in July 1992.

Shonin, Georgi (1935–97) Soviet cosmonaut whose one flight on *Soyuz 6* with Valeri Kubasov in October 1969 was part of the record (at the time) of three spacecraft (*Soyuz 6*, *7*, and *8*) and seven cosmonauts in Earth orbit at the same time. The cosmonauts manoeuvred their vehicles in relation to one another and conducted medical, scientific, and technical experiments. They landed in numerical order one day apart after each had spent 118 hours in space. Shonin was selected as a cosmonaut in 1960 and left the programme in 1979.

shooting star Another name for a *meteor.

short module A designation used for the core segment of the *Spacelab* when it is flown without its experiment segment on the space shuttle. This reduces the length to half, with the core segment being 2.7 m long and 4 m wide. It is pressurized and contains experiment racks, the control centre rack, and a transfer tunnel to the shuttle's mid-deck.

Shriver, Loren James (1944–) US astronaut. His three space shuttle flights include the April 1990 mission made by *Discovery* to deploy the **Hubble Space Telescope*. Shriver was selected as an astronaut in 1978, becoming deputy director of Launch and Payload Processing at NASA's Kennedy Space Center in 1997. He logged more than 386 hours in space before leaving NASA.

Shuang Cheng Tzu (or Jiuquan Satellite Launch Centre) Chinese launch centre located in the Gobi Desert. It is used to launch satellites into low Earth orbit and was the facility that launched China's first astronaut on 15 October 2003.

shutdown The moment that a rocket engine stops firing. Shutdown for the three main engines of the space shuttle occurs about nine minutes after lift-off.

shuttle mission simulator (SMS) NASA's primary system for training space shuttle crews. The US$100 million simulator system, located in Building 5 at the *Johnson Space Center, was completed in 1977. It has two shuttle cockpits equipped with identical controls, displays, and consoles. One, the motion-base crew station, has positions for the commander and pilot. The other, the fixed-base crew station, is configured for the commander, pilot, mission specialist, and payload operator. SMS is the only high-fidelity simulator capable of training for all possible phases of a mission, including the launch, ascent, abort, orbit, rendezvous, docking, payload handling, undocking, deorbit, entry, approach, landing, and rollout. The system can duplicate the main engine and solid rocket boosters, with computer-generated sound simulations.

(Image © NASA)

The shuttle mission simulator (SMS) at the Johnson Space Center. Pilot Frederick Sturckow was accompanied by four fellow astronauts and two cosmonauts on STS-105 to the *International Space Station* in August 2001.

S

shuttle orbiter medical system (SOMS) A name given to two kits of medical supplies carried aboard a space shuttle. The **medications and bandage kit** (MBK) and **emergency medical kit** (EMK) can be used for both minor illnesses and injuries, and for stabilizing more severe conditions.

The MBK contains wound dressings as well as medical treatments in the form of pills, capsules, and suppositories. The EMK includes diagnostic instruments, injectable medications, instruments for minor surgery, and a microbiological test.

Shuttle Radar Topography Mission (SRTM) An international mission in February 2000 to obtain the most complete and accurate high-resolution digital topographic database of the Earth. NASA and the National Imagery and Mapping Agency led the project. The SRTM was launched aboard the space shuttle *Endeavour* and accomplished its mission during ten days of operation, covering 80% of the Earth's land mass. The mission used radar interferometry, two radar images taken from slightly different locations so these differences can be calculated to show surface elevation or change. The main radar bounced pulses off the Earth that were received by two antenna, one in the shuttle payload bay and a second extended 60 m out of the bay.

http://www.jpl.nasa.gov/srtm/ Regularly updated with images from the Shuttle Radar Topography Mission (SRTM) and synthetic views of the Earth created with data from the mission, this site has information about the SRTM, its radar instrument and radar interferometry, and its data products.

shuttle training aircraft (STA) NASA's four Gulfstream II business jets, used to train space shuttle pilots in the skills of approach and landing. Pilots assigned to a shuttle mission receive about 100 hours of STA training, which is equal to 600 shuttle approaches. Each of the aircraft has been modified to fly like a space shuttle during landing. The STA also simulates the shuttle's high speed and steep angle of landing by putting its engines in reverse thrust and lowering the main landing gear.

side hatch Abbreviation of *ingress-egress side hatch, the side opening on the mid-deck of a space shuttle.

sievert (symbol Sv) An SI unit of radiation dose equivalent. It replaces the rem (1 Sv equals 100 rem). Some types of radiation do more damage than others for the same absorbed dose—for example, an absorbed dose of alpha radiation causes 20 times as much biological damage as the same dose of beta radiation. The equivalent dose in sieverts is equal to the absorbed dose of radiation in grays multiplied by the relative biological effectiveness. Humans can absorb up to 0.25 Sv without immediate ill effects; 1 Sv may produce radiation sickness; and more than 8 Sv causes death.

signal processing centre (SPC) The centre at each of NASA's three *deep-space communications complexes (DSCCs). It is used by network monitor and control (NMC) technicians to operate the antenna, called a deep-space station (DSS), which tracks spacecraft.

SIM Planet Quest A NASA spacecraft scheduled for launch in 2011 to measure the positions and distances of stars throughout the Galaxy, and to observe nearby stars for Earth-sized planets. SIM (formerly known as the Space Interferometry Mission) will be able to determine these measurements several hundred times more accurately than previous missions by using optical interferometry that combines light from two or more telescopes as if they were one gigantic telescope mirror. It will also pioneer a technique of blocking out light from a bright star in order to take images of areas near it. The spacecraft will reach a distance from Earth of about 95 million km after 5.5 years, and its interferometer will take observations of the complete celestial sphere during that time.

Single Aperture Far-Infrared Observatory (SAFIR) A large infrared space telescope with a mirror 8 to 10 m in diameter, proposed by NASA as the successor to its **Spitzer Space Telescope* and ESA's **Herschel Space Observatory*. *SAFIR* will study the earliest phases of forming galaxies, stars, and planetary systems at wavelengths from 20 micrometres to one millimetre, where these objects are brightest. The combination of large mirror size and ultra-cold temperature will make *SAFIR* more than 1 000 times more sensitive than *Spitzer* or *Herschel*, and it will approach the ultimate sensitivity limits at far-infrared and sub-millimetre wavelengths. 'Single Aperture' refers to the telescope's single primary mirror, as distinct from multi-mirror interferometry missions. Launch is not expected until around 2015–20.

single-stage-to-orbit A space vehicle that can go into orbit, or launch a payload into orbit, using only one rocket stage. An example was the proposed DC-X, which reached the experimental stage as the *DC-XA reusable launch vehicle.

SIR Acronym for *spaceborne imaging radar.

Skylab A US space station, launched on 14 May 1973, made from the adapted upper stage of a Saturn V rocket. At 75 tonnes, it was the heaviest object ever put into space and had a length of 25.6 m. *Skylab* contained a workshop for carrying out experiments in weightlessness, an observatory for monitoring the Sun, and cameras for photographing the Earth's surface.

Damaged during launch, it had to be repaired by the first crew of astronauts. Three crews, each of three astronauts, occupied *Skylab* for periods of up to 84 days, at that time a record duration for human space flight. *Skylab* finally fell to Earth on 11 July 1979, dropping debris on Western Australia.

Slayton, Deke (born Donald Kent Slayton) (1924–93) US astronaut, one of the 'Original Seven' chosen by *NASA in 1959 for the *Mercury* series of flights. Grounded for health reasons, he became director of NASA's Flight Crew Operations Branch for ten years. In 1972 he was returned to flight status, and in 1975 he finally flew in space, then the oldest person to have done so at the age of 51.

Although chosen for the second *Mercury* orbital mission, he was grounded because of a minor heart irregularity. As NASA's flight-crew director, he was responsible for astronaut training and the selection of crews for various flights, including the *Apollo* Moon missions. He finally flew on the *Apollo–Soyuz* Test Project, in which a US *Apollo* spacecraft docked in orbit with a Soviet *Soyuz* spacecraft.

He subsequently left NASA to join a private company, *Space Services Inc., which unsuccessfully attempted to develop rockets for a commercial launch service.

SLC 37 Abbreviation for *Space Launch Complex 37.

sleep kit A kit kept on board a space shuttle, containing eye covers and ear plugs, to help astronauts sleep. It is stored in the individual crew member's clothing locker during launch and re-entry.

sleep restraint A NASA term for a sleeping bag used in a spacecraft. Located in the mid-deck in a space shuttle, the sleep restraints are each attached to a padded board to keep the sleeper from floating around the craft (due to effects of low level gravity or *microgravity). The narrow private bunk area also contains a reading light and ventilation ducts.

SM Abbreviation for *service module.

Small Astronomy Satellite (SAS) A series of NASA satellites dedicated to celestial *X-ray astronomy. The first was *Uhuru*, also known as *SAS-1*, launched in December 1970. *SAS-2* was launched in November 1972 to take the first detailed look at the gamma-ray sky; however, a power-supply failure terminated the mission in June 1973. *SAS-3* was launched in May 1975 and discovered the precise location of about 60 X-ray sources before the satellite's lifespan ended in April 1979.

Small Missions for Advanced Research and Technology (SMART) A series of small spacecraft of the European Space Agency (ESA), to test new technologies for future use on larger missions. *SMART-1* was launched on 27 September 2003 as the first ESA mission to the Moon, orbiting and conducting scientific observations. However, its main purpose is to test solar electric primary propulsion (SEPP), based on the use of xenon gas as a propellant and on power sourced from the spacecraft's solar arrays.

SMART-1 used the competing gravity of the Earth and the Moon to conserve the fuel supply, going into orbit around the Moon in November 2004 after a voyage lasting 14 months instead of a matter of a few days. The new propulsion system may be used by ESA's *BepiColombo mission to Mercury planned to launch in 2012.

SMART Acronym for *Small Missions for Advanced Research and Technology.

Smith, Michael John (1945–86) US astronaut. He served as commander in NASA's Shuttle Avionics Integration Laboratory before being assigned as pilot of the space shuttle *Challenger* flight that exploded soon after its launch in January 1986. Smith and his six fellow crew members were killed in the explosion. He was selected as an astronaut in 1980.

Smith, Steven Lee (1958–) US astronaut. He has made four space flights and has performed seven space walks totalling 49 hours, the second-highest number of space walk hours in NASA history. Two space walks occurred during the April 2002 flight of *Atlantis* to the *International Space Station* (*ISS*) when a truss was installed. Smith joined NASA in 1989 as a payload officer and was selected as an astronaut in 1992. He is currently NASA's manager of the Automated Transfer Vehicle (ATV) Launch Package for the *ISS*.

SMS Abbreviation for *shuttle mission simulator.

Snapshot A NASA spacecraft with a nuclear power source, launched into Earth orbit on 3 April 1965. Its reactor supplied electrical power to a 1-kg ion engine. Weighing 440 kg, the craft was on an experimental mission for the US Air Force and US Army.

'Snoopy' helmet A nickname for the *communications carrier assembly, the soft cap worn by NASA astronauts under their bubble *space helmet. The name is taken from the character Snoopy in the popular *Peanuts* comic strip, who was often depicted wearing such a cap.

SOE Acronym for *sequence of events.

soft landing A carefully controlled landing of a spacecraft to avoid damage to the vehicle and, if crewed, to prevent injury to the astronauts. On return to Earth, this is accomplished by splashdowns in oceans or by using parachutes on land. Soft landings are not needed by a space shuttle, which lands like an aircraft.

soft suit A normal spacesuit made of flexible materials, such as nylon, dacron, and *Nomex cloth. This is in contrast with suits having a hard shell with movable joints. The *Mercury*, *Gemini*, and *Apollo* suits were soft suits, and the space-shuttle suits are soft with some hard materials. *Hard suits are considered more durable for longer extravehicular activities.

SOHO Abbreviation for **Solar and Heliospheric Observatory*.

Sojourner The first Mars roving vehicle, landed by NASA on the planet on 4 July 1997 by its **Mars Pathfinder* space probe. The rover, about the size of a microwave oven, weighed 11 kg. NASA specialists controlled Sojourner from Earth, sending it down a ramp from the *Pathfinder* and over a Martian plain to take photographs and measure and analyse rocks for nearly three months.

Having landed on the US Independence Day, it was named after Sojourner Truth, a former slave who had campaigned for abolitionism and equal rights for women.

***Solar and Heliospheric Observatory* (*SOHO*)** A space probe built by the *European Space Agency and launched by NASA on 2 December 1995 to study the *solar wind of atomic particles streaming towards the Earth from the Sun. It also observes the Sun in ultraviolet and visible light, and measures slight oscillations on the Sun's surface that can reveal details of the structure of the Sun's interior.

The US$1.2 billion probe is positioned towards the Sun, 1.5 million km from Earth. *SOHO*'s hydrazine fuel froze in June 1998 causing contact with it to be lost. Ground control at the Goddard Space Flight Center finally regained command of *SOHO* in September 1998, when *SOHO* was turned so that its solar power arrays faced the Sun. *SOHO* has been in continuous use since that time. It broadcasts daily images of the Sun, which are accessible to the public via the

Internet. On 28 October 2003, it observed a spectacular solar flare that required two of its instruments to be temporarily shut down to avoid damage.

SOHO carries equipment for 11 separate experiments, including the study of the Sun's corona and measurement of its magnetic field and of solar winds. The coronal diagnostic spectrometer (CDS) detects radiation at extreme ultraviolet wavelengths and allows the study of the Sun's atmosphere. The Michelson Doppler imager (MDI) measures Doppler shifts in light wavelengths and can detect winds caused by convection beneath the Sun's surface. The extreme-ultraviolet imaging telescope (EIT) investigates the mechanisms that heat the Sun's corona. The large-angle spectroscopic coronagraph (LASCO) images the corona by detecting sunlight scattered by the coronal gases.

http://sohowww.nascom.nasa.gov/ News, images, and background on the mission of the *Solar and Heliospheric Observatory* (*SOHO*) to study the Sun's structure, its outer atmosphere, and the particles that make up the solar wind. The site has available for download the *SOHO* screen saver, which automatically updates your desktop with the latest images of the Sun.

***Solar, Anomalous, and Magnetospheric Particle Explorer* (*SAMPEX*)** A NASA satellite launched on 3 July 1992 into a polar orbit around the Earth to obtain data on energetic particles from the Sun and cosmic rays from deep space over a complete solar cycle. It obtained samples of particles from solar flares and measured the effects of high-speed electrons on the chemistry of the upper atmosphere.

solar cycle A variation of activity on the *Sun over an 11-year period indicated primarily by the number of *sunspots visible on its surface. The next period of maximum activity is expected round 2011.

solar flare A brilliant eruption on the Sun above a *sunspot, thought to be caused by release of magnetic energy. Flares reach maximum brightness within a few minutes, then fade away over about an hour. They eject a burst of atomic particles into space at up to 1 000 kps. When these particles reach Earth they can cause radio blackouts, disruptions of the Earth's magnetic field, and *aurora.

In 2003 astronomers using the *Solar and Heliospheric Observatory* (*SOHO*) observed the biggest solar flare ever recorded. Solar flare strength is given an 'X' designation ranging from a minimum X1 up to X20 (the latter being the magnitude of the previous largest recorded solar flare, in 2001). The 2003 flare was so powerful that it overloaded the measuring devices, and estimates of its magnitude placed it at around X28.

***Solar Max* (in full: *Solar Maximum Mission* (*SMM*))** A satellite launched by NASA in 1980 to study solar activity. It discovered that the Sun's luminosity increases slightly when sunspots are most numerous. It was repaired in orbit by astronauts from the space shuttle in 1984 and burned up in the Earth's atmosphere in 1989.

Solar Orbiter A proposed European Space Agency probe which will observe solar activity from distances as close as 45 solar radii, one-fifth the Earth's distance from the Sun. Launch will take place in 2013 or after. Gravity assists from Venus will help manoeuvre into a 150-day orbit at an inclination of up to 35 degrees to the Sun's equator from where it will be able to observe the polar regions and the side of the Sun not visible from Earth. At its closest, *Solar Orbiter*'s orbital motion will match the rotation speed of the Sun so it can make extended observations of the development of solar storms.

solar panel A frame containing solar cells that is attached to a spacecraft to turn the energy of sunlight into electrical power. Solar panels are usually mounted as opposite pairs, as on the **Skylab* space station and the **Hubble Space*

(Image © NASA)

Golden solar arrays are lit from behind in this dramatic view of the *Hubble Space Telescope* as it sits in the space shuttle cargo bay during Hubble Servicing Mission 3 (STS-103). The Earth's atmosphere forms a bright streak between the telescope and the right solar panel.

Telescope. Probes such as *Mariner 9* and *Viking* have four solar panels (two opposite pairs).

solar power satellite (SPS) A proposal to launch solar power plants (consisting of large sheets of solar cells on frames) into Earth orbit to convert sunlight to electricity and send it to ground receiving stations. The idea was first proposed in the USA in the 1970s but has yet to be realized. Critics of SPS say it could be used as a weapons system.

Japan announced in 2001 that it had plans to launch a large solar power station into geostationary orbit by 2040, to generate 1 million kilowatts per second.

solar radiation The radiation given off by the Sun, consisting mainly of visible light, ultraviolet radiation, and infrared radiation, although the whole spectrum of *electromagnetic waves is present, from radio waves to X-rays. High-energy charged particles, such as electrons, are also emitted, especially from solar *flares. When these reach the Earth, they cause magnetic storms (disruptions of the Earth's magnetic field), which interfere with radio communications.

solar sail A spacecraft sail that uses the pressure of solar radiation for high-speed propulsion. It is the equivalent of the wind propelling a sailboat. The *solar wind is especially valuable for long missions and is NASA's choice for its future *Interstellar Probe.

http://www.ugcs.caltech.edu/~diedrich/solarsails/ Solar sailing for beginners. As well as news of recent solar-sail research and experiments, these pages describe how they work, the physics of solar sails, and how they can be manoeuvred. Several solar-sail designs are presented, accompanied by interesting artists' renditions.

solar shield A component of a spacecraft that protects the craft against the heat and radiation of the Sun or shields a spacecraft's instruments, such as a telescope, from the Sun's bright light.

Solar System The *Sun (a star) and all the bodies orbiting it: the nine *planets (Mercury, Venus, Earth, Mars, Jupiter, Saturn, Uranus, Neptune, and Pluto), their moons, and smaller objects such as *asteroids and *comets. The Sun contains 99.86% of the mass of the Solar System. The planets orbit the Sun in elliptical paths, and in the same direction as the Sun itself rotates. The planets nearer the Sun have shorter orbital times than those further away since the distance they travel in each orbit is less.

The inner planets (Mercury, Venus, Earth, and Mars) have solid, rocky surfaces; relatively slow periods of rotation (Mercury takes 59 days to complete one rotation, Venus 243 days, Earth nearly 24 hours, and Mars 24.5 hours); very few natural *satellites; and diameters less than 13 000 km. Venus can be seen with the unaided eye, appearing in the evening as the brightest 'star' in the sky. In contrast, the outer planets (Jupiter, Saturn, Uranus, and Neptune) have denser, gaseous atmospheres composed mainly of hydrogen and helium; fast periods of rotation (Jupiter takes under 10 hours for one rotation, Saturn nearly 10.25 hours); and many natural satellites. Uranus, Neptune, and Pluto were discovered after the development of the telescope.

The Solar System gives every indication of being a strongly unified system having a common origin and development. It is isolated in space. All the planets go around the Sun in orbits that are nearly circular and coplanar, and in the same direction as the Sun itself rotates. Moreover this same pattern is continued in the regular system of satellites that accompany Jupiter, Saturn, and Uranus. It is thought to have formed by condensation from a cloud of gas and dust in space about 4.6 billion years ago.

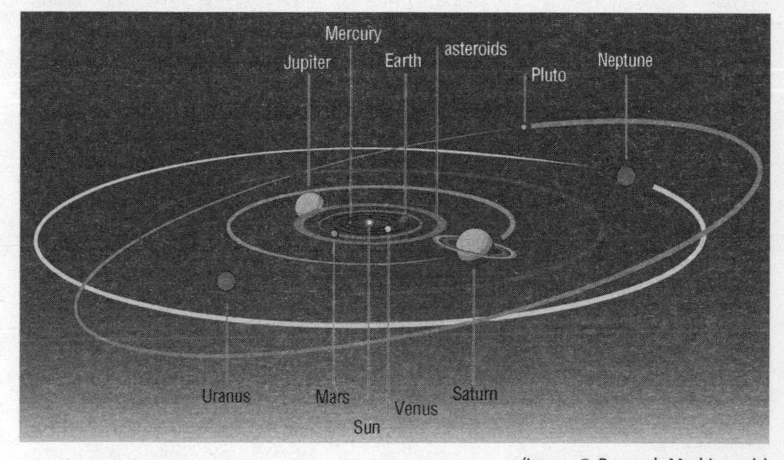

(Image © Research Machines plc)

Most of the objects in the Solar System lie close to the plane of the ecliptic. The planets are tiny compared to the Sun (not shown to scale). If the Sun were the size of a basketball, the planet closest to the Sun, Mercury, would be the size of a mustard seed 15 m from the Sun. The most distant planet, Pluto, would be a pinhead 1.6 km away from the Sun. The Earth, which is the third planet out from the Sun, would be the size of a pea 32 m from the Sun.

http://www.nineplanets.org/ Multimedia tour of the Solar System, with descriptions of each of the planets and major moons. In addition, there are appendices on such topics as astronomical names and how they are assigned, the origin of the Solar System, and hypothetical planets.

http://ic-www.arc.nasa.gov/ic/projects/bayes-group/Atlas/ Page dedicated to listing the best sources of high-resolution, often unprocessed images of each planet or moon in the Solar System.

http://www.lpi.usra.edu/research/outerp/moons.html Detailed and highly comprehensive page about the satellites of the outer planets. With links to separate pages for every satellite covered, this is more of a resource to dip into than read from start to finish. Images of each satellite are also provided, plus links to a special three-dimensional tour of the Solar System.

http://www.solarviews.com/eng/homepage.htm Educational tour of the Solar System. It contains information and statistics about the planets, moons, asteroids, comets, and meteorites found within the Solar System, and is supported by images.

solar-thermal-vacuum chamber (STV) A thermal-vacuum chamber that simulates strong sunlight, as well as the temperatures and vacuum experienced in space. The chamber is used to examine and adjust a spacecraft's thermal-control system. The STV at NASA's Johnson Space Center is among the world's largest, and the only one of its size that can simulate thermal conditions present near the planet Venus. The chamber is constructed of stainless steel and is 26 m tall and 8.2 m in diameter, with a door that is 7.6 m wide.

The **Cassini* spacecraft was tested there for 18 days while suspended from the chamber's walls by steel cables. This was under a solar simulator and in a temperature of −185 °C.

solar wind A stream of atomic particles, mostly protons and electrons, from the Sun's corona, flowing outwards at speeds of between 300 kps/200 mps and 1 000 kps.

The fastest streams come from 'holes' in the Sun's corona that lie over areas where no surface activity occurs. The solar wind pushes the gas of comets' tails away from the Sun, and 'gusts' in the solar wind cause geomagnetic disturbances and aurorae on Earth.

Solid Controlled Orbital Utility Test rocket (Scout) NASA's first solid-fuel rocket. A series of Scout rockets was launched during the period 1960 to 1993, carrying a total of 92 satellites into orbit for NASA, the US Air Force, US Navy, and several other nations. Scout rockets also made 19 sub-orbital flights to investigate potential problems for crewed spacecraft, such as *re-entry heat and radio blackouts.

The four-stage Scout was approximately 23 m long, with a launch weight of 21 329 kg. It was assembled with payloads in a horizontal position, a feature that distinguished it from the larger expendable rockets subsequently built by NASA.

solid fuel A rocket fuel that is in a solid state. It is mixed with an oxidizer to create a solid propellant like a plastic cake. Solid-fuel rockets are more reliable because they have fewer working parts than liquid-propellant rockets. However, they can not be turned on and off whenever desired. The fuel, known as the *grain, is shaped around the inside of the rocket.

The space shuttle uses solid rocket boosters (SRBs) during its launch, the largest solid-fuel motors ever flown. Their aluminium fuel and ammonium perchlorate oxidizer are bound together by a polymer. Each SRB weighs approximately 57 tonnes on the launch pad, with 85% of that being the solid fuel.

solid-state imaging subsystem (SSI) A spacecraft subsystem composed of a telescope, image sensor, and electronics. The SSI, which weighs 29 kg, was designed and assembled at NASA's Jet Propulsion Laboratory, California. It has a wavelength range that extends from the visible into the near-infrared. The SSI flew on the **Galileo* probe, using the same type of telescope that had been aboard the **Voyager* probes; it was focused on infinity.

solid-state recorder A machine that records data digitally without the use of moving parts. Installed on spacecraft to replace tape data storage, solid-state recorders provide increased flexibility by allowing multi-tasking: they can record two data streams at the same time, allowing both the science and engineering data streams to be captured on a single recorder.

On the **Hubble Space Telescope* the solid-state recorder is about the same size and shape as the reel-to-reel recorder, but can hold 12 gigabits of data, compared to only 1.2 gigabits on the tape. NASA first tested the solid-state recorder during a ten-day space shuttle mission (STS-95) in October 1998. Solid-state recorders have since been used in all spacecraft.

Solovyov, Anatoli Yakovlevich (1938–) Soviet cosmonaut. He flew five missions to the *Mir* space station between 1988 and 1997, including 179 days on *Mir* from February to August 1990, and a launch aboard the space shuttle *Atlantis* in June 1995. He made a record 14 space walks, accumulating over 60 hours of extravehicular activity experience. Solovyov was selected as a cosmonaut in 1979.

SOMS Acronym for *shuttle orbiter medical system.

Sonny Carter Training Facility (SCTF; formerly Weightless Environment Training Facility (WETF)) A NASA facility that includes the *Neutral Buoyancy Laboratory, located in Building 29 at the Johnson Space Center.

sounding rocket A small research rocket that ascends to the Earth's upper atmosphere and proceeds in a *sub-orbital trajectory, studying the atmosphere, the planet, or outer space. The Viking rocket, first flown in 1949, was sometimes used as a sounding rocket and in 1954 took the first photographs of the Earth from space.

sound-level meter A battery-run meter used in a space shuttle or space station to measure noise levels in the cabin. The flight crew of a shuttle has the responsibility of taking readings at various locations in the equipment and crew compartments. The results are logged into a record book and may also be recorded on tape.

southern lights The common name for the *aurora australis, coloured light in southern skies.

***Soyuz* (Russian: 'union')** A Soviet (later Russian) series of spacecraft, capable of carrying up to three cosmonauts. It is the longest serving crewed spacecraft in the world. *Soyuz* spacecraft consist of three parts: a rear section containing engines; the central crew compartment; and a forward compartment for working and living space. Although the craft were originally used for independent space fight, from 1998 the *Soyuz* ferried crews and components to the **International Space Station* (*ISS*), scheduled for completion in 2006. When NASA grounded its space shuttles after the 2003 *Columbia* space shuttle disaster, the *Soyuz* craft assumed all of the flights to the *ISS* until the shuttles' projected return in 2005.

Soyuz 1 crashed on its first flight in April 1967, killing the lone pilot, Vladimir *Komarov. Yet in 1968 *Soyuz 3* had the first crewed rendezvous and possible docking by a cosmonaut; in 1969 the flights of *Soyuz 6*, *Soyuz 7*, and *Soyuz 8* marked the first time three spacecraft and seven astronauts had been put into Earth orbit simultaneously; and in 1971 *Soyuz 11* linked up with the first space

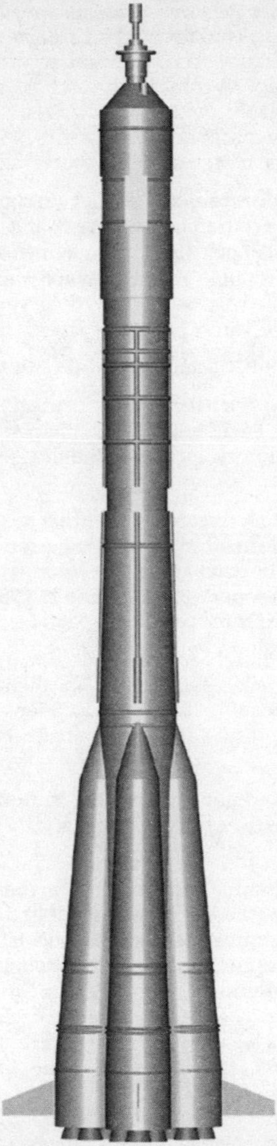

(Image © Research Machines Plc)

The Soyuz rocket is used to launch the crewed *Soyuz* spacecraft, which delivers cosmonauts to the *International Space Station*.

station, *Salyut 1* (*see* SALYUT), although three cosmonauts died on re-entry due to loss of pressure in the spacecraft caused by a faulty valve. In 1975 the *Apollo–Soyuz* Test Project resulted in a successful docking of the two spacecraft in orbit.

space (or outer space) The void that exists beyond Earth's atmosphere. Above 120 km, very little atmosphere remains, so objects can continue to move quickly without extra energy. The space between the planets is not entirely empty, but filled with the tenuous gas of the *solar wind as well as dust.

http://image.gsfc.nasa.gov/poetry//ask/askmag.html Well-organized NASA site with answers to questions about astronomy and space. The site endeavours to send a reply to any question about space and astronomy within three days. There are useful links to question archives, general files, and other space-related sites.

http://www.space.com/ Latest news on space exploration, plus space quizzes, games, and other fun stuff. The site also includes a children's page with activities.

space adaptation syndrome Alternative term for *space sickness.

space bike Informal term for *manned manoeuvring unit.

spaceborne imaging radar (SIR) An instrument that takes remote-sensing radar images from space. It is generally used to study the Earth's environment. SIR-C, one version built for the space shuttle in 2001, measures the Earth's surface at three different wavelengths, compared to one used by the European Space Agency's *ERS-1* satellite. SIR-C is used in combination with a synthetic aperture radar to measure such phenomena as deforestation, desertification, soil moisture-retention, and ocean dynamics.

SIR was developed by NASA's *Jet Propulsion Laboratory, and was first launched on 12 November 1981 as Shuttle Imaging Radar-A aboard the **Columbia* space shuttle.

http://www.jpl.nasa.gov/radar/sircxsar/ Presents recently released radar images obtained using SIR-C/X-Band SAR and processed at the Jet Propulsion Laboratory. The site has radar images organized by category (for example, archaeology, ecology); there are small browsable images for viewing online and high-resolution versions for a much closer look.

spacecraft Any device sent into space from Earth. Although rockets are space vehicles, normal spacecraft have payloads to accomplish missions as well as their own command systems, and supporting ground crews. Spacecraft, which can be crewed or uncrewed, include satellites, space probes, space stations, and space shuttles.

Spacecraft Assembly Facility (SAF) A NASA building in which instruments and software are integrated in a spacecraft and tested. This is done at the *Jet Propulsion Laboratory in Building 179 which has two *clean rooms. Spacecraft that have been assembled and tested there include **Galileo*, the **Hubble Space Telescope*, and the **Voyager* and **Viking* probes.

spacecraft bus A framework of a spacecraft's structure. The bus provides stability for modules sensitive to lift-off strain and vibrations, and houses instruments and equipment, including gyroscopes, cables, and plumbing. It also has attachment points for antennae, *booms, and other components; and there are points for holding and moving the spacecraft during construction, testing, transporting, and launch.

spacecraft central computer A single computer on a spacecraft that has overall control of the vehicle's activity. On some types of spacecraft it is called the **command and data-handling subsystem** (CDHS). This central computer is responsible for such important tasks as *telemetry (transmission to a distant receiving station), the interpretation of commands from Earth, fault protection, and *safing measures.

spacecraft clock (SCLK) A counter on a spacecraft that records the passing of time during a mission. It regulates nearly all activities, and many commands uplinked to the spacecraft are set for execution on counts of the spacecraft clock. In telemetry, the SCLK (pronounced 'sklock') helps in the processing, storage, retrieval, distribution, and analysis of data. SCLK is the best on-board estimate of *spacecraft event time (SCET) but is not as constant as SCET.

spacecraft event time (SCET) The *Universal Time Coordinated (UTC) measured on board a spacecraft. It is equivalent to *transmission time (TRM) added to *one-way light time (OWLT). SCET added to OWLT is equal to *earth-received time (ERT).

spacecraft position The position of a spacecraft orbiting the Earth or travelling through interplanetary space. Both positions are expressed in angles. The position of an Earth satellite is measured by the angles of the *deep-space stations (antennae) pointing at it. The location of a vehicle in distant space can be determined by *precision ranging using the angular measurements of two widely separated deep-space stations or by the process of *very long-baseline interferometry. The measurements are accurate to thousandths of a degree.

spacecraft team The ground team that has responsibility for a spacecraft and its operations, beginning from when the mission is planned, with the team discussing the vehicle and its instruments with contractors.

The spacecraft team is part of the flight team, which operates under the mission control team, and it also draws up the spacecraft event file (SEF) that is used to compile a mission's *integrated sequence of events (ISOE).

NASA's **Mars Climate Orbiter* space probe, launched on 11 December 1998, was lost because its spacecraft team and mission navigation team did not notice that different measurements (imperial and metric) were being used.

space debris A defunct spacecraft (or pieces of spacecraft) that remain in orbit after the end of a mission. This space junk ranges from lens covers to spent upper stages (*see* MULTISTAGE ROCKET) of craft and decommissioned satellites. There is concern that high-speed collisions between such objects will eventually produce even more debris, as well as posing a potential hazard to new spacecraft and to astronauts' spacesuits during extravehicular activities. In 1996, the antenna of a French satellite, *Cerise*, was severed by a piece of an Ariane 4 third-stage rocket.

Only 700 of the more than 8 500 spacecraft in orbit are operational. Fragments often come from explosions of upper stages and spacecraft, adding to some 120 000 pieces of space debris larger than 1 cm.

The European Space Agency (ESA) coordinates research into space debris through the Mission Analysis Section of its *European Space Operations Centre (ESOC) at Darmstadt, Germany. As well as assessing the risk of collisions between operational spacecraft and human-made or natural objects, ESOC provides forecasts of re-entry times and locations for space debris.

Space Exploration Initiative (SEI) An initiative announced in 1989 by US president George Bush that set a long-term goal for NASA to develop a permanent base on the Moon and a crewed mission to the planet Mars in the early 21st century. A presidential policy directive of 13 March 1992 designated NASA as the main SEI implementing agency, and NASA established an Office of Exploration to develop the plans. The directive gave supporting roles to the US Department of Energy and US Department of Defense. His son, President George W Bush, restated this resolve in January 2004, proposing a crewed return to the Moon by 2020 and a crew to Mars later this century.

Space Flight Operations Facility (SFOF) The operations control centre for the *deep-space network, based at NASA's *Jet Propulsion Laboratory, California. SFOF issues commands and communications during, for example, a mission's *cruise phase and *encounter phase. It is also home to the Flight System Testbed used to design and build smaller exploratory spacecraft.

Space Flight Tracking and Data Network (STDN) A NASA ground network that provides tracking and data acquisition to flights of the space shuttles and expendable launch vehicles (ELVs). STDN ground stations are located at Merritt Island, Florida; Bermuda; and Dakar, Senegal. They also provide preflight, launch, and early orbit communications to the shuttle.

space food The special food for astronauts on a mission. It can be fresh, irradiated, freeze-dried, or dehydrated. Early flights carried soft food in tubes or as tablets. Modern supplies on the space shuttle and *International Space Station* include about 100 choices of food and 20 drinks. NASA's condiments include salt, pepper, ketchup, mayonnaise, mustard, and taco sauce. Vitamins are also available, as are sweets and chewing gum.

http://spacelink.nasa.gov/NASA. Projects/Human.Exploration.and. Development.of.Space/Living.and.Working.In.Space/Space.Food/.index. html Articles on food in space. Documents describe how food is prepared for flights, list the food on various NASA missions, and provide a history of space food in the online book *Space Food and Nutrition*.

Space Foundation A non-profit US organization established in 1983 to promote an understanding of the practical and theoretical utilization of space. A major project since the 1980s has been to build and test an experimental solar sail. The Space Foundation supports space activities and education, hosting the annual National Space Symposium and International Space Symposium. Each year it presents four awards, including a Space Achievement Award (won in 2004 by the Ariane 4 launch team) and an Education

Achievement Award, as well as inducting people, organizations, and companies into its Space Foundation Hall of Fame. The foundation's members include space professionals and the general public. It is located in South Pasadena, California.

Space Frontier Foundation A US organization that promotes the large-scale permanent human settlement of space. Its members include scientists and engineers, political and media professionals, and entrepreneurs. Established in 1988, the foundation works for a 'citizen's space policy' and the opening of space to everyone through free markets and free enterprise.

In 1994, the foundation lobbied to privatize the **International Space Station* as 'Alpha Town' to maximize the economic activity there. In July 2004, the organization held its fifth annual symposium in Las Vegas, Nevada, 'Return to the Moon V', promoting lunar settlement.

http://www.space-frontier.org/ Presents the aims and activities of the Space Frontier Foundation. The site is frequently updated with news of the society's drive to 'open up the space frontier'. There are lots of interesting and well-illustrated articles, the best of which focus upon near-Earth objects and the future exploration of the Moon.

Space Grant (in full NASA National Space Grant College and Fellowship Program) A NASA educational organization, established in 1988, that funds research projects through a national network of 52 university-based state consortia (including the District of Columbia and the Commonwealth of Puerto Rico). The network also includes over 820 affiliated businesses, state and local government agencies, and non-profit organizations. Space Grant has awarded more than 12 000 US citizens with tuition assistance in science, engineering, and related fields.

***Space Habitation Module* (*Spacehab*)** A space research laboratory on a space shuttle that doubles the living space and quadruples the room for payload hardware. The module is 4 m in diameter and 2.7 m long, with a payload capacity of 2 160 kg. The modules are built by SPACEHAB, Inc. of Arlington, Virginia, who pay NASA to launch them. The company then leases experiment space to NASA, governmental agencies, private industry, universities, and research institutions. The first *Spacehab* was launched on 21 June 1993 aboard the space shuttle *Endeavour* and carried 22 experiments. The second *Spacehab* was launched on 3 February 1994 aboard the space shuttle *Discovery*, carrying 13 experiments that included crystal growth research to develop improved semiconductors. The third *Spacehab* flight was on 3 February 1995 on board *Discovery*. The latter *Spacehab* was an improved version with a camera to assist the mission's rendezvous docking with the Russian space station *Mir*.

space helmet The protective headgear worn by astronauts, especially for extravehicular activities (EVAs).

Helmets are so critical that the *Mercury* helmets accounted for half of the spacesuit cost. These were made of fibreglass and had a Plexiglas visor. The communication wiring entered from the rear. *Gemini* helmets included two additional visors, one for protection against heat and ultraviolet light and one protecting against impact. *Apollo* astronauts wore a bubble-shaped helmet of polycarbonate material and a soft *'Snoopy' helmet beneath it to hold the communications gear.

For *Apollo 17*, the last Moon landing, Velcro pads were positioned inside the helmets so astronauts could scratch their noses against them. This was retained for the polycarbonate space shuttle EVA helmets, along with a device to clear blocked ears. Shuttle space walkers also have a miniature television camera and lighting mounted on a visor assembly fitting over the helmet.

Spacelab A small space station built by the European Space Agency (ESA) which was carried in the cargo bay of the US space shuttle, returning to Earth with the shuttle. *Spacelab* consisted of a pressurized module in which astronauts worked, and a series of pallets, open to the vacuum of space, on which equipment was mounted. *Spacelab* was used for astronomy, Earth observation, and experiments utilizing the conditions of weightlessness and vacuum in orbit.

The first *Spacelab* mission lasted from 28 November to 8 December 1983. Three other missions were flown between 1983 and 1985. The first crewed mission managed by a nation other than the USA or USSR was a German *Spacelab* mission, **D1 Spacelab**. It carried a record crew of eight, and lasted from 30 October to 6 November 1985 aboard the space shuttle *Challenger*. *Spacelab* life sciences missions were flown in June 1991 and October 1993; a life science and microgravity mission was flown in June–July 1996. *Spacelab* was used in 25 shuttle flights, but was finally decommissioned in 1998.

http://science.ksc.nasa.gov/shuttle/technology/sts-newsref/spacelab.html#spacelab Space shuttle reference manual with a chapter describing in detail the make-up and operation of the *Spacelab* pressurized experiment module.

Space Launch Complex 37 SLC 37 A launch pad at the *Kennedy Space Center from which uncrewed **Apollo* project missions were launched, as well as other spacecraft. The complex was built in 1963 and there were eight *Apollo* launches from 1964 to 1968. SLC 37 was deactivated in 1972, and the Boeing Company converted the 53-ha site in 2000 into a launch complex for its Delta IV launch vehicles. The new facility includes a 101-m mobile service tower (MST) and a six-storey Horizontal Integration Facility (HIF) of 6 975 sq m for rocket assembly.

space law The body of law governing space-related activities. This is generally part of international law, although national space laws have been passed in more than 20 individual countries, including the USA, UK, and Russia. The major laws for space exploration have been developed by the United Nations (UN), including five international treaties and five sets of principles. The treaties include the *Moon treaty (never ratified), the *Outer Space Treaty, Liability Convention, Rescue Agreement for astronauts, and Registration Convention for objects launched into space. The UN defines astronauts as 'envoys of mankind in space' who should receive all possible assistance in the event of accident, distress, or emergency landing. Another

UN law covers international liability for damages caused by spacecraft on Earth or to other spacecraft.

http://www.permanent.com/archimedes/LawLibrary.html Simple site with the text of international agreements, policy information, and essays. Downloadable Adobe documents contain discussion of various aspects of space law, including arms control, space tourism, and remote sensing.

space legs A term used by astronauts for moving skilfully in microgravity. Those new to the conditions usually need time to find their 'space legs'. Early efforts at manoeuvring are often awkward and too rapid, leading to collisions and the displacement of small items that then float in the spacecraft cabin.

space medicine The medical specialty concentrating on the alleviation of the physiological effects encountered by astronauts in the microgravity atmosphere of space. Branches include *space sickness, radiation, and changes in bone and muscle and the cardiovascular immunology systems. Psychological problems, such as coping with isolation in space, are also addressed.

NASA has a medical science division that supports missions and conducts research, a medical operations branch responsible for operational space medicine (OSM) for crews, and a flight medicine clinic (FMC) for astronauts and former astronauts.

space motion sickness (SMS) Alternative name for *space sickness, the feeling of nausea experienced by many astronauts during their first few days in space.

Space Physics Data Facility (SPDF) A facility at NASA's *Goddard Space Flight Center that enhances the scientific return from NASA programmes in the space physics sciences. These include solar physics, cosmic and heliospheric physics, magnetospheric physics, and ionospheric, thermospheric, and mesospheric physics. SPDF ensures that NASA's space physics is accessible, communicates significant discoveries from the space physics programmes, and performs space physics research. Many of these programmes were formerly handled by the *National Space Science Data Center (NSSDC), but were taken over by SPDF in 1992.

http://spdf.gsfc.nasa.gov//spdf.html Designed to provide researchers with space physics data, this site has lots of information about space weather and how it is monitored. The Space Physics Data Facility's 'space weather resource page' makes it easy to pick out missions targeted at specific solar, near-Earth space, or atmospheric phenomena.

spaceport A spacecraft launching location, for example the *Kennedy Space Center (KSC). Towns around launching areas are often called spaceports, for example Titusville, Florida, near KSC is referred to as a spaceport.

space probe Any instrumented object sent beyond Earth to collect data from other parts of the Solar System and from deep space. The first probe was the Soviet *Luna 1*, which flew past the Moon in 1959. The first successful planetary probe was the US *Mariner 2*, which flew past Venus in 1962. The first probe to orbit another planet was *Mariner 9*, which went into orbit around

Mars in November 1971. The first space probe to leave the Solar System was *Pioneer 10* in 1983.

space programme The systematic exploration and utilization of space, such as the development and launch of a *space probe or use of the *space shuttle.

http://www.solarviews.com/solar/eng/history.htm Who was the first woman astronaut? Find out at this site, which also contains information on the history of rocketry and chronologies of exploration by the USA, Russia, Japan, and Europe.

space qualified A term indicating that a component is satisfactorily manufactured to a specified standard and can be launched as part of a spacecraft. In the USA, items are certified as space qualified by NASA, which approves suppliers for its contractors. Before export licenses can be given, US space-qualified components must be approved by either the Commerce Department or State Department.

space race The intense competition between the USA and USSR to become the leader in space activities from the late 1940s until the USSR's collapse in 1991. Rivalry began after World War II when both the USA and the USSR set up programmes for German rocket scientists who had helped develop the V-2 military rocket in the 1940s. The race was officially initiated with the Soviet launch of its *Sputnik* spacecraft in 1958. The USSR achieved the early triumphs in space, including the first spacecraft to reach the Moon (*Luna 2* in 1959), the first person in space (Yuri *Gagarin in 1961), and the first space walk (Aleksei *Leonov in 1965). The greatest prize, however, went to the USA for landing the first astronauts on the Moon (Neil *Armstrong and Buzz *Aldrin in 1968). Although the Soviets continued in their successes, such as achieving residential endurance records on the *Mir* space station and the robotic return of lunar surface samples, the demise of communism (leading to the break up of the USSR in 1991 and accompanying economic collapse) devastated the space programme. The newly independent Russia subsequently received financial support from the West for its continuation of the programme. The space race ended with the emergence of joint projects, such as the **Apollo–Soyuz* Test Project and the **International Space Station*.

Space Radar Laboratory (SRL) A suite of instruments mounted in the payload bay of the **Endeavour* space shuttle and operated by the crew on two topographical Earth-mapping missions, SRL-1 and SRL-2, launched as STS-59 on 9 April 1994 and STS-68 on 30 September 1994. SRL contained the US *spaceborne imaging radar SIR-C, and the German–Italian X-band *synthetic aperture radar. The 11-day missions each imaged over 400 target sites covering the equivalent of 20% of the Earth's surface. The topographical elevation images were derived by measuring the reflection of radar pulses (emitted from the shuttle) back from the Earth's surface.

Space Science Board (SSB) A board of the National Academy of Sciences in the USA that investigates the scientific aspects of human space exploration. It was established in 1958 approximately six months after the USSR launched its *Sputnik* spacecraft. The SSB advises NASA, the US Department of Defense, and other bodies on the aspects and problems of crewed space flights and space stations, and interplanetary probes. The organization did much to convince NASA to focus on the planet Mars as the main element of the US planetary exploration programme.

Space Science Data Operations Office (SSDOO) A NASA office responsible for the project management of certain missions and for the development and operation of data and information systems that support the processing, management, archiving, and distribution of data concerning space physics, astrophysics, and planetary missions. SSDOO is part of NASA's Space Sciences Directorate that conducts space-science projects and research at the *Goddard Space Flight Center.

Space Services Inc. (SSI) A US company in Houston, Texas, that in 1982 conducted the first successful privately-funded commercial rocket launch with its *Conestoga rocket. SSI was established by a Texas group that included the former astronaut Deke Slayton, who was its president until his death in 1993. The company constructed its launch pad on Matagorda Island near Corpus Christi, Texas, but its first rocket, the liquid-fuelled Percheron, exploded during a test firing of its engine in 1981. The Conestoga was then developed using solid fuel and made a 10-minute sub-orbital flight in September 1982. It failed, however, to launch NASA's Commercial Experiment Transporter (COMET) in 1995. The company had been bought in 1990 by EER Systems which was then merged in 2003 into Government Services Inc.

SpaceShipOne The three-person spacecraft which on 4 October 2004 won the *Ansari X Prize by becoming the first privately manned vehicle to exceed an altitude of 100 km twice within a fortnight. The craft, built by Scaled Composites of Mojave, California, was dropped from a White Knight carrier aircraft and used its own rocket engine to boost itself on a sub-orbital trajectory, landing back on a runway at Mojave Airport. The first flight, with pilot Mike Melvill, took place on 29 September 2004 and reached 102 km while the second, on 4 October 2004 with pilot Brian Binnie, reached 112 km. On each flight, only the pilot and ballast representing two passengers was on board.

http://www.scaled.com/projects/tierone/041004_spaceshipone_x-prize_flight_2.html

space shuttle (in full: space shuttle orbiter) A reusable crewed spacecraft developed by NASA to reduce the cost of using space for commercial, scientific, and military purposes. The orbiter, the part that goes into space, is 37.2 m long and weighs 68 tonnes. The first shuttle, **Columbia*, was launched on 12 April 1981. At the end of each mission, the space shuttle can be flown back to Earth to land on a special runway 4.5 km long and 91 m wide, and is then available for reuse.

Four space shuttles were built initially: *Columbia*, **Challenger*, first launched on 4 April 1983, **Discovery* on 30 August 1984, and **Atlantis* on 3 October 1985. The rocket boosters were redesigned after *Challenger* was destroyed in a mid-air explosion in 1986, killing its seven astronauts. Flights resumed in September 1988. A replacement orbiter, **Endeavour*, was built, making its first flight in May 1992. The break-up of *Columbia* on re-entry on 1 February 2003 killed all seven astronauts aboard. NASA then halted all flights and established the

Columbia Accident Investigation Board which recommended 15 safety changes. The first renewed launch of a shuttle, *Discovery*, took place on 26 July 2005. When the flights were halted, there had been 113 shuttle flights.

The USSR produced **Buran*, a shuttle of similar size and appearance to the US one. It was launched on its first and only flight, without a crew, by the *Energiya rocket on 15 November 1988. In June 2001, it was announced that *Buran* was being prepared for relaunch to transport space tourists.

Although most of the shuttle's cargoes are uncrewed, two to eight crew members occupy the orbiter's nose section, and missions last up to 18 days, although 30-day missions may be possible with the addition of more fuel cells to provide power. In its cargo bay, the orbiter can carry up to 29 tonnes of satellites, scientific equipment, **Spacelab*, or military payloads. At launch, the shuttle's three main engines are fed with liquid fuel from a cylindrical tank attached to the orbiter; this tank is discarded shortly before the shuttle reaches orbit. Two additional solid-fuel boosters provide the main thrust for launch, but are jettisoned after two minutes.

The aftermath of the *Columbia* disaster was an increased focus on astronaut safety and a decrease in the number of missions that space shuttles would have to fly. One high-profile programme that was abandoned on safety grounds was the proposed 2006 mission to upgrade and carry out maintenance on the *Hubble Space Telescope*.

http://spaceflight.nasa.gov/shuttle/index.html Official NASA site for all shuttle missions. There is extensive technical and non-technical information, both textual and graphic. Questions can be sent to shuttle crew members during missions. There is an extensive list of Frequently Asked Questions. There are helpful links to related sites and even a plain-English explanation of NASA's bewildering jargon and acronyms.

http://southport.jpl.nasa.gov/pic.html Small but spectacular collection of radar images taken from the space shuttle *Endeavour* in 1994 and offered by NASA. The collection also includes 3D images, videos, and animations. The high-technology enthusiast will also find ample details on the imaging radar system used for the project.

space shuttle main engine (SSME) One of the three main engines that power a space shuttle into orbit. Each reusable engine weighs around 3 150 kg and is 4.3 m long and 2.28 m in diameter. The engines burn for 8.5 minutes during lift-off and ascent, consuming more than 1.9 million L of fuel (liquid hydrogen and liquid oxygen) in the immense external tank. The engines shut down just before the shuttle goes into orbit. The SSME was developed at NASA's *Marshall Space Flight Center, where it was modified in 1988 and 1995. A further upgrade was completed in 2001.

space sickness (or space adaptation syndrome) The feeling of nausea, sometimes accompanied by vomiting, experienced by about 40% of all astronauts during their first few days in space. It is akin to travel sickness, and is thought to be caused by confusion of the body's balancing mechanism, located in the inner ear, by weightlessness. The sensation passes after a few days as the body adapts.

space simulator A module or laboratory on Earth that artificially imitates the conditions found in space. Devices can replicate *microgravity or simulate other conditions in a spacecraft. Microgravity simulators include the *drop tower and NASA's *Sonny Carter Training Facility. Command module and lunar module simulators were used by the *Apollo* astronauts prior to their missions in the 1960s and 1970s.

space station Any large structure designed for human occupation in space for extended periods of time. Space stations are used for carrying out astronomical observations and surveys of Earth, as well as for biological studies and the processing of materials in weightlessness. The first space station was the Soviet **Salyut 1*, launched in 1971. In 1973, NASA launched **Skylab*. The core of the Soviet space station **Mir* was launched in 1986. In 1998, the first component of the **International Space Station*, being constructed by the USA, Russia, and 14 other nations, was launched. The first crew arrived at *ISS* in November 2000 under the command of NASA astronaut Bill Shepherd. By 17 September 2001, six habitable modules had been added to the *ISS*.

spacesuit A protective suit worn by astronauts and cosmonauts during launch and in space. During launch it acts as a high-altitude pressure suit and also guards against depressurization in the vacuum of space. A spacesuit has to be worn outside the spacecraft during *extravehicular activity. It provides an insulated, air-conditioned cocoon in which a person can live and work for hours at a time while outside the spacecraft. Inside the suit is a cooling garment that keeps the body at a comfortable temperature even during vigorous work. The suit provides air to breathe and removes exhaled carbon dioxide and moisture. The suit's outer layers insulate the occupant from the extremes of hot and cold in space (−150 °C in the shade to 180 °C in sunlight), and from the impact of small meteorites.

types of spacesuit

Astronauts and cosmonauts flying *Gemini*, *Mercury*, *Vostok*, and *Voskhod* spacecraft 1961–6 wore suits based largely on pressure suits developed for high-performance jet aircraft. NASA's *Apollo* spacesuit was called the state-of-the-art suit. The advanced design suits used on the Moon had 21 layers that successfully protected astronauts against temperature, radiation, and the absence of pressure. 60 suits were produced at a cost of $90 million. NASA chose the name of extravehicular mobility unit (EMU) for the complete *Apollo* lunar suit system, which included a portable life-support system (PLLS). The suits had a more reliable (British-made) cooling system than the earlier *Mercury* and *Gemini* spacesuits because of the greater work required on the lunar surface. The astronauts' lunar gloves were made of a metal-woven fabric with strong nylon-covered fingertips, and later models added a slip-resistant silicon coating. Lunar boots had several layers, including one of Teflon-coated *Beta cloth, and the soles were of moulded silicon rubber. The first hard boots were later replaced by a more flexible version.

The EMUs worn by *Apollo* command module pilots were lighter versions with fewer layers. For the *Apollo–Soyuz* Test Project in 1975, astronauts wore a modified *Apollo* spacesuit, with a new cover layer added.

After the *Challenger* accident in 1986, NASA created a new pressure suit to allow astronauts to escape by parachute in the event of an emergency during launch or before landing but only at certain altitudes. The bright-orange one-piece suit, resembling but not the same as the suit worn by the STS-1–4 space shuttle crews, has an oxygen supply, parachute, life raft, life preserver, and additional survival equipment.

For space walks from the space shuttle, astronauts wear a two-piece EMU which connects at the waist and which incorporates a PLSS backpack and a chest-mounted emergency oxygen supply and control panel. The suit resembles the *Apollo* Moon-walking suit in design but incorporates many improvements to aid mobility. Early shuttle astronauts wore manned manoeuvring units (MMUs), large backpacks powered by pressurized nitrogen gas. These were later replaced by smaller simplified aid for extravehicular activity rescue (SAFER) units, which are safety devices to ensure that if an astronaut loses tether contact with the orbiter, he or she can manoeuvre back to the ship.

Space Task Group (STG) A task force assembled by NASA in its first week of existence. It was set up to develop the *Mercury* spacecraft project. The task group at *Langley Research Center in Hampton, Virginia, consisted of 35 engineers. Specifications for the *Mercury* spacecraft were drawn up in November 1958 and NASA announced the project on 17 December (the US aviation pioneers Orville and Wilbur Wright made their first flight on the same date in 1903). By 1961, the STG had some 750 employees, and it was transferred to Houston, Texas, as the Manned Spacecraft Center, which later became the *Johnson Space Center.

***Space Technology Research Vehicle* (*STRV*)** Either of two microsatellites developed by the UK Defence Research Agency at Farnborough, Hampshire, to test various spacecraft systems and technologies. Each satellite weighed 50 kg. They were launched in June 1994 by the *Ariane rocket from Kourou, French Guiana. The *STRVs* carried instruments to measure radiation, particles, and electrostatic charging. Aboard one vehicle was a *cold-ion detector (CID) to monitor the activity of ions and their effects on electronic systems, as the spacecraft passed through the Earth's radiation belts.

Space Telescope Science Institute (STScI) The NASA establishment founded in 1981 at Johns Hopkins University, Baltimore, Maryland, which operates the *Hubble Space Telescope* on behalf of its own researchers and other astronomers worldwide. STScI is developing the **James Webb Space Telescope*, *Hubble*'s successor, which it will also operate when it is launched.

space tracking and data network (STDN) NASA's early system of tracking satellites in Earth orbit. STDN, operated by the *Goddard Space Flight Center (GSFC), established 14 telecommunications ground stations around the world for tracking, although this covered only 14% of a satellite's orbit. Data collected by the stations was relayed to Goddard. STDN was superseded by the *Tracking and Data-Relay Satellite (TDRS) system that provides a coverage of 85–100%.

space transportation system (STS) NASA's official name for its space-shuttle system. Each shuttle flight is given an STS number. The first orbital shuttle flight, by *Columbia* in April 1981, was designated STS-1, and the *Endeavour* mission to the **International Space Station* in April 2001 was STS-100.

space travel Any voyage in space. Space travel normally refers to crewed journeys and is often used to describe space trips by ordinary citizens. The first space tourist was Dennis Tito, a 60-year-old financier from California, who paid US$20 million to the Russians to visit the *International Space Station* during April 2001. Civilians have visited space since the 1980s, including three US Congressmen, a Japanese journalist, a Russian bureaucrat, and a Ukrainian. NASA's 'citizen-passenger programme' ended after Christa McAuliffe, a teacher, died in the *Challenger* space shuttle disaster in 1986.

Among companies developing plans for hotels in space are DaimlerChrysler, Hilton, and Shimizu. In 1998, Buzz *Aldrin, the second person to walk on the Moon (1969), established the non-profit ShareSpace Foundation to promote space tourism. Space Adventures, the US company that helped Dennis Tito broker his Russian flight, has already booked more than 100 reservations for sub-orbital flights, with the first scheduled for 2005 (although the vehicle has yet to be built).

space walk Alternative term for *extravehicular activity, the movement and work carried out by an astronaut outside a spacecraft.

SPACEWARN A monthly bulletin published by NASA's *National Space Science Data Center to provide information on satellites and space probes. Issued on the first day of each month, the bulletin has a listing of the month's launches as well as details of those from the previous month. *SPACEWARN* was first published in 1991.

http://nssdc.gsfc.nasa.gov/spacewarn/ Monthly bulletins (dating back to 1991) containing details of rocket launchings of satellites and space probes. The announcements include spacecraft names, launch dates and times, and interesting comments about the launch and the spacecraft.

Spartan 201 A series of small NASA satellites that study the Sun, being deployed and retrieved by a space shuttle. Spartan 201-1 flew in April 1993, Spartan 201-2 in September 1994, Spartan 201-3 in September 1995, and Spartan 201-5 in November 1998. In November 1997, Spartan 201-4 failed to deploy. Each of the satellites uses telescopes to study the Sun's extremely hot *corona and its expansion into the *solar wind. The data provides new understanding about the source of energy that heats the corona and accelerates the solar wind particles.

SPC Abbreviation for *signal processing centre.

specific impulse (ISP) The amount of rocket thrust produced by a certain weight of fuel. It is calculated by dividing the rocket's thrust by its flow of fuel. This is expressed in seconds, because the thrust is measured in pounds and the fuel flow in pounds/seconds, so the pounds cancel each other out. The ISP of liquid hydrogen is 420 seconds. A higher ISP means a better performing rocket.

spectrometer In physics and astronomy, instrument used to study the composition of light emitted by a source. The range, or *spectrum, of wavelengths emitted by a source depends upon its constituent elements, and may be used to determine its chemical composition.

The simpler forms of spectrometer analyse only visible light. A **collimator** receives the incoming rays and produces a parallel beam, which is then split into a spectrum by either a diffraction grating or a prism mounted on a turntable. As the turntable is rotated each of the constituent colours of the beam may be seen through a **telescope**, and the angle at which each has been deviated may be measured on a circular scale. From this information the wavelengths of the colours of light can be calculated. Spectrometers are used in astronomy to study the electromagnetic radiation emitted by stars and other celestial bodies. The spectral information gained may be used to determine their chemical composition, or to measure the *redshift of light associated with the expansion of the universe and thereby calculate the speed with which distant stars are moving away from the Earth.

spectroscopic binary A *binary star in which two stars are so close together that they cannot be seen separately, but their individual light spectra can be distinguished by a spectroscope.

As the two stars revolve around their mutual centre of mass, they alternately approach and recede from the observer, resulting in a periodic Doppler shift (*see* DOPPLER EFFECT) in the lines of their spectra. In about one case in six, the component stars are sufficiently similar in brightness for the spectra of both to appear, giving a double-line spectroscopic binary. The line-of-sight velocity of the brighter star, or of each star in a double-line spectroscopic binary, is measured from the Doppler shift at various stages of the orbital period. Analysis of these velocity curves gives a lower limit of the combined mass of a single-line binary, or of the mass of each component of a double-line binary.

spectrum (plural: spectra) The pattern of frequencies or wavelengths obtained when electromagnetic radiations are separated into their constituent parts. Visible light is part of the *electromagnetic spectrum and most sources emit waves over a range of wavelengths that can be broken up or 'dispersed'; white light can be separated (for example, using a triangular prism) into red, orange, yellow, green, blue, indigo, and violet. The visible spectrum was first studied by English physicist Isaac Newton, who showed in 1666 how white light could be broken up into different colours.

There are many types of spectra, both emission and absorption, for radiation and particles, used in spectroscopy. An incandescent body gives rise to a **continuous spectrum** where the dispersed radiation is distributed uninterruptedly over a range of wavelengths. A gaseous element gives a **line spectrum**—one or more bright discrete lines at characteristic wavelengths. Molecular gases give **band spectra** in which there are groups of closely-packed lines. In an **absorption spectrum** dark lines or spaces replace the characteristic bright lines of the absorbing medium. The **mass spectrum** of an element is obtained from a mass spectrometer and shows the relative proportions of its constituent isotopes.

speed brake Any device used to slow a spacecraft. The space shuttle's speed brake is made up of two identical rudder halves that can be opened into a wedge to slow it for landing.

speed of light The speed at which light and other *electromagnetic waves travel in a vacuum. Its value is 299 792 458 m per second but for most calculations 3×10^8 m s^{-1} (300 million metres per second) suffices. In glass the speed of light is two-thirds of its speed in air, about 200 million metres per second. The speed of light is the highest speed possible, according to the theory of *relativity, and its value is independent of the motion of its source and of the observer. It is impossible to accelerate any material body to this speed because it would require an infinite amount of energy.

spinning solid upper stage A former name for the *payload assist module (PAM), a rocket that establishes a satellite's orbit.

spin stabilization The act of spinning a spacecraft in order to stabilize its attitude in orbit. The vehicle is normally spun around its ***X** axis, the longest one.

spin table A device on a space vehicle used to deploy a geosynchronous or geostationary satellite. On the space shuttle, the satellite is located in the cargo bay on the large round spin table that turns at 50 revolutions a minute. The shuttle is positioned at the proper attitude, the satellite is spun, pyrotechnic charges release clamps holding it on the table, and four coiled springs then boost the satellite into space.

spiral galaxy One of the main classes of galaxy in the Hubble classification comprising up to 30% of known galaxies. Spiral galaxies are characterized by a central bulge surrounded by a flattened disc containing (normally) two spiral arms composed of hot young stars and clouds of dust and gas. In about half of spiral galaxies (barred spirals) the arms originate at the ends of a bar across the central bulge. The bar is not a rigid object but consists of stars in motion about the centre of the galaxy.

Spirit One of NASA's twin *Mars Exploration rovers, launched on 10 June 2003. It landed on 4 January 2004 in the crater Gusev, which was once filled with a lake.

Spitzer, Lyman (1914–97) US astrophysicist. From 1947 to 1979, he was chairman of the Department of Astrophysical Sciences at Princeton University, New Jersey, and director of the Princeton Observatory. Under Spitzer's direction, a group of Princeton scientists developed the 81-cm telescope for the **Copernicus* ultraviolet satellite. Spitzer later steered the development and refurbishment of the **Hubble Space Telescope*.

***Spitzer Space Telescope* (*SST*)** A NASA space observatory launched on 25 August 2003 and first named the Space Infrared Telescope Facility. A major step in its *Origins Program, *SST* is a cryogenically-cooled (using liquid helium) infrared observatory that can investigate both the Solar System and distant regions of the Universe. Offering an improved capability over existing observatories, it consists of an 85-cm telescope and three science instruments for imaging and spectroscopy. It has a projected lifespan of up to five years or more. The *SST* Science Center is located at the *California Institute of Technology.

http://spitzer.caltech.edu/ Documents the history, design, and results of the *Spitzer Space Telescope*. This large site has interesting tutorials, science and technology information, and a spacecraft image gallery.

splashdown The landing of a space vehicle in an ocean. Splashdowns in the Pacific Ocean, with recovery by the US Navy, have been used by NASA since space exploration began, including the return of the *Apollo 11* crew who made the first lunar landing.

SPOT Acronym for **Satellite Pour l'Observation de la Terre*.

Spring, Sherwood Clark (1944–) US astronaut. He was responsible for deploying three communications satellites from the space shuttle *Atlantis*, launched in November 1985. He was selected as an astronaut in 1980 and retired from NASA in 1988. He then directed the Army Space Program Office in Washington, DC. Spring retired from the US Army in 1994 and became an aerospace consultant with The Application Science Corporation (TASC) in Reston, Virginia.

Springer, Robert Clyde (1942–) US astronaut. His two flights included the March 1989 mission of the space shuttle *Discovery* to deploy a *Tracking and Data-Relay Satellite. Springer was selected as an astronaut in 1980 and left NASA in 1990.

SPS Abbrevation for either *service propulsion system or *solar power satellite.

***Sputnik* (Russian: 'fellow traveller')** A series of ten Soviet Earth-orbiting satellites launched from 1957 by R-7 rockets (see illus. on p. 322). *Sputnik 1* was the first artificial satellite, launched on 4 October 1957. It weighed 84 kg, with a 58 cm diameter, and carried only a simple radio transmitter, which allowed scientists to track the spacecraft as it orbited Earth. It burned up in the atmosphere 92 days later. The *Sputnik* research team was headed by Sergei *Korolev. *Sputniks* were superseded in the early 1960s by the **Cosmos* series.

Sputnik 2, launched on 3 November 1957, weighed about 500 kg and had on board the dog Laika, the first living creature in space. Later *Sputniks* were test flights of the **Vostok* spacecraft.

SRTM Abbreviation for *Shuttle Radar Topography Mission.

SSB Abbreviation for *Space Science Board.

SSDOO Abbreviation for *Space Science Data Operations Office.

SSI Abbreviation for either *Space Services Inc. or *solid-state imaging subsystem.

SSME Abbreviation for *space shuttle main engine.

Image © Research Machies Plc)

The Semiorka R-7 rocket used to launch the Soviet *Sputnik* satellites.

SSP Abbreviation for *surface science package.

STA Acronym for *shuttle training aircraft.

stabilization control system (SCS) A NASA system introduced for the *Mercury* spacecraft to point its space capsule in the correct direction.

The SCS controlled the spacecraft's attitude by thrusters consisting of jets of hydrogen peroxide. An astronaut could operate the system with a control stick or put it into the automatic mode that could find the horizon using gyroscopes and infrared heat detectors.

stabilizing fin (or stabilizer) A vertical or horizontal surface used to steady the path of a craft flying in the Earth's atmosphere. Any vehicle flying in the atmosphere, including a rocket or a spaceplane, requires stabilization and uses some form of fin or fins.

Stafford, Thomas Patten (1930–) US astronaut. His flights include commanding the *Apollo 10* mission in May 1969 that orbited the Moon, preparing the way for the *Apollo 11* landing that followed two months later. He was also commander of the *Apollo–Soyuz* Test Project in 1975. Stafford was selected as an astronaut in 1962.

He left NASA in 1975 to assume command of the Air Force Flight Test Center. In 1990 and 1991 Stafford headed a Synthesis Group, organized by NASA and the White House, to recommend future strategy in space exploration.

stage zero (T-zero) A term describing the beginning of the first stage of the propulsion process of a rocket or launcher. T-0 is often used in countdowns to indicate the moment of lift-off. Some rockets take off at engine ignition, as in the case of hypergolic propellants which ignite spontaneously on contact, while some lift off a few seconds after engine ignition, allowing the engine to build up to full thrust; in each case this is termed T-0. However, T-0 is also sometimes and confusingly used to mark the point of engine ignition, and actual lift-off may therefore occur at T + a number of seconds.

staging The moment when one rocket stage has finished burning and the next stage begins. Staging of the space shuttle's solid rocket boosters occurs two minutes after lift-off.

standard time The four standard times officially used in the USA to record space shuttle launches and landings. Those at the *Kennedy Space Center in Florida occur during Eastern Standard Time (EST) from 2 a.m. on the last Sunday in October until 2 a.m. on the first Sunday in April. They happen during Eastern Daylight Time (EDT) the rest of the year. Landings at *Edwards Air Force Base in California occur during Pacific Standard Time (PST) from 2 a.m. on the last Sunday in October until 2 a.m. the first Sunday in April. The rest of the year they happen during Pacific Daylight Time (PDT).

EST is five hours behind *Universal Time Coordinated (UTC); PST is three hours behind EST and eight hours behind UTC.

star A luminous globe of gas, mainly hydrogen and helium, which produces its own heat and light by nuclear reactions. Although stars shine for a very long time—many billions of years—they change in appearance at different stages in their lives (they are said to have a 'life cycle'). Stars seen at night belong to our galaxy, the Milky Way. The Sun is the nearest star to Earth; other stars in the Milky Way are large distances away.

The smallest mass possible for a star is about 8% that of the Sun (80 times that of *Jupiter), otherwise nuclear reactions do not occur. Objects with less than this critical mass shine only dimly, and are termed **brown dwarfs**.

origin

Stars are born when nebulae (giant clouds of dust and gas) contract under the influence of gravity. These clouds consist mainly of hydrogen and helium, with traces of other elements and dust grains. The temperature and pressure in its core rises as the star grows smaller and denser.

At first the temperature of the star scarcely rises, as dust grains radiate away much of the heat, but as it grows denser less of the heat generated can escape, and it gradually warms up. At about 10 million °C the temperature is hot enough for a nuclear reaction to begin, and hydrogen nuclei fuse to form helium nuclei; vast amounts of energy are released, contraction stops, and the star begins to shine.

main-sequence stars

Stars at this stage are called main-sequence stars. When all the hydrogen at the core of a main-sequence star has been converted into helium, the star swells to become a **red giant**, about 100 times its previous size and with a cooler, redder surface.

white dwarfs
What happens next depends on the mass of the star. If this is less than 1.2 that of the Sun, the star's outer layers drift off into space to form a planetary nebula, and its core collapses in on itself to form a small and very dense body called a white dwarf. Eventually the white dwarf fades away, leaving a non-luminous **dark body**.

supernovae
If the mass is greater than about eight times that of the Sun, the star does not end as a white dwarf but passes through its life cycle quickly, becoming a red *supergiant. The star eventually explodes into a brilliant **supernova**. Part of the core remaining after the explosion may collapse to form a small superdense star, consisting almost entirely of neutrons and therefore called a **neutron star**. Neutron stars, also called *pulsars, spin very quickly, giving off pulses of radio waves.

black holes
If the collapsing core of the supernova has a mass more than three times that of the Sun it does not form a neutron star; instead it forms a *black hole, a region so dense that its gravity not only draws in all nearby matter but also all radiation, including its own light, as the velocity of escape from its surface exceeds that of light.

starburst galaxy A spiral galaxy that appears unusually bright in the infrared part of the spectrum due to a recent burst of star formation, possibly triggered by the gravitational influence of a nearby companion galaxy.

Star City A Russian compound that is home to the *Gagarin Cosmonaut Training Centre. It is located in the countryside northeast of Moscow. Cosmonauts and their families live at Star City, and NASA has its own facilities and housing for its astronauts who train there. Star City was originally a highly restricted military base, and military sentries still guard its entrance. Since the fall of communism and the collapse of the USSR in 1991, however, there are conducted tours of the centre and members of the public are able to experience mock astronaut training in the simulators.

star cluster A group of related stars, usually held together by gravity. Members of a star cluster are thought to form together from one large cloud of gas in space.

Open clusters such as the Pleiades contain from a dozen to many hundreds of young stars, loosely scattered over several light years. Globular clusters are larger and much more densely packed, containing perhaps 10 000–1 000 000 stars.

Stardust A NASA probe to obtain a sample of dust and gas from the head of a comet. Launched on 7 February 1999, *Stardust* first flew past asteroid Annefrank in November 2002 before passing through the head of Comet Wild 2 on 2 January 2004, within 240 km of its nucleus, photographing its irregular surface and capturing cometary particles on the surface of a paddle coated with a low-density glass foam called aerogel. *Stardust* is due to return to Earth with its samples in January 2006.

http://stardust.jpl.nasa.gov/ News updates on the *Stardust* mission with photographs and animations from its encounter with Comet Wild 2.

star scanner A spacecraft instrument that scans the sky for known bright stars in order to control where the vehicle is pointing. On-board software systems use this to determine the orientation. When stars are not visible enough, the scanner can be set for guidance by its gyroscopes.

Starsem A joint European-Russian company established in 1996 to market commercial launches of **Soyuz* spacecraft. Starsem oversees the marketing and management of commercial launch services, including mission preparations at the *Baikonur Cosmodrome in Kazakhstan, payload deployment, and satellite tracking. Its headquarters are in Paris, France.

The company launched the European Space Agency's (ESA's) *Mars Express probe on 2 June 2003. In March 2004 it signed a contract with *Arianespace and the ESA to launch two experimental *Galileo* satellites on *Soyuz* rockets, the first by 2005.

Starsem is composed of four space organizations: the European Aeronautic Defence and Space Company (EADS), Arianespace, the Russian space agency *Rosaviakosmos, and the *Samara Space Centre in Russia.

star tracker A system used as part of a space shuttle's navigation system. Its two units are located outside the crew compartment, forward and to the left of the commander's window. It is normally mounted in the aft crew station, and only moved to near the commander's station for ascent and deorbit thrusting periods. The star tracker aligns the *inertia measurement unit (IMU) approximately every 12 hours by measuring the line-of-sight vector to at least two stars. If the IMU alignment error is more than 1.4°, the star tracker is not able to correct it until the *crewman optical alignment sight corrects it to within 1.4°.

The star tracker system also provides angular data from the shuttle to a target, as for close or rendezvous operations with a target.

Each of the units has a door that protects it during the shuttle's ascent and re-entry but is open when the shuttle is in orbit.

state vector The readings of the velocity and three-dimensional position of a spacecraft at a specific time. The data assist in navigation.

State vectors can also be used to calculate Keplerian elements, which describe the shape and size of a spacecraft's orbital ellipse.

static firing The test-firing of a rocket engine on the ground. This can be done vertically or horizontally, with the rocket attached to the ground.

station keeping (or keeping station) The act of keeping a spacecraft in its proper attitude by using its *thrusters. Station keeping is required to keep a vehicle in geostationary orbit, so that it seems to be motionless above the Earth.

STDN Abbreviation for either *Space Flight Tracking and Data Network or *Space Tracking and Data Network.

steering thruster The *thruster used to steer a spacecraft in a certain direction. The Ariane 5 rocket had six small steering thrusters on its upper stage to keep it aimed correctly during the launch.

Two *Mir* cosmonauts had to end their extravehicular activity on 6 April 1998 when a steering thruster, used to point the space station towards the Sun, ran out of fuel, and they had to return to activate a replacement engine.

Stewart, Robert Lee (1942–) US astronaut. During the *Challenger* space shuttle mission in February 1984, he and Bruce *McCandless conducted the first untethered *extravehicular activity. He also flew on the first mission of *Atlantis*, in October 1985. Stewart was selected as an astronaut in 1979 and left NASA in 1986.

STG Abbreviation for *Space Task Group.

Still-Kilrain, Susan (born Susan Still) (1961–) US astronaut. Her two space shuttle flights were as the pilot of the **Microgravity Science Laboratory* in April and July 1997, both aboard *Columbia*. The first flight was aborted after more than 95 hours because of problems with a fuel cell. The second lasted for more than 376 hours and involved experiments on materials and combustion in microgravity. Still-Kilrain was selected as an astronaut in 1995. After her flights, she was a specialist for the space shuttle in the Office of Legislative Affairs at NASA's headquarters in Washington, DC. She retired from NASA in December 2002 to return to the US Navy.

stinger A retrieving device used by NASA space shuttle astronauts during an extravehicular activity to recover a satellite in need of maintenance work or repairs. The astronaut floats out with the stinger and inserts it into the satellite's rocket nozzle, then releases a lever to expand toggle fingers that engage the satellite. The stinger is then shortened with a crank until its ring presses tightly against the satellite. An astronaut in the shuttle manoeuvres the craft's robotic Canadarm until it attaches to a *grappling pin on the stinger and brings the satellite to the shuttle for the necessary work.

storage subsystem A spacecraft's on-board computer memory system. Data is held for a specific amount of time before it is overwritten. This enables a ground station to command a spacecraft to replay a specific segment if reception was interrupted by meteorological conditions or other station problems.

stowage The placing of personal, scientific, and other items in *modular lockers, as on space stations and space shuttles. Crew members on a shuttle carry out stowage activities before re-entry into the Earth's atmosphere and landing.

Stratospheric Aerosol and Gas Experiment (SAGE) A series of three Sun photometer instruments launched by NASA to investigate aerosol, ozone, and nitrogen dioxide levels in the Earth's atmosphere. SAGE I was launched in February 1979 and collected solar radiance data during a period of nearly three years. This was combined with meteorological data from the *National Oceanic and Atmospheric Administration to yield altitude profiles of aerosol extinction and ozone and nitrogen dioxide concentrations. SAGE II was launched in October 1984 to investigate the same targets and water vapour. It revealed that the eruption of Mt Pinatubo in the Philippines caused a loss of nitrogen dioxide in the stratosphere, and it increased knowledge of the Antarctic ozone hole. Sage III, with expanded wavelength coverage, was scheduled for launch in December 2001 aboard a Russian spacecraft and in 2004 to the **International Space Station.*

Strekalov, Gennady Mikhailovich (1940–) Russian cosmonaut. He served as flight engineer on four flights, including *Soyuz* T-11, a joint Soviet–Indian mission in April 1984 that carried out scientific research aboard the *Salyut 7* space station; and on *Mir* August–December 1990 and March–July 1995. He logged more than 153 days in space. Strekalov then joined the *Energiya Rocket and Space Complex, based at the mission control centre, working on flights of spacecraft belonging to the Academy of Sciences. He was selected as a cosmonaut in 1974.

Strelka One of two dogs launched aboard the Soviet spacecraft *Korabl Sputnik 2* in 1960. *See* BELKA AND STRELKA.

structural model The full-sized non-flight structural prototype of a spacecraft, built to assist in its final design and to aid decisions as to the location of equipment and instruments. It is superseded by the *engineering model in the development of the flight project.

STRV Abbreviation for *Space Technology Research Vehicle.

STS Abbreviation for *space transportation system.

STScI Acronym for *Space Telescope Science Institute.

Sturckow, Frederick Wilford (1961–) US astronaut. He flew with the first **International Space Station (ISS)* assembly mission in December 1998, on board the space shuttle *Endeavour*. In August 2001, he flew on the *Discovery* mission to deliver the third resident crew to the *ISS*. He was assigned as crew commander of a 2004 flight that was postponed after the *Columbia* tragedy the previous year. He became deputy of the Shuttle Operations Branch in the Astronaut Office and leads the Kennedy Space Center Operations Support team. Sturckow was selected as an astronaut in 1994.

STV Abbreviation for *solar-thermal-vacuum chamber.

***STW/China* (or *DHF-2*)** China's first experimental communications satellite, launched on 8 April 1984, making China the fifth nation to develop, build, and launch a geostationary satellite.

***Submillimetre Wave Astronomy Satellite* (*SWAS*)** A NASA satellite launched on 6 December 1998 to study the chemical composition of interstellar gas clouds and the processes that lead to the formation of stars and planets. It carries a telescope with an elliptical mirror 0.55 m by 0.71 m and detectors which are sensitive to the submillimetre spectral lines emitted by water molecules, oxygen, carbon, and carbon monoxide. *SWAS* used its

S

instruments to map giant molecular clouds in the Galaxy's spiral arms, the sites of current star formation.

sub-orbital The flight or path of a rocket or spacecraft that does not follow Earth orbit. The first US astronaut travelled on a sub-orbital flight: Alan Shepard was launched in May 1961 to an altitude of more than 161 km, and after a 15-minute flight landed 483 km from Cape Canaveral, Florida.

subsystem In computing, hardware and/or software that performs a specific function within a larger system. Silicon Graphics, for example, uses subsystems to perform the many calculations needed for computer animation.

On spacecraft, subsystems control such vital areas as communications, electrical power, and the life-support system on crewed flights.

Suisei (Japanese: 'comet') Japan's second deep-space probe. The first was Sukigake, identical except for the scientific instruments on board. Launched in August 1985 to carry out a fly-by of Halley's Comet, Suisei had its closest approach to the comet in March 1986, detecting cometary water, carbon monoxide, and carbon dioxide ions.

The mission was intended to include an encounter with the Giacobini–Zinner comet in 1998, but Suisei exhausted its on-board fuel supply in February 1991 and Earth contact with the craft was lost.

Sullivan, Kathryn Dwyer (1951–) US astronaut. Her three space shuttle flights include the *Discovery* mission in April 1990 to deploy the **Hubble Space Telescope*. In 1984 she was the first US woman to make a space walk. She logged more than 532 hours in space. Sullivan was selected as an astronaut in 1978 and left NASA in 1992 to become the chief scientist at the National Oceanic and Atmospheric Administration.

Sun The *star at the centre of our Solar System. It is about 5 billion years old, with a predicted lifetime of 10 billion years; its diameter is 1.4 million km, its temperature at the surface (the *photosphere) is about 5 800 K/5 530 °C, and at the centre 15 million K/about 15 million °C. It is composed of about 70% hydrogen and 30% helium, with other elements making up less than 1%. The Sun's energy is generated by nuclear fusion reactions that turn hydrogen into helium, producing large amounts of light and heat that sustain life on Earth.

Space probes to study the Sun have included NASA's series of *Orbiting Solar Observatory satellites, launched between 1963 and 1975, the **Ulysses* space probe, launched in 1990, the Japanese **Yohkoh*, launched in 1991 and the ESA–NASA **Solar and Heliospheric Observatory* (*SOHO*) launched in 1995.

At the end of its life, it will expand to become a *red giant the size of Mars's orbit, then shrink to become a *white dwarf. The Sun is about 149 million km from Earth (the closest star to Earth), with light and heat taking about seven minutes to reach Earth. The Sun spins on its axis every 25 days near its equator, but more slowly towards its poles. Its rotation can be followed by watching the passage of dark *sunspots (cooler regions of about 3 600 K/3 300 °C) across its disc. Sometimes bright eruptions called *flares occur near sunspots. Above the Sun's photosphere (its visible surface which emits light

S

and heat) lies a layer of thinner gas called the *chromosphere, visible only by means of special instruments or at eclipses. Tongues of gas called *prominences extend from the chromosphere into the corona, a halo of hot, tenuous gas surrounding the Sun. Gas boiling from the corona streams outwards through the Solar System, forming the *solar wind. Activity on the Sun, including sunspots, flares, and prominences, waxes and wanes during the **solar cycle**, which peaks every 11 years or so, and seems to be connected with the solar magnetic field.

http://umbra.nascom.nasa.gov/sdac.html Home page of the Solar Data Analysis Center at NASA's Goddard Space Flight Center. The site includes information about, and images of, the Sun, taken from space and also from ground-based observation posts. There are also numerous links to related pages.

http://www.solarviews.com/eng/sun.htm All you ever wanted to know about our closest star, including cross-sections, photographs, a history of exploration, animations of eclipses, and much more. The user can also take a multimedia tour of the Sun or find out what the weather is like on the Sun.

http://seds.lpl.arizona.edu/nineplanets/nineplanets/sol.html Clear and concise information about the Sun. There are key facts and a summary of what we still have to learn about the Sun. The site features a very good resource list, with links to many other excellent Sun-related sites.

sunseeker A common name for a photoelectric device on a spacecraft that keeps a spectrograph or other instrument always pointing towards the Sun.

Sun sensor A light-sensitive device that orients a spacecraft towards the Sun. It is part of the *attitude control system.

When 'Sun acquisition' occurs, the device's sensing voltage changes from low to high. The *TOPEX/Poseidon spacecraft, launched on 10 August 1992 to measure ocean surfaces, carried two Sun sensors to ensure that the vehicle's altimeter antenna pointed straight down to the sea.

sunshade A protective covering on a space vehicle to shield it from the Sun's rays. The first crew to visit the *Skylab* space station in 1973 discovered unbearable temperatures within because the heat shield had been damaged. Temperatures on the outer surface reached more than 150 °C. Two astronauts had to extend a parasol-like sunshade over the space vehicle to cool it down enough to be habitable.

sunspot A dark patch on the surface of the Sun, actually an area of cooler gas, thought to be caused by strong magnetic fields that block the outward flow of heat to the Sun's surface. Sunspots consist of a dark central **umbra**, about 4 000 K (3 700 °C), and a lighter surrounding **penumbra**, about 5 500 K (5 200 °C). They last from several days to over a month, ranging in size from 2 000 km to groups stretching for over 100 000 km.

Sunspots are more common during active periods in the Sun's magnetic cycle, when they are sometimes accompanied by nearby *flares. The number of sunspots visible at a given time varies from none to over 100, in a cycle averaging 11 years. There was a lull in sunspot activity, known as the Maunder minimum, 1645–1715, that coincided with a cold spell in Europe.

http://www.exploratorium.edu/sunspots/ Well-written and easily understandable explanation of sunspots. There is a high-resolution image of a group of sunspots and a link for further sunspot pictures. There is information on how to observe sunspots and the Sun's cycle of magnetic activity. For more detailed information, there are links to solar observatories.

Sun-synchronous orbit The orbit of an artificial satellite that always keeps the same relative orientation to the Sun. It is mostly used for Earth-observation missions, since the orbital plane ensures that the satellite will repeatedly pass over the same area of the Earth during the same conditions of sunlight. This is a *low Earth orbit that is nearly a *polar orbit, inclined about 98° to the Equator.

An example of a non-Earth spacecraft in a Sun-synchronous orbit is the **Mars Global Surveyor*. The satellite is in a 2 p.m. Mars Local Time orbit to take advantage of well-placed shadows for the best views of the planet.

supergiant The largest and most luminous type of star known, with a diameter of up to 1 000 times that of the Sun and an apparent magnitude of between 0.4 and 1.3. Supergiants are likely to become *supernovae.

Super Guppy A cargo aircraft used by NASA to transport huge delicate space components and equipment. The aircraft has a 7.6-m diameter fuselage to handle oversized loads and a unique fold-away nose that opens 110 degrees for cargo loading. It is more than 43 m long with a wingspan of more than 47 m.

The first version, the Pregnant Guppy 377PG, was developed in 1962 by the California-based Aero Spacelines. The Pregnant Guppy replaced the only other means to move *Apollo* rocket stages from California to Florida, a slow ship-journey through the Panama Canal. The current aircraft, the Super Guppy 77SGT-F, has been used to transport components for the **International Space Station*.

superior planet A planet that is farther away from the Sun than the Earth is: that is, Mars, Jupiter, Saturn, Uranus, Neptune, and Pluto.

supernova The explosive death of a star, which temporarily attains a brightness of 100 million Suns or more, so that it can shine as brilliantly as a small galaxy for a few days or weeks. Very approximately, it is thought that a supernova explodes in a large galaxy about once every 100 years. Many supernovae—astronomers estimate around 50%—remain undetected because of obscuring by interstellar dust.

The name 'supernova' was coined in 1934 by Swiss astronomer Fritz Zwicky and German-born US astronomer Walter Baade. Zwicky was also responsible for the division of supernovae into types I and II.

Type I supernovae are thought to occur in *binary star systems, in which gas from one star falls on to a *white dwarf, causing it to explode.

Type II supernovae occur in stars ten or more times as massive as the Sun, which suffer runaway internal nuclear reactions at the ends of their lives, leading to explosions. These are thought to leave behind neutron stars and *black holes. Gas ejected by such an explosion causes an expanding radio source, such as the Crab nebula. Supernovae are thought to be the main source of elements heavier than hydrogen and helium.

The first supernova was recorded (although not identified as such at the time) in AD 185 in China. The last supernova was seen in our Galaxy in 1604, but many others have been seen since in other galaxies. In 1987 a supernova visible to the unaided eye occurred in the Large Magellanic Cloud, a small neighbouring galaxy.

supernova remnant (SNR) The glowing remains of a star that has been destroyed in a *supernova explosion. The brightest and most famous example is the Crab nebula.

Super Weight Improvement Program (SWIP) A NASA programme during the 1960s to reduce the weight of the *Apollo* lunar module. SWIP was created by Joseph Gavin, vice-president of the module project at *Grumman Aerospace, which originally built the craft. The design changes included a reduction of the number of windows and removal of the astronauts' seats, with the crew strapped into standing positions. When the redesigned lunar module landed on the Moon, it weighed just over 14 850 kg.

surface insulation A part of the space shuttle's thermal protection system (TPS) to protect the vehicle's surface from heat caused by atmospheric friction during re-entry. Besides *thermal tiles, some surface insulation is composed of a felt reusable surface insulation (FRSI) of coated *Nomex cloth to protect the upper fuselage and an advanced flexible reusable surface insulation (AFRSI), consisting of a quilted blanket of silica fibre, for some other areas.

Surface science package (SSP) An instrument placed in the **Huggens* probe which measured and relayed data from its landing point on Titan. Different sensors were able to measure a variety of physical and chemical properties on the surface of the moon.

Surveyor A series of seven US crewless Moon landers launched 1966–68 that paved the way for the **Apollo* project. Each craft was equipped with television cameras and a scoop for examining lunar soil. *Surveyor 1* landed in the Ocean of Storms in June 1966, becoming the first US spacecraft to make a soft landing on the Moon. With the exceptions of *Surveyor 2* and *Surveyor 4*, all the landings were successful. *Surveyor 3*, which landed in the Ocean of Storms in April 1967, was visited by the crew of *Apollo 12* in November 1969.

sustainer A special engine that keeps a rocket firing after the boosters or a stage have fallen away. The Atlas rocket had a 27 143-kg thrust sustainer.

Suzaku A joint Japanese–US X-ray astronomy satellite, launched on 10 July 2005. It observes at a wide range of energetic wavelengths, ranging from soft X-rays to gamma rays (0.2–700 keV), with three instruments: an X-ray micro-calorimeter (X-ray Spectrometer, on XRS), four X-ray CCDs (the X-ray Imaging Spectrometers, or XISs), and a hard X-ray detector, or HXD. *Suzaku* has four foil X-ray telescopes (XRTs) focusing X-rays onto each of the four XISs, along with a fifth XRT used with the XRS.

Swarm A group of three satellites that are to study the Earth's magnetic field. Scheduled for launch in 2009, *Swarm* is an Earth Explorer Opportunity Mission of the European Space Agency (ESA). It will conduct the most extensive survey ever of the Earth's geomagnetic field, to aid scientific understanding of the Earth's interior and climate. The three satellites are to be released from a single launcher, with one of them flying at 530 km above the other two flying at 450 km. They will take high-precision and high-resolution measurements of the magnetic field's strength, direction, and variation, along with navigation, accelerometer, and electric field measurements.

SWAS Abbreviation for **Submillimetre Wave Astronomy Satellite*.

SWG Abbreviation for *Science Working Group.

Swift A NASA satellite to detect and study gamma-ray bursts which was launched on 20 November 2004. It carries three instruments to observe gamma-ray bursts and their afterglows caused as the shockwave from the burst hits interstellar gas at gamma-ray, X-ray, ultraviolet, and optical wavelengths. *Swift*'s wide-angle Burst Alert Telescope (BAT) detects gamma-ray bursts and immediately notifies astronomers on the ground. The spacecraft then aligns itself on the burst to study it in more detail with its narrow-field X-ray Telescope (XRT) and Ultraviolet/Optical Telescope (UVOT) to image and take spectra of the afterglow, while ground-based astronomers simultaneously point their own telescopes at it.

http://swift.gsfc.nasa.gov/docs/swift/swiftsc.html Keep up to date with research on gamma-ray bursts and watch *Swift* in action on this site.

swing arm A rotating arm on an *umbilical tower that connects to different levels of a space vehicle on the launch pad. Astronauts use a swing arm to enter their spacecraft, and technicians use the device to ready the launch vehicle. The arms swing away just before lift-off.

swing-by The path of a spacecraft as it flies past another planetary body and its course is changed and speed increased by the gravitational forces of this body. Swing-by differs from a *fly-by, which is a flight past a planetary body without any significant effect on the flight path or speed. Swing-by is often described as a gravity-assist manoeuvre.

synchronous orbit Another term for *geostationary orbit.

synchronous rotation Another name for *captured rotation.

Syncom An early series of US geosynchronous communication satellites. *Syncom 3* broadcast the opening of the Olympic Games from Japan in August 1964.

synthetic aperture radar (SAR) An instrument used for broad-area imaging at high resolutions. SAR is not exclusively used in space, and is employed extensively by aircraft for environmental monitoring, Earth resource mapping, and military systems. It measures the reflection of radar pulses (emitted from the shuttle or aircraft) back from the Earth's surface. The use of radar pulses means that the mapping can take place regardless of atmospheric or light conditions. The X-band synthetic aperture radar (X-SAR) was developed by the Agenzia Spaziale Italiana (ASI; Italian Space Agency) and the Deutsches Zentrum für Luft- und Raumfahrt (DLR; German Space Agency) for use in space as part of the *Space Radar Laboratory.

synthetic vision A computer vision system used instead of direct sight by astronauts for some work in space. The Zarya and Unity modules of the **International Space Station*, for instance, were connected using synthetic vision.

T An abbreviation of 'time' for launching a spacecraft. The countdown to launch is expressed as 'T minus' hours, minutes, or seconds. During a space shuttle launch, the crew begin pre-flight systems checks at T minus 1 hour, 30 minutes. Just before launch, at T minus 2 seconds, the shuttle's computers check that the engines have reached their correct power.

TAEM Abbreviation for *terminal area energy management.

Taikonaut The name used in the west for a Chinese astronaut. It comes from the Chinese word 'taikong' meaning space or cosmos. The official Chinese name is yuhangyuan, meaning 'travellers of the Universe'.

tandem staging A launch system that uses rockets in stacked stages, rather than attached to the side. Each stage falls away after its fuel is burned. Examples include NASA's three-stage *Saturn rocket and China's two-stage *Long March rocket.

Tanegashima Space Centre (TNSC) A Japanese launch site established in 1969, when the original National Space Development Agency of Japan (NASDA) was formed. Located in the south of Kagoshima Prefecture, along the southeast coast of Tanegashima, it is the largest space-development facility in Japan and is known as the most beautiful rocket-launch complex in the world. Tanegashima launches satellites for weather, broadcasting, and communications, while scientific satellites and probes are launched from *Uchinoura Space Centre.

Facilities at Tanegashima include the Takesaki Range (for small rockets), the Osaki Range (for J-I and H-IIA launch vehicles), the Masuda Tracking and Communication Station, the Nogi Radar Station, and the Uchugaoka Radar Station. There are also facilities for test firings of liquid and solid-fuel rocket engines.

Tanner, Joseph Richard (1950–) US astronaut. His three space shuttle flights include the *Discovery* mission in February 1997 to service the **Hubble Space Telescope*. During the mission he made two space walks. He logged more than 742 hours in space and 33 hours of space walks. He was assigned to a 2004 flight that was postponed following the *Columbia* space shuttle disaster the previous year. Tanner joined NASA's Johnson Space Center in 1984 as an aerospace engineer and research pilot. He was selected as an astronaut in 1992.

target vehicle A space vehicle designated for approach by another. The target vehicle may be a satellite being retrieved for repairs, or a spacecraft used in a docking manoeuvre. In July 1966, the *Gemini 10* astronauts made the first orbital docking, connecting to an *Agena rocket target vehicle and using its engines for propulsion. The mission proved that a space rendezvous and docking were possible.

Taurus rocket A four-stage launch vehicle developed by the US *Orbital Sciences Corporation. It is able to put satellites weighing up to 1 350 kg into a low Earth orbit, and those of up to 360 kg into a *geosynchronous transfer orbit. By 2004, the Taurus had been launched six times, carrying a total of 11 satellites into orbit.

TCM Abbreviation for *trajectory correction manoeuvre.

TD-1 A satellite launched in 1972 by the European Space Agency (ESA) to make the first ultraviolet survey of the sky. 'TD-1' is also the designation for a North Korean two-stage launch vehicle.

TDRS Abbreviation for *Tracking and Data-Relay Satellite.

Technology Transfer Programme (TTP) A European Space Agency (ESA) programme created in 1990 to transfer space technology to industrial and commercial sectors. In the past decade, the programme has made more than 150 successful transfers which helped create or save 2 500 jobs in Europe and create 20 new companies. In 2004, it had a portfolio of 450 space technologies available. Examples of transferred technology included special spacesuit materials being used for tracksuits, and new coatings for bearings in a rocket launcher being used on car wheels. The transfers created new businesses and saved companies millions of euros in research, development, maintenance, and operation costs.

Teflon cloth A cloth made of a tough translucent synthetic resin. It has been used in spacesuits since the *Apollo* programme. The space shuttle suits incorporate Gore-Tex, a new version of *Teflon cloth, that is stretched into a thin membrane. It is five times more resistant to abrasions and also more resistant to heat and cold.

TEI Abbreviation for *trans-Earth injection.

telemetry A measurement at a distance, in particular the systems by which information is obtained and sent back by instruments on board a spacecraft.

telemetry system (TLM) system used to *downlink information from a spacecraft, such as science data and the status of the craft, for example its temperature and pressure. Ground stations, such as NASA's *deep-space network, deliver the data to the flight project. It is then usually displayed, stored, distributed, analysed, and published.

telescope An optical instrument that magnifies images of faint and distant objects. Telescopes in space have been used to study infrared, ultraviolet, and X-ray radiation that does not penetrate the atmosphere, but carries much information about the births, lives, and deaths of stars and galaxies. *See also* RADIO TELESCOPE.

http://galileo.rice.edu/science.html Early history of the telescope, documenting its first astronomical use by Italian astronomer Galileo and English astronomer Thomas Harriot. There are photographs of Galileo's and English physicist Isaac Newton's telescopes and interesting drawings and engravings from old manuscripts.

Television Infrared Observation Satellite (Tiros) A series of US weather satellites controlled by the *National Oceanic and Atmospheric Administration. *Tiros 1*, launched in April 1960, was the world's first weather satellite. It failed after 2.5 months, but in that time succeeded in sending back about 23 000 photographs from around the Earth. *Tiros 9*, launched into polar orbit in January 1965, recorded the first complete picture of the Earth's weather in 480 pictures taken during nine orbits. Two upgraded polar orbiters of the *Tiros-N* type now operate together, insuring that the weather data for any region of the Earth are no more than six hours old. The main instrument on each is the *advanced very high-resolution radiometer (AVHRR).

Tele-X A Swedish telecommunications satellite launched on 2 April 1989 by an Ariane rocket and taken out of operation on 16 January 1998. It was managed by the Swedish Space Corporation. Its coverage was primarily for Sweden, Norway, and Finland. *Tele-X* weighed 1.3 tonnes, measured 1.7 × 2.4 × 2.4 m, and had a solar array of 19 m.

Telstar A US communications satellite, launched on 10 July 1962, which relayed the first live television transmissions between the USA and Europe. *Telstar* orbited the Earth every 2.63 hours, and unlike later geostationary satellites was only usable when in line-of-sight of two tracking stations.

***Tenma* (Japanese: 'heavenly horse')** The second Japanese *X-ray astronomy satellite. Launched in February 1983, its original name *Astro B* was changed to *Tenma*, after the constellation Pegasus. Before ending its mission in November 1985, the satellite carried out the first sensitive measurements of the 'iron line' region of the electromagnetic spectrum, where iron emission lines can be detected in the radiation from *pulsars, *binary stars, and hot *plasma.

ten-minute hold A crucial break near the end of the countdown to launch a space shuttle. The hold is to provide time to check that all systems are synchronized and functioning properly and to receive a final weather clearance. The countdown is then resumed at *T minus 9 minutes.

Tereshkova, Valentina Vladimirovna (1937–) Soviet cosmonaut, the first woman to fly in space. She was the solo pilot of a three-day flight 16–19 June 1963 in *Vostok 6*, orbiting the Earth 48 times. Her flight on the last of the *Vostok* series was her only experience in space.

Tereshkova was born in Maslennikovo in western Russia, the daughter of a tractor driver. She married the cosmonaut Andrian Nikolayev, in 1963. Their daughter, Elena, born a year later, was the first child of parents who had both been in space.

terminal area energy management (TAEM) A computer program on a space shuttle that makes the last adjustments of speed and altitude as the spacecraft commences landing. Beginning about six minutes before touchdown, the TAEM constantly checks the shuttle's position, attitude, velocity, and descent angle, and automatically guides the craft into its final approach to the runway.

terminator The line across the Earth (or other planet) that separates day and night (or dark and light). A space shuttle passes a terminator every 45 minutes, so that it experiences 16 sunrises and 16 sunsets every 24 hours.

Terra A NASA satellite launched on 18 December 2001 as the flagship for its *Earth-observing system. It began collecting data on 24 February 2000 and has an expected lifespan of 15 years. *Terra* weighs 5 190 kg and is 6.8 m long. Its instruments include the moderate-resolution imaging spectroradiometer (MODIS), which uses a radiometer to measure biological and physical processes on land and sea; measurements of pollution in the troposphere (MOPITT), provided by the Canadian Space Agency to measure carbon monoxide and methane in the troposphere with a spectroscope; and Japan's advanced spaceborne thermal-emission and reflection radiometer (ASTER) that uses three sensors for images of land, water, ice, and clouds.

http://terra.nasa.gov/ Attractive site presenting images and results from the instruments carried by the *Terra* Earth-observation satellite. There are galleries presenting the best recent images, as well as background information on climate and land-surface monitoring.

terraforming The hypothetical idea of creating an Earthlike environment on another planet. Some scientists believe that astronauts could create an oxygenated atmosphere that would make Mars and other planets inhabitable.

http://library.thinkquest.org/10274/ View of what a colony on Mars might look like in 2050. Navigate around the fictional Martian colony, Koinae, learning about daily life, the climate, terrain, and research projects on Mars.

http://www.users.globalnet.co.uk/~mfogg/ Good selection of essays and articles about terraforming the planets. The site has a gallery with space art depicting terraformed planets, as well as links to terraforming resources on the Internet and a comprehensive bibliography.

terrestrial planet Any of the four small, rocky inner *planets of the Solar System: *Mercury, *Venus, *Earth, and *Mars. The *Moon is sometimes also included, although it is a satellite of the Earth and not strictly a planet.

Terrestrial Planet Finder (TPF) A proposed NASA mission in which a pair of space telescopes will study planetary formation around other stars as well as extrasolar planets themselves. The TPF mission would involve a coronagraph operating at visible wavelengths, to be launched around 2014, and a long-baseline interferometer operating in the infrared, for launch before 2020. These observatories will measure the size, temperature, and orbits of planets as small as the Earth in distant solar systems, while spectroscopic observations will reveal the composition of their atmospheres to find whether a planet currently supports life or someday could do so. TPF's extremely high resolution will also allow it to image targets of astrophysical interest, such as the disks surrounding the central black holes of galaxies.

tether A cable that connects an astronaut to the spacecraft during an *extravehicular activity. From a space shuttle, an astronaut embarking on a space walk will first reach outside the secure air lock to pull in the safety tether and attach it to a ring on his or her spacesuit. A ring at the other end of the tether slides on a line leading through the cargo bay into space. If there is some distance to go, as when retrieving a satellite, the astronaut unhooks the tether and uses a *manned manoeuvring unit (MMU).

tethered satellite system (TSS) A joint project of NASA and the Agenzia Spaziale Italiana (ASI; Italian Space Agency) to demonstrate the use of long tethers in space. Tests have been conducted to show how satellites on tethers can be deployed and controlled, and how they can generate electrical power. The *TSS-1* Italian satellite was launched aboard the space shuttle *Atlantis* on 31 July 1992. The satellite was connected to the shuttle by a conductive tether. The electric potential was created as the tether passed through the Earth's magnetic field, the satellite charging positively and the shuttle negatively. The two vehicles separated by 167 m, but the mission failed to achieve the full extension of 20 km. *TSS-1R* (the 'R' stands for 'reflight') was launched on *Columbia* on 22 February 1996 with the same goals. The satellite was lost when the tether broke, partly due to a protruding bolt, after being extended to more than 19 km.

Thagard, Norman Earl (1943–) US astronaut and physician. He logged 140 days in space on five shuttle flights, including the May 1989 mission of *Atlantis* to deploy the **Magellan* spacecraft. He was also the cosmonaut-researcher for the Russian *Mir 18* mission when he was launched from the *Baikonur Cosmodrome in March 1995 for a 115-day flight that involved 28 experiments. Thagard was selected as an astronaut in 1978 and retired in 1996.

thermal tile A tile used to protect a spacecraft from frictional heat during re-entry. A space shuttle's thermal protection system (TPS) has a layer of more than 30 000 thermal tiles fitted by hand and glued to its aluminium underbelly to absorb and dissipate the heat. The silica-fibre tiles are coated with silica glass. Reinforced carbon–carbon tiles are placed on the leading edges of the wings, where they must withstand re-entry temperatures of up to 1 378 °C.

Thiele, Gerhard Julius Paul (1953–) German astronaut with the European Space Agency (ESA). He flew on the *Endeavour* space shuttle flight in February 2000 for the *Shuttle Radar Topography Mission, logging over 268 hours in space. Thiele was selected to train as an astronaut at the German Aerospace Research Establishment in 1988. In 1992 he joined NASA's Johnson Space Center (JSC) to train as a payload specialist. He trained for a second time at the JSC in 1996, as a mission specialist, and joined the ESA in 1998.

Thirsk, Robert Brent (1953–) Canadian astronaut and physician. He flew with the *Columbia* space shuttle mission in June 1996 that carried out 43 experiments in life and materials sciences. Thirsk was selected in 1983 as a Canadian astronaut and led an international research team to study the effect of weightlessness on the heart and blood vessels. He also headed a programme to design and test an antigravity spacesuit. Thirsk was assigned to NASA's Johnson Space Center in 1998 to train as a mission specialist and went on to become a CapCom (spacecraft communicator) for the **International Space Station*.

t

Thomas, Andrew Sydney Withiel (1951–) US astronaut. His three missions include 141 days in space as a crew member aboard the Russian space station *Mir*, transported there on the *Endeavour* space shuttle in January 1998. He also flew on *Discovery* in March 2001 on the mission that took the second resident crew to the **International Space Station*, and undertook a space walk lasting more than six hours. His fourth mission was aboard *Discovery* in July–August 2005, the first space shuttle launch since the *Columbia* disaster of 2003. Thomas joined NASA's Jet Propulsion Laboratory in 1989 and was selected as an astronaut in 1992.

Thomas, Donald Alan (1955–) US astronaut. He logged more than 1040 hours during his four space shuttle flights, which included the 15-day *Columbia* mission in July 1994 that conducted *microgravity experiments. He was assigned to the sixth resident crew for the **International Space Station* in November 2002 but was held back because of medical issues involved in the long-duration mission. Thomas worked at the Lockheed Engineering and Sciences Company, where he reviewed space shuttle payloads, before joining NASA's Johnson Space Center as a materials engineer. He was selected as an astronaut in 1990.

Thornton, Kathryn Ryan Cordell (1952–) US astronaut. She logged more than 975 hours in space during her four flights, including the first flight of the space shuttle *Endeavour* in May 1992. She also logged more than 21 hours of extravehicular activity. Thornton was selected as an astronaut in 1984 and left NASA in 1996.

Thornton, William Edgar (1929–) US scientist-astronaut and physician. He flew on the space shuttle *Challenger* in August 1983 and April 1985, studying the effects of weightlessness on the human body on the first flight, and on monkeys and rodents on the second (the first animal payload in a crewed flight). He developed the treadmill used for in-flight exercise on shuttles. He served with the US Air Force, working in space medicine research, before being selected as a scientist-astronaut in 1967. He left NASA in 1994.

Thor rocket A rocket developed by the US Air Force as an intermediate-range ballistic missile (IRBM). Its first launch was on 20 September 1957. A year later, it was combined with the second stage of a Vanguard rocket to form the Thor-Able launch vehicle used to launch the **Pioneer* probe to the Moon. In 1959 it was combined with an upper stage Agena rocket, and launched a **Discoverer* satellite. A year later, the first Thor Delta was launched and became known as the Delta, flying more than 200 times until the Delta III was introduced in 1998.

three-axis stabilization A control of the attitude of a spacecraft by stabilization of its X, Y, and Z axes. This can be done using small thrusters to rock the craft slowly back and forth, as with the two *Voyager* space probes since 1977. Another method is the use of *momentum wheels, which steady a spacecraft even further, necessary when instruments are making observations.

three-way communications A mode of spacecraft communications that uses two ground stations. One station receives a downlink from the craft, and the other station provides an uplink. For example, a three-way situation would

t

occur if a spacecraft approaching Australia sent telemetry to a *deep-space network (DSN) station in that country, while still receiving an uplink from a station in the USA.

throat A narrow passage on a rocket engine for exhaust gases. It is located between the combustion chamber and the nozzle.

thrust The propulsive pressure exerted by, for example, a jet or rocket engine.

thrust chamber Alternative name for the *combustion chamber in a liquid-fuel rocket.

thruster A small solid or liquid propellant rocket engine fired in short bursts to alter the orientation of a spacecraft in relation to *pitch, *yaw, and *roll, in order to maintain *attitude control.

thrust level The maximum thrust power of a rocket. A space shuttle's two solid rocket boosters each have a thrust level of about 1 497 tonnes. NASA's three-stage *Delta rocket, by contrast, has a thrust level of 78 tonnes for the first stage, 3.43 tonnes for the second, and 1.25 tonnes for the third.

thrust vector control (TVC) A rocket thrust control system on the space shuttle. During a launch, the system controls the attitude and trajectory by swivelling the rocket nozzles to direct the thrust. This involves the three main engines and two solid rocket boosters during lift-off and the first-stage ascent, and then only the main engines during the second-stage ascent. TVC is part of the shuttle's flight control system (FCS).

Thuot, Pierre Joseph (1955–) US astronaut. During his three space shuttle flights he logged more than 17 hours of extravehicular activity. On the first flight of *Endeavour* in May 1992 he undertook three space walks, including the first three-person group walk, the longest space walk to date at 8.5 hours. Thuot was selected as an astronaut in 1985 and left NASA in 1995.

Tidbinbilla A space tracking station in Australia, just south of Canberra, part of NASA's *deep-space network. It provides tracking facilities and command transmissions in support of NASA's crewed and uncrewed spacecraft.

TIG Acronym for *time of ignition.

time of ignition (TIG) The moment when a rocket engine fires. During a space shuttle launch, this happens at 'T minus 5 seconds' (*see* *T) for the three main engines. TIGs are also carefully scheduled for orbit manoeuvres, such as a rendezvous, and for slowing the shuttle so it can drop out of orbit to return. During the latter, the crew's preparations for a deorbit begin at 'TIG minus four hours'.

Tiros Abbreviation for **Television Infrared Observation Satellite*.

Titan The largest moon of the planet Saturn, with a diameter of 5 150 km and a mean distance from Saturn of 1 222 000 km. It was discovered in 1655 by Dutch mathematician and astronomer Christiaan Huygens, and is the second-largest moon in the Solar System (only Ganymede, of Jupiter, is larger).

Titan is the only moon in the Solar System with a substantial atmosphere (mostly nitrogen), topped with smoggy orange clouds that obscure the surface, which may be covered with liquid ethane lakes. Its surface atmospheric pressure is greater than Earth's. Radar signals suggest that Titan has dry land as The well as oceans (among the planets, only Earth has both in the Solar System).
Huygens probe carried aboard NASA's **Cassini* spacecraft landed on Titan on 14 January 2005.

Titan rocket A family of US space rockets, developed from the Titan intercontinental missile. Two-stage Titan rockets launched the **Gemini* project crewed missions. More powerful Titans, with additional stages and strap-on boosters, such as the Titan–Centaur, were used to launch spy satellites and

(Image © NASA)
The Titan rocket that put *Gemini 12* astronauts James Lovell and Buzz Aldrin into orbit on 11 November 1966. Among their tasks was docking with the Agena target vehicle launched an hour previously.

space probes, including the **Viking* and **Voyager* probes, **Mars Observer*, and **Cassini*.

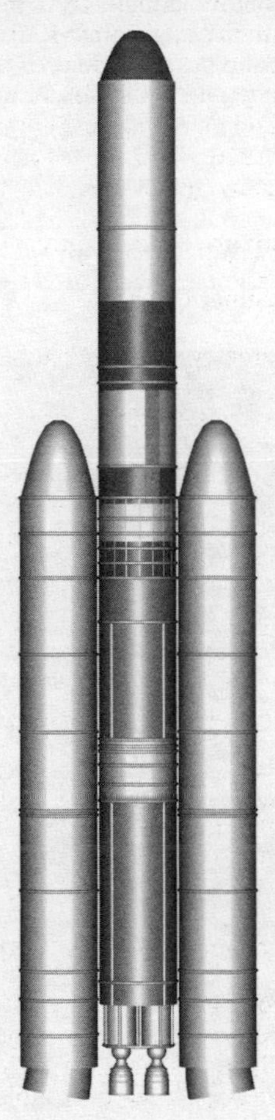

(Image © Research machines plc)
The Lockheed–Martin Titan rocket. A Titan 4B was launched in October 2001, carrying an undisclosed cargo, likely to be some form of spy satellite.

Titov, Gherman Stepanovich (1935–2000) Soviet cosmonaut. He flew the world's second crewed spacecraft in orbit, *Vostok 2*, in August 1961, the first flight of more than one day (it lasted 25 hours, 18 minutes). During the 17.5 orbits of the Earth, he flew the spacecraft manually. After his flight, he was assigned to the experimental *Spiral* spaceplane project that was eventually cancelled. Titov was selected as an cosmonaut in 1960 and resigned in 1970.

Titov, Vladimir Georgievich (1947–) Russian cosmonaut. In September 1983, he survived an emergency abort when a fire consumed the launch vehicle less than a minute before launch. The vehicle exploded seconds after his *Soyuz* module was automatically separated. His missions include a record stay of 365 days on the Russian space station *Mir* (December 1987–December 1988). He also made two flights on NASA space shuttles to dock with *Mir*: on *Discovery* in February 1995 and on *Atlantis* in September 1997. Titov was selected as a cosmonaut in 1976. He retired as a cosmonaut and went on to work for the *Boeing Company in Moscow.

TKSC abbreviation for *Tsukuba Space Centre.

TLI Abbreviation for *translunar injection.

TLM Abbreviation for *telemetry system.

TOA Abbreviation for *top of atmosphere.

Tognini, Michel (1949–) French astronaut with the European Space Agency. He flew aboard the *Soyuz* mission launched in July 1992, spending 14 days conducting joint Soviet–French scientific experiments; and aboard the space shuttle *Columbia* in July 1999, which deployed the **Chandra X-Ray Observatory*. He was selected as a French astronaut in 1985. He trained at the Gagarin Cosmonaut Training Centre in 1986 and NASA's Johnson Space Center in 1995. He became a CapCom (spacecraft communicator) for the *International Space Station* and on 1 May 2003 was also appointed as head of the Astronaut Division of the European Space Agency.

Tokarev, Valery Ivanovich (1952–) Russian cosmonaut. He flew aboard the space shuttle *Discovery* in May 1999, delivering supplies to the **International Space Station* (*ISS*) in preparation for the arrival of the first resident crew. He was selected as a cosmonaut in 1987, originally to fly the **Buran* space shuttle. He was assigned to the backup crew of the tenth resident crew of the *ISS*.

TOMS Acronym for *total ozone mapping spectrometer.

TOPEX/Poseidon A mission launched in August 1992 to map the surface topography of the Earth's oceans. It is a joint project involving NASA and the *Centre National d'Etudes Spatiales (CNES), the French space agency, which launched the craft aboard an Ariane 4 rocket. Able to reveal minute differences in ocean height, the spacecraft has monitored the effects of ocean currents on global climate change and investigated such phenomena as the El Niño climate changes. The data provided by TOPEX/Poseidon has also improved the understanding of the Earth's gravity field and produced the most accurate global maps of tides to date, as well as the first global views of seasonal changes in ocean currents.

http://topex-www.jpl.nasa.gov/ Scientific data and educational resources relating to the TOPEX/Poseidon mission. The latest research results are reported and an image gallery displays TOPEX/Poseidon images, pictures of the spacecraft, and mission posters.

top of atmosphere (TOA) The highest region of the Earth's atmosphere. Spacecraft often use sensing instruments to measure radiation at this level. For example, the satellites in NASA's Pathfinder programme employ them to carry out a global study of layered cloud systems and their influence on TOA radiation.

total ozone mapping spectrometer (TOMS) An instrument developed by NASA to measure ozone indirectly by monitoring ultraviolet light. It observes backscattered ultraviolet radiation, which is solar radiation that has reached the Earth's lower atmosphere and been scattered by clouds and air molecules back through the stratosphere. TOMS measures the total amount of ozone in a column of air from the Earth's surface to the top of the atmosphere.

In 1996, the first instruments were launched on Japan's **Advanced Earth-Observing Satellite* and on the *Earth Probe TOMS* satellite. Three others have since flown. TOMS is part of NASA's long-term 'Mission to Planet Earth' to study the planet as a global environmental system.

http://toms.gsfc.nasa.gov/index.html Presents the science of the total ozone mapping spectrometer and provides information about the spacecraft on which it is carried. There are tutorials, regular news updates, and a multimedia gallery with still images, charts, and animations.

touchdown The instant when a spacecraft lands on the Earth, another planet, or a moon. A space shuttle travels at 370 kph when its wheels have achieved touchdown.

TPF Abbreviation for *Terrestrial Planet Finder.

TRACE Acronym for **Transition Region and Coronal Explorer*.

tracking The monitoring the path of a spacecraft from tracking stations on the ground. It is carried out by radio frequencies and radar. An example is NASA's *deep-space network for space probes. Systems that use satellites to track other spacecraft are NASA's *Tracking and Data-Relay Satellite system and the European Space Agency's *Data-Relay and Technology Mission.

***Tracking and Data-Relay Satellite* (*TDRS*)** One of the nine satellites launched into geosynchronous orbit by NASA to track other spacecraft in low Earth orbits and relay their communications. The first of the second-generation satellites was *TDRS-H* launched in 2000, followed by *TDRS-I* and *TDRS-J* in 2002. These include the space shuttle, the *Hubble Space Telescope*, and the Landsat satellite series. Each *TDRS* weighs 2 270 kg and measures 17.4 m from the edges of the opposite solar panels. One *TDRS* was lost in the *Challenger* space shuttle disaster in 1986. The satellites are part of the Tracking and Data-Relay Satellite System (TDRSS) that includes ground stations. Each *TDRS* is controlled from the White Sands Test Facility in New Mexico by the Mission Operations and Data Systems Directorate of the *Goddard Space Flight Center.

tracking station A ground station that confirms the location of a spacecraft during its mission, using radar or other means, and keeps in radio contact. A tracking station is normally part of a network, such as the European Space Agency's *ESTRACK.

tracking system (TRK) A system used to determine a spacecraft's location, including its distance, velocity, and trajectory. NASA's *deep-space network (DSN) uses such methods as Doppler, ranging, and antenna control. Distance is determined by uplinking data that the spacecraft echoes. Measuring the Doppler shift change of frequency on the vehicle's downlink carrier provides the velocity. Doppler and range measurements together provide the trajectory. Navigation files are used by the DSN to predict where to point their antennae.

tracking, telemetry, and control (TT+C) The use of voice, data, and radar equipment to monitor the flight path and operational stages of a space flight, including the launch, the deployment of the payload, the bringing of the payload into operational status, and the control of mission systems. TT+C uses traditional and digital voice communications, 'bent pipe' data flow where real-time data is fed along electronic pipelines via various fixed and transportable stations located around the world, and tracking and data-relay satellites augmenting the ground stations.

trajectory The path of a spacecraft, rocket, or other object in space or the Earth's atmosphere. A space probe travelling to a planet is, at the same time, in orbit around the Sun, so its trajectory can be plotted as the part of its solar orbit from the launch to the target body.

trajectory correction manoeuvre (TCM) The adjustment in a spacecraft's trajectory by firing its thrusters. This normally involves a change in velocity of a few metres per second. The manoeuvre must first be determined by calculating the directional change needed to establish the desired trajectory. As propellant is limited, only very small corrections can be made.

trans-Earth injection (TEI) A NASA term for a spacecraft leaving its orbit around the Moon for a trajectory that returns it to Earth. **Apollo* project astronauts had to burn the *service propulsion system engine on their command and service module for 2.5 minutes to break free of lunar gravity

transfer orbit An elliptical path followed by a spacecraft moving from one orbit to another, designed to save fuel although at the expense of a longer journey time.

Space probes travel to the planets on transfer orbits. A probe aimed at Venus has to be 'slowed down' relative to the Earth, so that it enters an elliptical transfer orbit with its perigee (point of closest approach to the Sun) at the same distance as the orbit of Venus; towards Mars, the vehicle has to be 'speeded up' relative to the Earth, so that it reaches its apogee (furthest point from the Sun) at the same distance as the orbit of Mars.

Geostationary transfer orbit is the highly elliptical path followed by satellites to be placed in *geostationary orbit around the Earth (an orbit coincident with Earth's rotation). A small rocket is fired at the transfer orbit's apogee to place the satellite in geostationary orbit.

Transit The world's first operational satellite navigation system. It was developed in the early 1960s by the Johns Hopkins University in Baltimore, Maryland, for the navigation of US Navy ballistic missile submarines. The system was operated by the Naval Space Operations Center at Point Mugu, California. The first Transit satellite, also called the *Navy Navigation Satellite*, was launched unsuccessfully in 1959, but *Transit 5A1* became the first operational navigation satellite on 19 December 1962. The system was made available to civilians in 1967 and the last satellite was launched in 1988, making a total of six in polar orbits. The Transit navigation system was terminated in 1996.

***Transition Region and Coronal Explorer* (*TRACE*)** The NASA satellite used to examine the Sun's corona and the so-called transition region between the upper chromosphere and corona where temperatures rise rapidly. *TRACE*, launched on 2 April 1998, carried a telescope of 0.3-m aperture which imaged the motions of hot gas in the Sun's outer layers at a range of ultraviolet wavelengths. These gas motions are controlled by complex magnetic fields, and it is interactions in the magnetic fields that give rise to solar flares and large-scale mass ejections.

translation The movement of a spacecraft in any one direction along a straight line. This can include reversing.

translunar injection (TLI) The moment when a spacecraft leaves Earth orbit to begin a trajectory to the Moon. The *Apollo* spacecraft required a speed of nearly 40 225 kph to achieve TLI.

transmission time (TRM) The time taken for an uplink of data from a ground station to a spacecraft, measured in *Universal Time Coordinated (UTC).

transponder An electronic device on board a spacecraft, often used to combine the communications transmitter and receiver.

Treschev, Sergei Yevgenyevich (1958–) Russian cosmonaut. He was a foreman and engineer at the *Energiya Rocket and Space Complex from 1984 to 1986, supporting crew training aboard the Russian space station *Mir*. Treschev was selected as a cosmonaut in 1992, and flew on *Endeavour* on 5 June 2002 to the **International Space Station* as a member of the fifth resident crew. He remained there for six months, performing one space walk to install various devices, including a frame to house components for future space walks.

Triton The largest of Neptune's moons. It has a diameter of 2 700 km, and orbits Neptune every 5.88 days in a retrograde (east to west) direction at a distance of 354 000 km. It takes the same time to rotate about its own axis as it does to make one revolution of Neptune.

It is slightly larger than the planet Pluto, which it is thought to resemble in composition and appearance. Probably Triton was formerly a separate body like Pluto but was captured by Neptune. Triton was discovered in 1846 by English astronomer William Lassell (1799–1880) only weeks after the discovery of Neptune. Triton's surface, as revealed by the *Voyager 2* space probe, has a temperature of 38 K (−235 °C), making it the coldest known place in the Solar System. It is covered with frozen nitrogen and methane, some of which evaporates to form a tenuous atmosphere with a pressure only 0.00001 that of the Earth at sea level. Triton has a pink south polar cap, probably coloured by the effects of solar radiation on methane ice. Dark streaks on Triton are thought to be formed by geysers of liquid nitrogen. The surface has few impact craters (the largest is the Mazomba, with a diameter of 27 km), indicating that many of the earlier craters have been erased by the erupting and freezing of water (cryovulcanism).

TRK Abbreviation for *tracking system.

TRM Abbreviation for *transmission time.

Truly, Richard Harrison (1937–) US astronaut and NASA administrator. He flew with the November 1981 mission of the space shuttle *Columbia*, the first reuse of a shuttle and the first to carry a scientific payload. He also flew on the *Challenger* mission in 1983 that accomplished the first night-time launch (30 August) and landing (5 September). He was selected as an astronaut in 1965 and left NASA in 1983.

Tryggvason, Bjarni Valdimar (1945–) Canadian astronaut. He was a payload specialist aboard the *Discovery* space shuttle mission in August 1997 to test the microgravity vibration isolation mount (MIM), on which he worked as its principal investigator. He was selected as a Canadian astronaut in 1983 and assigned to NASA's Johnson Space Center in 1998 to train for a future shuttle flight. He went on to perform training and technical duties at the Johnson Space Center and handles microgravity programmes at the Canadian Space Agency.

Tsiolkovsky (or Tsiolkovskii), Konstantin Eduardovich (1857–1935) Russian scientist who developed the theory of space flight. He published the first practical paper on astronautics in 1903, dealing with space travel by rockets using liquid propellants, such as liquid oxygen.

Tsiolkovsky was born in the Spassk district and had little formal education; he was deaf from the age of ten. He never actually constructed a rocket, but his theories and designs were fundamental in helping to establish the reality of space flight.

In 1883 Tsiolkovsky proved that it is feasible for a rocket-propelled craft to travel through the vacuum of space. He calculated that in order to achieve flight into space, speeds of 11.26 kps or 40 232 kph would be needed—the escape velocity for Earth. Known solid fuels were too heavy, so Tsiolkovsky worked out how to use liquid fuels. He also suggested the 'piggyback' or step principle, with one rocket on top of another. When the lower one was expended, it could be jettisoned (reducing the weight) while the next one fired and took over.

http://www.informatics.org/museum/tsiol.html Illustrated account of the life and work of Konstantin Tsiolkovsky. The site is that of the Konstantin E Tsiolkovsky State Museum of the History of Cosmonautics in Kaluga City, Russia. There is discussion of Tsiolkovsky's many works—primarily his far-sighted ideas for space travel, but also his philosophy. The pages contain lots of drawings from Tsiolkovsky's manuscripts.

TSS Abbreviation for *tethered satellite system.

TsSKB-Progress Alternative name for the *Samara Space Centre.

Tsukuba Space Centre (TKSC) A major facility of the Japan Aerospace Exploration Agency (JAXA) located in Tsukuba Science City north of Tokyo. Opened in 1972, it has been responsible for the development of Japanese satellites and rockets as well as tracking and controlling them after launch. Currently it is developing and testing the Japanese Experiment Module (JEM) Kibo for the International Space Station. TKSC also undertakes astronaut training. Japan's Office of Space Flight and Operations, the Office of Space Application, the Institute of Space Technology and Aeronautics, and part of the Institute of Space and Astronautical Science are located at the TKSC.

T-38 jet trainer A jet aircraft developed by US manufacturing company Northrop and used by NASA to prepare astronauts for space flight. Its high-g and low-g manoeuvres accustom astronauts to the conditions occurring during the ascent of a launch vehicles, the free-fall atmosphere of space, and re-entry. The T-38 flights are taken about three days before launch. The US Air Force's first T-38A 'Talon' jet trainer flew on 10 April 1959 and entered service on 17 March 1961 as the force's first supersonic trainer.

TTP Abbreviation for *Technology Transfer Programme.

turbo code An advanced type of coding system to control errors in spacecraft transmissions.

TVC Abbreviation for *thrust vector control.

twang A descriptive term used by space shuttle astronauts for the tension and bend experienced by the shuttle for six seconds before lift-off. This is because the power of the firing rocket engines strains the shuttle while it remains bolted to the launch pad.

TWNC Abbreviation for *two-way non-coherent mode.

two-way non-coherent mode (TWNC) A means of sending data from a spacecraft using an on-board oscillator to generate a downlink frequency. Normally, a coherent (consistent) mode is used for communications, whereby a spacecraft generates a downlink coherent to a strong uplink generated from a ground station. In contrast, in the event that a spacecraft receives a corrupted uplink, the TWNC (pronounced 'twink') creates its downlink signal using the on-board oscillator. For example, TWNC is used when a spacecraft is 'hidden' behind a planet in relation to the position of the ground station.

Tyurin, Mikhail Vladislavovich (1960–) Russian cosmonaut. He was part of the third resident crew on the **International Space Station*, arriving on *Discovery* in August 2001 for four months. From 1984 Tyurin was an engineer with the *Energiya Rocket and Space Complex, researching the psychological aspects of training cosmonauts to manually control a spacecraft. He was selected as a cosmonaut in 1993.

UARS Abbreviation for **Upper-Atmosphere Research Satellite.*

Uchinoura Space Centre (USC) A launch site of the Japan Aerospace Exploration Agency (JAXA), founded in 1962. It is located in Uchinoura on the east of Ohsumi Peninsula, Kagoshima Prefecture, and was originally known as Kagoshima Space Centre. It was re-named in October 2003 when JAXA came into being. USC launches sounding rockets and scientific satellites, and also manages tracking and data. More than 360 rockets as well as 23 satellites and probes have been launched from here, including *Akebono, Yohkoh, Geotail, HALCA, Nozomi,* and *Hayabusa.*

UFO Acronym for *unidentified flying object.

***Uhuru* (or *SAS-1*; Swahili: 'freedom')** The first satellite mission dedicated totally to celestial X-ray astronomy. It was the first of NASA's *Small Astronomy Satellite series, but launched from the San Marco platform in Kenya in December 1970 on the seventh anniversary of Kenyan independence. Before its lifespan ended in March 1973, *Uhuru* discovered the diffuse X-ray emission from clusters of galaxies, and detected 339 X-ray sources from supernova remnants, binaries, and galaxies.

UKISC Acronym for *United Kingdom Industrial Space Committee.

ULDB Abbreviation for *Ultra-Long Duration Balloon.

ullage rocket A spacecraft's small auxiliary rocket that fires in space to shift fuel into the bottom of partly empty tanks of larger rockets about to be fired. The empty space in a propellant tank is called ullage.

Ultra-Long Duration Balloon (ULDB) A NASA project to develop giant helium-filled balloons able to support scientific observations on the edge of space. The balloons are designed to operate above 99% of the Earth's atmosphere, for up to 100 days. The prototype carries a trans-iron galactic element recorder (TIGER), a payload weighing 990 kg, to study outer space. The ULDB is the largest single-chamber, high-pressure balloon ever flown, with a diameter of 59 m and a height of 35 m.

The first two ULDB launches, from Australia in February and March 2001, were unsuccessful. During the first, the balloon developed a leak in its thin plastic skin after a few hours and crashed when brought down by remote control. The second was forced down by shifting high-altitude winds less than 24 hours into its flight. NASA is planning future ULDB missions to study the Sun and find new planets. In March 2003, an ULDB was launched in Australia and terminated after 12 hours due to gas loss.

ultra-stable oscillator (USO) A device carried on some spacecraft to provide a more stable *downlink frequency. In normal conditions, a spacecraft receives a stable *uplink frequency and then uses it to generate a stable downlink frequency. The USO is used when an uplink is impossible, as when the spacecraft passes behind the Moon or a planet. Low-mass oscillators are also used for this purpose, but they are affected by temperature changes and are not highly stable. The USO, however, is kept in a temperature-controlled unit.

ultraviolet astronomy The study of cosmic ultraviolet emissions using artificial satellites. The USA launched a series of satellites for this purpose, receiving the first useful data in 1968. Only a tiny percentage of solar ultraviolet radiation penetrates the atmosphere, this being the less dangerous longer-wavelength ultraviolet radiation. The dangerous shorter-wavelength radiation is absorbed by gases in the ozone layer high in the Earth's upper atmosphere.

The US *Orbiting Astronomical Observatory (OAO) satellites provided scientists with a great deal of information regarding cosmic ultraviolet emissions. *OAO-1*, launched in 1966, failed after only three days, although *OAO-2*, put into orbit in 1968, operated for four years instead of the intended one year, and carried out the first ultraviolet observations of a supernova and also of Uranus. *OAO-3* (*Copernicus*), launched in 1972, continued transmissions into the 1980s and discovered many new ultraviolet sources. The *International Ultraviolet Explorer*, which was launched in January 1978 and ceased operation in September 1996, observed all the main objects in the Solar System (including Halley's comet), stars, galaxies, and the interstellar medium. Since then, observations have moved into the extreme ultraviolet with **ROSAT*, the **Extreme Ultraviolet Explorer*, and the **Far-Ultraviolet Spectroscopic Explorer*.

ultraviolet radiation The light rays invisible to the human eye, of wavelengths from about 4×10^{-4} to 5×10^{-6} millimeters (where the X-ray range begins). Physiologically, they are important but also dangerous, causing the formation of vitamin D in the skin and producing sunburn in excess.

Levels of ultraviolet radiation have risen an average of 6.8% a decade in the northern hemisphere and 9.9% in the southern hemisphere 1972–96, according to data gathered by the *total ozone mapping spectrometer on the *Nimbus 7* satellite.

Ulysses A space probe to study the Sun's poles, launched in 1990 by a US space shuttle. It is a joint project by NASA and the European Space Agency. In February 1992, the gravity of Jupiter swung *Ulysses* on to a path that looped it first under the Sun's south pole in 1994 and then over its north pole in 1995 to study the Sun and solar wind at latitudes not observable from the Earth. *Ulysses* continued orbiting the Sun, overflying the south pole for a second time in 2000 and the north pole in 2001.

http://ulysses.jpl.nasa.gov/ Information and news of the *Ulysses* spacecraft. Instrument payload and scientific objectives are described in detail, as well as the mission time line and flight plan. There are useful links to relevant solar astronomy sites.

umbilical cord (or umbilical) A flexible tube, pipe, or cable that supplies crucial elements, such as oxygen, information, rocket fuel, and electrical supplies. During *extravehicular activity, an umbilical cord runs from the spacecraft to the astronaut's *chest pack, supplying high-pressure oxygen. This cord also contains radio wires, a hose that delivers nitrogen to the *manned manoeuvring unit (MMU) for the jet gun, and a strong nylon tether to keep the umbilical cord from pulling loose.

umbilical tower (formerly launch umbilical tower (LUT); or gantry; or service tower) A tall fixed scaffold framework on a launch pad that is used to assemble and service a rocket on its launch pad. It supplies the rocket with fuel, electricity, and other reserves. The structure includes cranes, lifts, girders, ladders, cables, and floodlights. *Booms, or swing arms, are used by technicians to service various parts of the launch vehicle. These arms swing away at lift-off. The umbilical tower of the Saturn V rocket was 116 m high, and its width was 18 × 34 m at the base, tapering to 12 × 12 m.

umbra The central region of a shadow that is totally dark because no light reaches it, and from which no part of the light source can be seen (compare *penumbra). In astronomy, it is a region of the Earth from which a complete *eclipse of the Sun can be seen.

undershoot boundary The lower edge of a spacecraft's re-entry corridor. If the vehicle enters below this level, the high atmospheric density will lead to overheating.

unidentified flying object (UFO) Any light or object seen in the sky of which the immediate identity is not apparent. Despite unsubstantiated claims, there is no evidence that UFOs are alien spacecraft. On investigation, the vast majority of sightings turn out to have been of natural or identifiable objects, notably bright stars and planets, meteors, aircraft, and satellites, or to have been perpetrated by pranksters. The term **flying saucer** was coined in 1947.

In 1968, the US Air Force sponsored a scientific investigation of UFOs, and came to the conclusion that there was no extraterrestrial cause of these phenomena. Nevertheless, thousands of people in different parts of the world claim to have seen something they believed to be an alien craft. In the vast majority of cases, such objects are described as either cigar-shaped or saucer-shaped, resembling illustrations in old science fiction magazines.

http://www.bufora.org.uk/ British UFO Research Organization home page. Featuring news, selected articles, and events lists, as well as a research and investigation section, this site is ideal for anyone waiting for a close encounter.

http://www.pbs.org/wgbh/nova/aliens/alienhome.html Companion to the US Public Broadcasting Service television programme *Nova*, this page takes a serious approach to alien abduction. It is divided into two sections: the believers and the sceptics. The debate includes renowned scientists, as well as abductees, who weigh in with their opinions on alien kidnappings. An interview with an artist abductee includes sketches of the aliens he claims to have come into contact with. The late astronomer Carl Sagan also states his position and there is a special feature on the physical evidence. Is there any? Find out here.

United Kingdom Industrial Space Committee (UKISC) A principal trade association of the UK space industry. It represents the interests of its member companies (which total about 20), providing a forum for policy and supporting the UK government on space issues.

UKISC works with the *British National Space Centre, briefing government officials on space matters, and providing support to the Parliamentary Space Committee.

United States Geological Survey (USGS) An independent US government body concerned with natural resources. Established in 1879, it celebrated its 125th anniversary in 2004. It has worked with NASA on using satellites for remote-sensing data, such as the USGS images taken from **Landsat 7* showing the eruption of Mount Etna, Sicily, in July 2001. The organization maintains the National Satellite Land Remote-Sensing Data Archive with its collection of images spanning nearly 40 years. USGS also works with NASA on AmericaView, a programme to expand remote-sensing education and promote the availability of natural-science data. *Apollo* astronauts were given training in geology by USGS prior to their explorations of the Moon in the 1960s and early 1970s.

In 2004, USGS had 10 000 scientists, technicians, and support staff in some 400 offices. There are branches in every US state, and in several other nations. Its budget is more than US$1 billion a year.

http://wwwflag.wr.usgs.gov/USGSFlag/Space/ Information on US Geological Survey mapping and imagery of the terrestrial planets and satellites. There are facilities to make custom image maps online, access to IAU approved gazetteers of planetary nomenclature, and many other useful resources.

United States Microgravity Laboratory (USMG) One of two NASA scientific laboratories launched aboard its **Spacelab* module to study the effects of *microgravity (weak gravity causing weightlessness in space). Each USMG carried 14 experiments. Launched in the space shuttle *Columbia*, USMG-1 was launched on 25 June 1992 for a 14-day mission and USMG-2 on 20 October 1995 for a 16-day mission. Among the experiments on both flights were those to grow plants and protein crystals and to study how liquids react under microgravity conditions.

Unity The first US module to be delivered to the **International Space Station* (*ISS*); also known as Node 1. The Unity Node is a cylinder 5.5 m long and 4.6 m in diameter. It is attached to the station's Zarya Control Module and acts as connecting passageway to living and work areas of the *ISS*. Unity was delivered by the space shuttle *Endeavour* on the STS-88 mission in December 1998.

Universal Time (UT) Another name for Greenwich Mean Time. It is based on the rotation of the Earth, which is not quite constant. Since 1972, UT has been replaced by *Universal Time Coordinated (UTC), which is based on uniform atomic time.

Universal Time Coordinated (UTC; or Coordinated Universal Time) A replacement adopted in 1972 for *Universal Time, maintained to scientific standards by atomic clocks. UTC adds or subtracts leap seconds twice a year to adjust for irregularities in the Earth's rotation. UTC is used by NASA's *deep-space network for events recorded by when transmission of the event is received (*see* EARTH-RECEIVED TIME). UTC clocks are also installed on board spacecraft, including space shuttles. However, the times of most events during a flight are measured by *mission elapsed time (MET) clocks.

Universe All of space and its contents, the study of which is called *cosmology. The Universe is thought to be around 13–14 billion years old, and is mostly empty space, dotted with galaxies for as far as telescopes can see. These galaxies are moving further apart as the universe expands. The standard theory of how the Universe came into being and evolved is the *Big Bang theory of an expanding Universe originating in a single explosive event.

The most distant detected galaxies and *quasars lie 10 billion light years or more from Earth. Apart from those galaxies within the *Local Group, all the galaxies we see display *redshifts in their spectra, indicating that they are moving away from us. The further we look into space, the greater are the observed redshifts, which implies that the more distant galaxies are receding at ever greater speeds.

This observation led to the theory of an expanding Universe, first proposed in 1929 by US astronomer Edwin *Hubble, and to Hubble's law, which states that the speed with which one galaxy moves away from another is proportional to its distance from it. Current data suggest that the galaxies are moving apart at a rate of 50–100 kps for every million *parsecs of distance (one parsec equals 3×10^{13} km).

http://sln.fi.edu/planets/planets.html Site designed for teachers and students, with pages on 'Space Science Fact'—the universe as humans know it today—and 'Space Science Fiction'—the universe as humans imagine it might be. Features include planetary fact sheets, information about planets outside the Solar System, virtual trips to black holes and neutron stars, space quotes, and a course in spaceship design.

http://www.windows.ucar.edu/ Dramatic site containing lots of information about the Universe. It contains sections on the Earth and the Solar System, plus myths about the Universe. The articles within each section are available to read at three levels: beginner, intermediate, and advanced. There is also a 'Kid's Space' section that contains games and an 'Ask a Scientist' section.

Universities Space Research Association (USRA) A US non-profit private organization through which universities cooperate with each other, the government, and other organizations to further space science and technology, and to promote education in those areas. It also maintains a system of science councils of scientific experts who guide specific areas of research. It receives grants and contracts from NASA for most of its activities.

USRA, established in 1969, is under the auspices of the National Academy of Sciences. It began with 49 college and university members, and by 2001 had 88 members, including two in Britain, two in Canada, and two in Israel.

USRA also administers the Lunar and Planetary Institute (LPI) at the USRA Center for Advanced Space Studies in Houston, Texas.

uplink A radio signal transmitted from a ground station to a spacecraft.

upper atmosphere The area of the Earth's atmosphere beginning at an altitude of about 16 km. The upper atmosphere is above the troposphere, and includes the stratosphere, mesosphere, and thermosphere. Many satellites orbit in this area, including the *Upper-Atmosphere Research Satellite* at 585 km.

***Upper-Atmosphere Research Satellite* (*UARS*)** NASA's orbiting observatory to investigate the upper atmosphere's structure and behaviour. It was launched in September 1991 aboard the space shuttle *Discovery*. UARS is the first satellite to provide systematic and comprehensive data on the stratosphere, mesosphere, and lower thermosphere. The information will increase understanding of the ways in which human activities and natural events affect these areas. In 2004, seven of its ten instruments were still working.

Among the accomplishments of the spacecraft's ten science instruments have been the infrared mapping of the effects of aerosols, the first global maps of chlorofluorocarbons (CFCs) from space, and the first direct measurement of winds from space.

Uranus The seventh planet from the Sun, discovered by German-born British astronomer William *Herschel in 1781. It is twice as far out as the sixth planet, Saturn. Uranus has a mass 14.5 times that of Earth. The spin axis of Uranus is tilted at 98°, so that each pole points towards the Sun in turn as the planet moves around its orbit, giving extreme seasons.

mean distance from the Sun 2.9 billion km

equatorial diameter 51 100 km

rotation period 17 hours 12 minutes

year 84 Earth years

atmosphere deep atmosphere composed mainly of hydrogen and helium

surface composed primarily of rock and various ices with only about 15% hydrogen and helium, but may also contain heavier elements, which might account for Uranus's mean density being higher than that of Saturn

satellites over 20 known moons, the largest of which is Titania, 1 580 km in diameter

rings 11 rings, composed of rock and dust, around the planet's equator, were detected by the US space probe *Voyager 2* in 1977. The rings are charcoal black and may be debris of former 'moonlets' that have broken up. The ring furthest from the planet centre (51 000 km), Epsilon, is 100 km at its widest point

Uranus has a peculiar magnetic field, in that it is tilted at 60° to the axis of spin, and is displaced about a third of the way from the planet's centre to its surface. Uranus spins from east to west, the opposite of the other planets, with the exception of Venus and Pluto. The rotation rate of the atmosphere varies with latitude, from about 16 hours in mid-southern latitudes to longer than 17 hours at the equator. *Voyager 2* reached Uranus in January 1986 but the planet's clouds proved to be bland and almost featureless.

http://www.solarviews.com/solar/eng/uranus.htm Did you know that Uranus is tipped on its side? Find out more about Uranus, its rings, and its moons at this site. Also included are a table of statistics about the planet, photographs, and animations of its rotation.

Usachev, Yury Vladimirovich (1957–) Russian cosmonaut. He spent two long residencies on the Russian space station *Mir* as a board engineer (January–July 1994 and February–September 1996). In May 2000, he flew on the *Atlantis* space shuttle mission to construct the **International Space Station* (*ISS*), and became a member of its second resident crew in March 2001, transported to the *ISS* on *Discovery* for a four-month stay. He has logged over 670 days in space and taken six space walks. Usachev worked for the *Energiya Rocket and Space Complex, involved in *extravehicular activity training and space construction techniques. He was select as a cosmonaut in 1989.

USC Abbreviation for *Uchinoura Space Centre.

USGS Abbreviation for *United States Geological Survey.

USMG Abbreviation for *United States Microgravity Laboratory.

USO Acronym for *ultra-stable oscillator.

US Space and Rocket Center A commercial centre in Huntsville, Alabama, that claims to have the greatest collection of rocket and space memorabilia in the world. It includes a space museum and simulators: visitors can experience a force of 3 g in a g-force accelerator and sit in an *Apollo* command module. The centre also has the non-profit Space Camp with programmes for young people and adults, including a simulated shuttle mission. Another part of the centre is the Aviation Challenge offering fighter-pilot experience by means of simulators.

UT Abbreviation for *Universal Time.

UTC Abbreviation for coordinated universal time, an alternative name for *universal time coodinated, the standard measurement of time.

VAB Acronym for *Vehicle Assembly Building.

vacuum In general, a region completely empty of matter; in physics, any enclosure in which the gas pressure is considerably less than atmospheric pressure (101 325 pascals).

VAFB Abbreviation for *Vandenberg Air Force Base.

Valles Marineris A complex system of rift valleys and canyons on the surface of *Mars. It was first delineated by the *Mariner 9* spacecraft that surveyed the planet from its orbit in 1971. It is 4 500 km long and up to 8 km deep. Part of the system is visible from Earth-bound telescopes as a dark streak known as Coprates.

Van Allen, James Alfred (1914–) US physicist whose instruments aboard the first US satellite *Explorer 1* in 1958 led to the discovery of the Van Allen belts, two zones of intense radiation around the Earth. He pioneered high-altitude research with rockets after World War II.

After the end of World War II, Van Allen began utilizing unused German V2 rockets to measure levels of cosmic radiation in the outer atmosphere, the data being radioed back to Earth. He then conceived of rocket-balloons (rockoons), which began to be used in 1952. They consisted of a small rocket that was lifted by means of a balloon into the stratosphere and then fired off.

Van Allen radiation belts Two zones of charged particles around the Earth's magnetosphere, discovered in 1958 by US physicist James Van Allen. The atomic particles come from the Earth's upper atmosphere and the *solar wind, and are trapped by the Earth's magnetic field. The inner belt lies 1 000–5 000 km above the Equator, and contains protons and electrons. The outer belt lies 15 000–25 000 km above the Equator, but is lower around the magnetic poles. It contains mostly electrons from the solar wind.

The Van Allen belts are hazardous to astronauts and interfere with electronic equipment on satellites.

Vandenberg Air Force Base (VAFB) A US Air Force base in California that is home to the Air Force Space Command. Vandenberg is the only US military base from which NASA and commercial satellites are launched into polar orbit.

The base, about 240 km northwest of Los Angeles, covers 39 893 ha and has about 3 000 buildings. The Command's 30th Space Wing operates the base, being responsible for all space and missile activities on the US West Coast, and the 30th Operational Group is responsible for the *Western Test Range. Vandenberg began in 1941 as the US Army's Camp Cooke and in 1957 was transferred to the air force to become the nation's first space and ballistic missiles operational and training base.

http://www.vandenberg.af.mil/ Details of forthcoming launches from Vandenberg. The site includes images and video clips of launches, and a comprehensive history of the air force base.

Vanguard An early series of US Earth-orbiting satellites and their associated rocket launcher. *Vanguard 1* was the second US satellite, launched on 17 March 1958 by the three-stage Vanguard rocket. Tracking of its orbit revealed that the Earth is slightly pear-shaped. The series ended in September 1959 with *Vanguard 3*.

van Hoften, James Dougal Adrianus (1944–) US astronaut. He flew on the *Challenger* mission in April 1984 to repair the Solar Max satellite. In August 1985, he flew on *Discovery* to deploy three communications satellites. He logged a total of 22 hours of extravehicular activity. Van Hoften was selected as an astronaut in 1978 and left NASA in 1986.

variable star A star whose brightness changes, either regularly or irregularly, over a period ranging from a few hours to months or years. The *Cepheid variables regularly expand and contract in size every few days or weeks.

Stars that change in size and brightness at less precise intervals include **long-period variables**, such as the red giant Mira in the constellation Cetus (period about 331 days), and **irregular variables**, such as some red supergiants.

Eruptive variables emit sudden outbursts of light. Some suffer flares on their surfaces, while others, such as a *nova, result from transfer of gas between a close pair of stars. A *supernova is the explosive death of a star. In an *eclipsing binary, the variation is due not to any change in the star itself but to the periodic eclipse of a star by a close companion. The different types of variability are closely related to different stages of stellar evolution.

Veach, Charles Lacy (1944–95) US astronaut. His two space shuttle flights include the *Discovery* mission in April 1991 during which he was the primary operator of the *Canadarm. He then worked as the lead astronaut with robotics for the **International Space Station*. Veach joined NASA's Johnson Space Center in 1982 as an engineer and research pilot. He was selected as an astronaut in 1984.

vectored thrust A rocket thrust directed by swivelling nozzles. This is used to launch a spacecraft into its proper trajectory, and later to correct its attitude in space.

Vega A pair of Soviet space probes launched towards *Venus in December 1984 that later visited Halley's Comet.

On passing Venus in June 1985 both spacecraft released landing modules. During the descent, each lander then released a 3.4-m sounding balloon that floated at a height of 54 km in the atmosphere of Venus, travelling 11 000 km in two days. Precise tracking of the two balloons by radio telescopes on Earth (using *very long-baseline interferometry to obtain high-resolution images) allowed astronomers to study the atmospheric circulation of Venus. In March 1986, the two *Vega* probes passed within 9 000 km of the nucleus of Halley's Comet.

Vehicle Assembly Building (VAB) A NASA building, one of the largest constructions in the world, used to support space shuttle operations. It has a total capacity of 3 664 883 cubic m, equal to 3.75 times that of the Empire State Building, New York, and was originally built to assemble the *Apollo*–Saturn vehicles. The VAB is 160 m tall, 218 m long, and 158 m wide. It has a low bay area for maintenance and overhaul of the shuttle's main engine.

(Image © NASA)

Within the Kennedy Space Center's gigantic Vehicle Assembly Building, space shuttle *Discovery* is mated with the external fuel tank and solid rocket booster assembly already mounted on the mobile launcher platform.

The building has four high bays, each with a door 139 m high. Bays 1 and 3 are used to integrate and stack the complete shuttle vehicle, and 2 and 4 to check out and store the external tank, which is then attached to the shuttle waiting in 1 or 3. The completed stack, with solid rocket boosters, is put onto a mobile launcher platform and taken from the VAB to the launch pad by the *crawler-transporter.

Velcro The trademark fastener with tiny loops and hooks. It is used in space to fasten down items that would otherwise float when in zero gravity. In a space shuttle, such things as checklists, food containers and trays, towels, and medical kits are secured with Velcro. It is also used on the *foot restraints that keep crew members anchored to the deck.

Venera A series of 16 Soviet space probes launched towards *Venus 1961–83.
Venera 4 (October 1967), and *Venera 5* and *Venera 6* (May 1969), entered the atmosphere of the planet but were destroyed by the intense heat and pressure before they reached the surface. *Venera 7* became the first spacecraft to land on Venus, on 15 December 1970, followed by *Venera 8* in July 1972, *Venera 9* and *Venera 10* in October 1975 (first pictures from the surface), *Venera 11* and *Venera 12* in December 1978, and *Venera 13* and *Venera 14* in March 1982 (first soil analysis). *Venera 15* and *Venera 16*, arriving in orbit in October 1983, made radar maps of the cloud-covered planet.

vent out Used to release air, gases, or liquids, such as fuel, from a spacecraft into space.

Venus The second planet from the Sun. It can approach Earth to within 38 million km, closer than any other planet. Its mass is 0.82 that of Earth. Venus rotates on its axis more slowly than any other planet, from east to west, the opposite direction to the other planets (except Uranus and Pluto).
mean distance from the Sun 108.2 million km
equatorial diameter 12 100 km
rotation period 243 Earth days
year 225 Earth days
atmosphere Venus is shrouded by clouds of sulphuric acid droplets that sweep across the planet from east to west every four days. The atmosphere is almost entirely carbon dioxide, which traps the Sun's heat by the greenhouse effect and raises the planet's surface temperature to 480 °C, with an atmospheric pressure of 90 times that at the surface of the Earth
surface consists mainly of silicate rock and may have an interior structure similar to that of Earth: an iron-nickel core, a mantle composed of more mafic rocks (rocks made of one or more ferromagnesian, dark-coloured minerals), and a thin siliceous outer crust. The surface is dotted with deep impact craters. Some of Venus's volcanoes may still be active
satellites no moons
The first artificial object to hit another planet was the Soviet probe *Venera 3*, which crashed on Venus on 1 March 1966. Later **Venera* probes parachuted down through the atmosphere and landed successfully on its surface, analysing surface material and sending back information and pictures. In December 1978, NASA's **Pioneer Venus* probe went into orbit around the planet and mapped most of its surface by radar, which penetrates clouds. In 1992 the

(Image © NASA)
Global view of Venus compiled from images acquired with the synthetic aperture radar carried by the *Magellan* spacecraft, with data gaps filled using earlier *Pioneer Venus* data. The image, centered on 0° longitude, shows the wide variety of exotic landforms discovered on Venus.

US space probe **Magellan* mapped 99% of the planet's surface to a resolution of 100 m.

The largest highland area is Aphrodite Terra near the equator, half the size of Africa. The highest mountains are on the northern highland region of Ishtar Terra, where the massif of Maxwell Montes rises to 10 600 m above the average surface level. The highland areas on Venus were formed by volcanoes.

Venus has an ion-packed tail 45 million km in length that stretches away from the Sun and is caused by the bombardment of the ions in Venus's upper atmosphere by the solar wind.

http://www.jpl.nasa.gov/magellan/images.html Library of images of Venus from NASA's *Magellan* Project home page. Each image has a full explanation of the subject matter.

http://www.solarviews.com/solar/eng/venus.htm All you ever wanted to know about the planet Venus can be found at this site, which includes a table of statistics, photographs of the planet, animations, information about its volcanic features and impact craters, plus a chronology of exploration.

Venus Express The European Space Agency probe to Venus, based on the same design as ESA's earlier **Mars Express* spacecraft. *Venus Express* will go into a highly elliptical polar orbit, ranging between 250 km and 66 000 km from Venus, from which it will study the planet's atmosphere and surface for a planned 500 Earth days (about two Venusian years). The Venus Monitoring Camera (VMC) will image the planet's clouds at optical and ultraviolet wavelengths, while other instruments plumb the atmosphere to elucidate its structure, circulation, and interaction with the surface. *Venus Express* is scheduled for launch in October 2005.

very long-baseline interferometry (VLBI) In radio astronomy, a method of obtaining high-resolution images of astronomical objects by combining simultaneous observations made by two or more radio telescopes thousands of kilometres apart.

The maximum resolution that can be achieved is proportional to the longest baseline in the array (the distance between any pair of telescopes), and inversely proportional to the radio wavelength being used.

Vesta The third-largest asteroid in the Solar System, 530 km in diameter, discovered in 1807 by German astronomer Heinrich Olbers. Vesta orbits the Sun at a distance of 353 million km, with a period (time taken to circle the Sun) of 3.63 years. It is the only asteroid that ever becomes bright enough to be seen without a telescope. Vesta has a shallow crater at its south pole indicating that a large chunk of rock has been dislodged at some point.

Viking Either of two US space probes to the planet Mars, each one consisting of an orbiter and a lander. They were launched on 20 August and 9 September 1975, and transmitted colour pictures and analysed the planet's soil.

Viking 1 carried life-detection laboratories and landed on Mars on 20 July 1976 to carry out detailed research. Designed to work for 90 days, the craft operated for 6.5 years, losing contact with Earth in November 1982. *Viking 2* landed on Mars on 3 September 1976 and functioned for 3.5 years.

VLBI Abbreviation for *very long-baseline interferometry.

Voenno-Kosmicheski Sily (or VKS; Russian: 'military space forces') Russia's military organization in charge of space activities. The VKS controls the *Plesetsk Cosmodrome launch facility, where many military satellites have been launched. The organization also shares control (with *Rosaviakosmos) of the *Baikonur Cosmodrome and the *Gagarin Cosmonaut Training Centre.

voice-operated relay (VOR) A communication device developed for the *Apollo* helmet. When an astronaut spoke into his microphone, the VOR sensor relayed it to either the intercom system or to the intercom and radio circuits.

von Braun, Wernher Magnus Maximilian (1912–1977) German rocket engineer responsible for Germany's rocket development programme in World War II (V1 and V2), who later worked for the space agency *NASA in the USA. He also invented the *Saturn rocket (Saturn V) that sent the **Apollo* project spacecraft to the Moon in 1969.

During the 1940s he was technical director of the research team at Peenemünde on the Baltic coast that produced the V1 (flying bomb) and supersonic V2 rockets. In the 1950s von Braun was part of the team that produced rockets for US satellites (the first, *Explorer 1*, was launched early 1958) and early space flights by astronauts.

Von Braun was born in Wirsitz (now in Poland) and studied at Berlin and in Switzerland at Zürich. In 1930 he joined a group of scientists who were experimenting with rockets and in 1938 became technical director of the Peenemünde military rocket establishment; he joined the Nazi Party 1940. In the last days of the war in 1945 von Braun and his staff, not wishing to be captured in the Soviet-occupied part of Germany, travelled to the West to surrender to US forces. Soon afterwards von Braun began work at the US Army Ordnance Corps testing grounds at White Sands, New Mexico. In 1952 he became technical director of the army's ballistic-missile programme. He held an administrative post at NASA 1970–2, and founded the National Space Society.

http://liftoff.msfc.nasa.gov/academy/history/vonBraun/vonBraun.html Biographical sketch of German rocket engineer Wernher von Braun. These pages document his early years, his life in Germany, his escape to the West, and his work at the Marshall Space Flight Center. There are photographs and links to several other pages with biographical information.

VOR Acronym for *voice-operated relay.

***Voskhod* (Russian: 'ascent')** Two Soviet multi-man spacecraft modified from the single-seat *Vostok*. The *Vostok* ejector seat was removed to make room for two or three cosmonauts, but the overall size of the craft was the same. *Voskhod 1*, launched on 12 October 1964, carried a crew of three on a day-long, 16-orbit flight. *Voskhod 2*, launched on 18 March 1965, carried only two cosmonauts. The space for the third was filled by an extendable airlock through which Aleksei *Leonov crawled to make the first-ever space walk. *Voskhod* was launched by the same rocket as *Vostok* but with an upgraded second stage.

Voss, James Shelton (1949–) US astronaut. His space flights include the *Atlantis* mission in May 2000 to construct the **International Space Station (ISS)* and install equipment and supplies. In March 2001, he became a member of the second resident crew on the *ISS*, transported there on the space shuttle *Discovery* for a four-month stay. He has logged 201 days in space with four space walks totalling over 22 hours. Voss joined NASA's Johnson Space Center in 1984 as a vehicle integration test engineer. He was selected as an astronaut in 1987. He went on to be the deputy for flight operations in the Mission Integration and Operations Office for the Space Station Program.

Voss, Janice Elaine (1956–) US astronaut. She has logged over 49 days in space on five shuttle flights, including the *Discovery* mission in February 1995 that docked with the Russian space station *Mir*, and the February 2000 flight of *Endeavour* on a radar mapping mission. Voss worked at NASA's Johnson Space Center on computer simulations from 1973–5. She became a crew trainer in 1977, joined the *Orbital Sciences Corporation in 1987, and was selected as an astronaut in 1990. She went on to join the Space Station Program Office.

***Vostok* (Russian: 'east')** A Soviet single-seater spacecraft consisting of a metal sphere 2.3 m in diameter in which the cosmonaut sat on an ejector seat. After re-entry of *Vostok* through the atmosphere, the cosmonaut ejected from the capsule and landed separately by parachute. *Vostok 1*, launched on 12 April

(Image © NASA)
Photo montage of the gas giants from the *Voyager* 'grand tour'. Bottom to top, with increasing distance from the inner Solar System, are Jupiter, Saturn, Uranus, and Neptune.

1961, carried the first human in space, Yuri *Gagarin, who made a single-orbit flight lasting 108 minutes. On *Vostok 2*, launched on 6 August 1961, Gherman *Titov made the first day-long space flight. *Vostok 3* was launched on 11 August 1962 carrying Andrian *Nikolayev, who was joined in orbit a day later by Pavel *Popovich aboard *Vostok 4*. On 14 June 1963 Valery *Bykovsky was launched in *Vostok 5* on a five-day flight, during which he made 81 orbits of the Earth, the longest-ever individual space flight. While *Vostok 5* was still in orbit, the first woman in space, Valentina *Tereshkova, was launched aboard *Vostok 6*, the last of the series. The *Vostok* was little more than an automatic capsule with a passenger, and the basic design is still used to carry payloads into Earth orbit that need to be recovered afterwards, such as biological specimens and reconnaissance cameras.

Voyager Either of two US space probes. *Voyager 1*, launched on 5 September 1977, passed the planet Jupiter in March 1979, and reached Saturn in November 1980. *Voyager 2* was launched earlier, on 20 August 1977, on a slower trajectory that took it past Jupiter in July 1979, Saturn in August 1981, Uranus in January 1986, and Neptune in August 1989. Like the **Pioneer* probes, the *Voyagers* are on their way out of the Solar System.

Their tasks include helping scientists to locate the heliopause, the boundary at which the influence of the Sun gives way to the forces exerted by other stars. Both *Voyagers* carry coded recordings called 'Sounds of Earth', intended to enlighten any other civilizations that might find them.

Voyager 2 was not intended to visit Uranus and Neptune, but scientists were able to reprogram its computer to take it past those planets. *Voyager 2* passed by Neptune at an altitude of 4 800 km; its radio signals took 4 hours 6 minutes to reach Earth.

Wakata, Koichi (1963–) Japanese astronaut. He was the first Japanese *mission specialist on a space shuttle mission when he flew on *Endeavour* in January 1996 to retrieve the Space Flyer Unit, an uncrewed space experiment platform, launched from Japan. He also flew on the *Discovery* mission in October 2000 to expand the **International Space Station*. Wakata was selected as an astronaut with the *National Space Development Agency in 1992 and assigned to NASA's Johnson Space Center for training in the same year.

Walheim, Rex Joseph (1962–) US astronaut. He flew on the *Atlantis* flight in April 2002 to the **International Space Station*, making two space walks to prepare the station for future space walks. He has logged over 259 days in space and over 14 hours of space walks. Walheim was a flight controller and operations engineer at NASA's Johnson Space Center from 1986 to 1989. He was selected as an astronaut in 1996.

Walker, David Mathieson (1944–2001) US astronaut. His four space shuttle flights include the *Discovery* space shuttle mission in May 1984 that deployed the **Magellan* spacecraft. From 1994 he chaired the Johnson Space Center safety review board. Walker was selected as an astronaut in 1978 and left NASA in 1996.

walking orbit (or precessing orbit) An orbit that drifts slowly in relationship to the surrounding fixed inertial space. This is a helpful occurrence for a spacecraft studying a planet's surface. A walking orbit around a body in space is caused by gravitational influences of other bodies, such as a moon or the Sun. It may also be caused by a planet that is not a perfect sphere. These factors can be configured into a spacecraft's trajectory in order to create a walking orbit.

Wallops Flight Facility A NASA principal facility for managing and implementing sub-orbital research programmes, and one of the world's oldest launching sites. Located on Wallops Island off the Virginia shore, it manages the *Spartan 201 programme and University Class Explorer missions for low-cost experiments. It is also responsible for orbital tracking projects and small payloads for the space shuttle, such as the *Getaway Special, and for payloads for sounding rocket and balloon programmes.

Established in 1945 by the *National Advisory Committee for Aeronautics, NASA's predecessor, Wallops Flight Facility and its 900 employees are now part of the *Goddard Space Flight Center. The facility is open to industry for commercial launches and for space and aeronautics research.

http://www.wff.nasa.gov/ Updated regularly with news of launches from the flight facility, this site also details the installation's technical capabilities and the role it plays in the support of NASA projects. There is information on a variety of experimental aircraft.

Walz, Carl Erwin (1955–) US astronaut. His four space shuttle flights include the *Atlantis* mission in September 1996 that docked with the Russian space station *Mir*. He flew on *Endeavour* to the *International Space Station* as a member of the fourth resident crew. His 196 days there set the US endurance record in space, along with fellow crew member Daniel Bursch. He has logged 231 days in space and is now serving in the Office of Space Science at NASA's headquarters in Washington, DC. Walz was selected as an astronaut in 1990.

waste management system (WMS) NASA's term for toilet facilities on a spacecraft or the method of waste disposal in a spacesuit. On the space shuttle, a mid-deck compartment holds the toilet, which uses airflow to collect waste into a sealed chamber. The early astronauts were supplied only with plastic bags and hoses. Spacesuit waste management systems have included a urine collection device and a faecal containment system.

WDC SI Abbreviation for *World Data Center for Satellite Information.

Webb, James (Edwin) (1907–1992) US business executive and government official. A Sperry Gyroscope executive (1936–41) and Marine Corps major, he served as US president Harry Truman's budget director (1946–9), balancing the budget before joining the state department (1949–52). As NASA director from 1961, he oversaw the *Mercury*, *Gemini*, and *Apollo* programmes that put US astronauts ahead of the Russians. Webb left NASA in 1968 to head a science foundation.

Weber, Mary Ellen (1962–) US astronaut. Her space shuttle flights include the *Discovery* mission in July 1995 to launch NASA's TDRS-G communications satellite, and the *Atlantis* mission in May 2000 to construct the **International Space Station*. She logged more than 450 hours in space. Weber was selected as an astronaut in 1992. She was the Legislative Affairs liaison with Congress at NASA Headquarters in Washington, DC, in 1995 and resigned from NASA in 2002. She went on to be the associate vice-president at the University of Texas Southwestern Medical Center in Dallas.

weightlessness The apparent loss in weight of a body in free fall (see illus. on p. 368). Astronauts in an orbiting spacecraft do not feel any weight because they are falling freely in the Earth's gravitational field (not because they are beyond the influence of Earth's gravity). The same phenomenon can be experienced in a falling lift or in an aircraft imitating the path of a freely falling object.

Weitz, Paul Joseph (1932–) US astronaut. He was the pilot of **Skylab* 2 that was launched as the first crewed mission of the spacecraft in June 1973. In April 1983, he flew as commander of the first *Challenger* space shuttle mission. Weitz was selected as an astronaut in 1966, and retired in 1988 to become deputy director of the Johnson Space Center. He retired from NASA in 1994.

Western Test Range A large area of the Pacific Ocean west of California that is used for US satellite launches and missile testing. All of the nation's polar satellites are launched here.

The range is operated by the US Air Force's 30th Range Squadron within its 30th Operational Group, and all launches are the responsibility of the 2nd Space Launch Squadron. The vast range begins at *Vandenberg Air Force Base

(Image © NASA)
Astronaut Janice Voss experiencing weightlessness aboard a DC-9 aircraft. Parabolic trajectories are flown by aircraft such as this and the notorious KC-135 'vomit comet', to produce weightless conditions for experimental and training purposes.

and extends westward over the ocean. It includes tracking systems along the Pacific Coast.

Wetherbee, James Donald (1952–) US astronaut. He became the first US astronaut to command five space missions. His six space shuttle flights include two dockings with the Russian space station *Mir* when he was mission commander: on *Discovery* in February 1995 and on *Atlantis* in September 1997. He also commanded the March 2001 *Discovery* mission that delivered the second resident crew to the **International Space Station*, and the *Endeavour* flight in November 2002 to deliver the sixth resident crew. By 2004, he had logged more than 1 592 hours in space. Wetherbee was selected as an astronaut in 1984.

wet-trash stowage compartment A compartment in a space shuttle to store any rubbish that may produce gas (termed 'wet trash'). The unit, which is under the mid-deck floor, has a capacity of 0.2 cu m. It holds disposable wet-trash bag liners that are vented overboard each day.

wet workshop NASA's original plan to transform a rocket fuel-tank in orbit into its **Skylab* space station. To reduce financial costs, it was envisaged that astronauts would be launched into space by the hydrogen tank. They would then empty the tank of remaining fuel, pump in oxygen, and equip the station. After several Moon missions were cancelled, however, a powerful Saturn V rocket became available to launch an already fitted-out *Skylab* into orbit.

Wheelock, Douglas Harry (1960–) US astronaut. He was selected as an astronaut in 1998 and assigned to the Space Station Operations Branch of NASA's Astronaut Office awaiting selection for a space shuttle flight. He became a Russian liaison for the Space Station Operations Branch, working in Moscow, then served as a crew support astronaut for the second and fourth resident crews of the *International Space Station*. Since 2002, he has been a CapCom (spacecraft communicator) in the Mission Control Center in Houston, Texas. Prior to his selection as an astronaut, Wheelock worked for NASA's Johnson Space Center as an engineer for space shuttle missions from 1996.

Whipple shield A device for protecting spacecraft against high-speed impacts by debris. A Whipple shield, also known as a bumper shield, consists of an outer skin of metal around a spacecraft which will vaporize dust particles before they reach the craft itself. The shield and spacecraft are separated by a gap called a standoff. In a 'stuffed' Whipple shield, the gap can be filled with blankets of ceramic cloth and other toughened materials as used in bullet-proof vests to increase protection against larger impacts. Such shields are named after the American astronomer and comet expert Fred L. Whipple (1906–2004) who invented the concept in 1946.

White, Edward Higgins, II (1930–67) US astronaut. While accompanying Gus Grissom on the *Gemini 4* mission, he became the first US astronaut to undertake a space walk (see illus. on p. 370). On 27 January 1967 at Cape Kennedy, White, Grissom, and Commander Roger Chaffee of the US Navy were killed in a flash fire that swept their capsule during training for the first mission of the **Apollo* project.

White was born in San Antonio, Texas, educated at the US Military Academy, and joined the US Air Force in 1952. He was the second person to walk in space, the first being the Soviet cosmonaut Alexei Leonov, on 18 March 1965.

white dwarf A small, hot *star, the last stage in the life of a star such as the Sun. White dwarfs make up 10% of the stars in the Galaxy; most have a mass 60% of that of the Sun, but only 1% of the Sun's diameter, similar in size to the Earth. Most have surface temperatures of 8 000 °C or more, hotter than the

(Image © NASA)

Edward White attached to his *Gemini* spacecraft by a 7.6-m umbilical line and tether, floats freely in space—the first US astronaut to do so. In his left hand is the hand held manoeuvring unit (HHSMU) with which he propelled himself.

Sun. However, being so small, their overall luminosities may be less than 1% of that of the Sun. The Milky Way contains an estimated 50 billion white dwarfs.

White dwarfs consist of degenerate matter in which gravity has packed the protons and electrons together as tightly as is physically possible, so that a spoonful of it weighs several tonnes. White dwarfs are thought to be the shrunken remains of stars that have exhausted their internal energy supplies. They slowly cool and fade over billions of years.

white room A sterile room where NASA astronauts wait to board a space shuttle for launching. It is located at the end of the *catwalk leading to the shuttle and positioned next to the latter's open hatch. In the white room, technicians in lint-free uniforms help the waiting crew members into emergency-escape harnesses and lead them into the shuttle (always in the order of commander, pilot, mission specialists, and payload specialists).

White Sands Test Facility (WSTF) The primary area, situated in southern New Mexico, for space shuttle pilots to practice landings in a shuttle-training aircraft. It also tests spacecraft propulsion.

wicket tab A device on spacecraft to help an astronaut activate controls when vision is poor. Tabs provide raised tactile information, such as where the required controls are located, their particular function, and the sequence of activation. In the space shuttle, wicket tabs are placed on controls that are difficult to see by seated crew members during the ascent and re-entry phases.

***Wide-Field Infrared Explorer* (*WIRE*)** A NASA spacecraft launched in March 1999 to study infrared emissions from starburst galaxies, where rapid star formation occurs, and protogalaxies, which are infant galaxies. *WIRE* was one of the first missions of NASA's *Origins Program and part of its Small Explorer Program. The original four-month mission was aborted because an electronics box was incorrectly designed and ejected the cover of its telescope early, causing frozen hydrogen (used to cool infrared detectors) to turn into gas when exposed to the Sun. NASA is now using *WIRE*'s *star tracker for long-term monitoring of bright stars to probe their structure and support the agency's planet-finding programme.

Wilcutt, Terrence Wade (1949–) US astronaut. He has logged over 1 007 hours in space on four shuttle flights, including two dockings with the Russian space station *Mir*: as pilot of *Atlantis* in September 1996, and as commander of *Endeavour* in January 1998. He also commanded the September 2000 mission of *Atlantis* when it prepared the **International Space Station* for its first resident crew. His 2004 shuttle flight was postponed following the *Columbia* space shuttle disaster the previous year. Wilcutt was selected as an astronaut in 1990.

***Wilkinson Microwave Anisotropy Probe* (*WMAP*)** A NASA probe designed to make a full-sky map of temperature fluctuations in the *cosmic background radiation with much higher resolution, sensitivity, and accuracy than NASA's 1992 **Cosmic Background Explorer* satellite. *WMAP* detects anisotropy (tiny fluctuations) in the temperature of the cosmic background radiation to provide a detailed picture of the early universe. Its scientific observation point lies about 1.5 million km from the Earth, in order to prevent interference from terrestrial or solar microwave emissions. *WMAP* was launched on 30 June 2001 for an 18-month mission. In 2002 it completed full sky scans in April and August, and the *WMAP* team extended its mission for several years.

Williams, Dave (born Dafydd Rhys Williams) (1954–) Canadian astronaut and physician. He was a member of the crew on the 16-day **Spacelab* flight aboard the space shuttle *Columbia* in April 1998. The mission investigated the effects of *microgravity on the brain and nervous system. He was selected as an astronaut by the Canadian Space Agency in 1992. In 1995 he was assigned for training at NASA's Johnson Space Center, and from 1998 to 2002 was director of its Space and Life Sciences Directorate. His 2004 shuttle flight was postponed because of the *Columbia* space shuttle disaster the previous year.

Williams, Donald Edward (1942–) US astronaut. He flew on the space shuttle *Discovery* in April 1985, whose crew included the first US senator in space, Jake *Garn; and on the October 1989 *Atlantis* mission that deployed the **Galileo* spacecraft. He was selected as an astronaut in 1978 and retired from NASA in 1990.

Williams, Jeffrey Nels (1958–) US astronaut. He flew with the May 2000 mission of the *Atlantis* space shuttle that delivered equipment and supplies to the **International Space Station (ISS)*. During the mission he undertook a space walk of more than six hours. He logged over 236 hours in space and over six hours on space walks. Williams was a US Army pilot assigned to NASA's Johnson Space Center from 1987 to 1992 and a test pilot from 1993. He was selected as an astronaut in 1996. He went on to train for a future flight to the *ISS*.

window of opportunity A term that usually refers to the time frame available to launch a spacecraft to successfully complete its mission. This may vary from a few minutes to more than a week. The length of a window of opportunity depends upon the relative position of a target planet, moon, or other celestial body. A window of opportunity is also calculated for landing a spacecraft, such as a returning space shuttle.

wind profiling The vertical images of wind movements created by spacecraft using three *Doppler radar. One radar points down vertically and two radar point at small angles to this. Wind profiles can also be taken on other planets. A prototype miniature Doppler *lidar (detection device based on a laser) for wind profiling on Mars was under development at NASA's Jet Propulsion Laboratory in 2001.

WIRE contraction of **Wide-Field Infrared Explorer*.

Wisoff, Peter Jeffrey Kelsay (1958–) US astronaut. His four space shuttle flights include the January 1997 mission of *Atlantis* that docked with the Russian space station *Mir*. In October 2000, he flew with the *Discovery* mission to expand the **International Space Station*. He has logged nearly 43 days in space and nearly 20 hours of *extravehicular activity during three space walks. Wisoff was selected as an astronaut in 1990.

WMAP Abbreviation for **Wilkinson Microwave Anisotropy Probe*.

WMS Abbreviation for *waste management system.

Wolf, David Alexander (1956–) US astronaut and physician. He flew on the space shuttle *Columbia* mission in October 1993 to conduct experiments on the effect of *microgravity on humans. He spent 119 days aboard the Russian space station *Mir* from September 1997 to January 1998. He then flew on *Atlantis* in October 2002 to the **International Space Station* on an assembly mission, making three space walks. He logged over 143 days in space and 28 hours on space walks. Wolf joined the Medical Sciences Division of NASA's *Johnson Space Center in 1983, and was selected as an astronaut in 1990. He has since left NASA.

Worden, Alfred Merrill (1932–) US astronaut. He was the command module pilot for *Apollo 15*, launched to the Moon in July 1971. He orbited the Moon for three days while David *Scott and Jim *Irwin explored the lunar surface. Worden mapped the Moon from his orbit and later took the first deep-space extravehicular activity. He was selected as an astronaut in 1966 and left NASA in 1975.

World Data Center for Satellite Information (WDC SI; formerly World Data Center-A for Rockets & Satellites) A NASA data archive and exchange centre in the *National Space Science Data Center (NSSDC) at the Goddard Space Flight Center. Its information, which is available to visitors during normal working hours, includes launches of satellites, space probes, and rockets, and descriptions of spacecraft and experiments. It also handles requests for NSSDC data from around the world and publishes the **SPACEWARN* bulletin.

World Space Week An annual international week celebrating the importance of space and its exploration. The United Nations General Assembly established the event and the dates of 4–10 October, and it coordinates the celebrations along with the non-profit organization Spaceweek International Association. Worldwide events are held to highlight the benefits of space exploration, maintain funding for space programmes, and promote institutions involved in space.

WSTF Abbreviation for *White Sands Test Facility.

X-20 Alternative name for **Dyna-Soar*.

X-43A NASA's hypersonic research aircraft that broke the world speed record in March 2004. It flew at Mach 5, just over seven times the speed of sound, or about 8 000 kph. The experimental craft, which is 3.56 m long, was launched from a converted B-52 bomber on top of a Pegasus booster rocket off the coast of southern California. It flew on its own power for six minutes, reaching an altitude of nearly 30 500 m.

The *X-43A*, known as a ramjet or scramjet, burns hydrogen mixed with oxygen from the atmosphere. Heavy fuel tanks are not needed. The first *X-43A* flight, on 2 June 2001, failed after its Pegasus rocket veered off course.

X axis An imaginary line that runs through a spacecraft from fore to aft.

XEUS A proposed European Space Agency observatory for X-ray astronomy, known in full as the X-ray Evolving Universe Spectrometer mission. It is the potential follow-on to **XMM-Newton* and is being jointly studied by ESA and the Japan Aerospace Exploration Agency (JAXA). As currently envisioned, *XEUS* will consist of two spacecraft, one containing the mirror and the other containing the detectors, flying in formation at the L2 *Lagrangian point of the Earth's orbit. *XEUS* would be launched around 2015, with a second mission utilizing a larger mirror some 10 years later.

XMM-Newton A European Space Agency (ESA) spacecraft launched in December 1999 to make simultaneous X-ray and optical observations of the sky. It carries three gold-coated imaging X-ray telescopes and an optical monitor, the first on an X-ray observatory. It has an expected lifetime of ten years and is the second cornerstone of ESA's *Horizon 2000 Science Programme.

X prize Former name of the *Ansari X prize.

X-ray astronomy The detection of X-rays from intensely hot gas in the universe. Such X-rays are prevented from reaching the Earth's surface by the atmosphere, so detectors must be placed in rockets and satellites. The first celestial X-ray source, Scorpius X-1, was discovered by a rocket flight in 1962.

Since 1970, special satellites have been put into orbit to study X-rays from the Sun, stars, and galaxies. These include the **Chandra X-Ray Observatory* and the **Röntgen Satellite*. Many X-ray sources are believed to be gas falling in to *neutron stars and *black holes.

http://www.xray.mpe.mpg.de/ Impressive X-ray resource at the Max Planck-Institut für Extraterrestrische Physik. There are details of X-ray missions, institute projects, and publications, and a well-organized image gallery.

http://sci.esa.int/xmm/ Covers all aspects of ESA's *XMM-Newton* Mission. There are regular news updates and superb new images returned by the spacecraft. There are pages profiling the instruments carried and an interactive display showing the sky at different wavelengths.

X-ray telescope A telescope designed to receive *electromagnetic waves in the X-ray part of the spectrum. X-rays cannot be focused by lenses or mirrors in the same way as visible light, and a variety of alternative techniques is used to form images. Because X-rays cannot penetrate the Earth's atmosphere, X-ray telescopes are mounted on *satellites, *rockets, or high-flying balloons.

X-series A NASA series of experimental aircraft and spaceplanes that acted as prototypes for operational vehicles. The series started in 1946, and the *X-1* flew 49 missions before breaking the sound barrier on 14 October 1947. In 1962, the *X-15* was the first plane to reach space, and was flown at a record atmospheric speed of Mach 6.7 in 1967. The *X-15* served as an important test bed for **Apollo* project and *space shuttle technology. The *X-23* and *X-24* acted as space shuttle aerodynamic prototypes in 1969–75. The *X-38* was a flight demonstrator model for the *International Space Station*'s *crew return vehicle, built in 1997. Most of the later X-series planes are for aeronautical use.

Yang Liwei (1965–) Chinese astronaut. The country's first person in space, Lt Col Yang was launched on 15 October 2003 aboard the *Shenzhou-5* spacecraft for a flight that lasted 21 hours and 23 minutes. His reentry capsule landed by parachute in Inner Mongolia.

Yang was born in Suizhong County in Liaoning Province in northeast China. He joined the Chinese People's Liberation Army (PLA) in 1983 and graduated from aviation college of the PLA air force in 1987. He became a pilot and in 1999 was selected with 13 others to train for the space flight at the Astronaut Training Base in Beijing and at the Jiuquan Launch Centre.

yaw The movement of the nose of a spacecraft or rocket from side to side.

Y axis An imaginary line that runs from side to side through a spacecraft.

***Yohkoh* (Japanese: 'sunbeam')** A Japanese satellite that studies solar flares and other energetic phenomena that take place on the Sun. It was launched in August 1991 from the Kagoshima Space Centre by Japan's Institute of Space and Astronautical Science. US and British scientists collaborated on its Bragg crystal spectrometer, and US scientists collaborated on the soft X-ray telescope. Other instruments are a wide-band spectrometer and hard X-ray telescope.

http://www.solar.isas.ac.jp/english/ Description of the *Yohkoh* spacecraft and its instrumentation. The site has spectacular *Yohkoh* SXT (soft X-ray telescope) images of the Sun—annotated to highlight features of special interest.

Young, John Watts (1930–) US astronaut. His first flight was on *Gemini 3* in 1965, followed by *Gemini 10* in 1966. In 1969 he flew the command module of *Apollo 10* and in 1972 commanded *Apollo 16*, which landed on the Moon. He and Charles *Duke collected 90 kg of Moon rocks. Young was commander of the first flight of a space shuttle, **Columbia*, in 1981, and commanded it on the ninth space-shuttle flight in 1983.

He was named chief of the Astronaut Office in 1975 and later became associate (technical) director of the *Johnson Space Center.

Born in San Francisco, California, in 1952 Young was awarded a Bachelor of Science degree in aeronautical engineering from the Georgia Institute of Technology. He was a test pilot with the US Navy before becoming an astronaut in 1962.

yuhangyuan Chinese name for astronauts. China's Xinhua News Agency announced in October 2000 that it would send robots on future Moon landings before yuhangyuan, who are 'extremely expensive resources'.

Yuri Gagarin Cosmonaut Training Centre Official name for the *Gagarin Cosmonaut Training Centre.

Zaletin, Sergei (1962–) Russian cosmonaut. He was assigned as the back-up crew commander for the 26th mission to the Russian space station *Mir* in 1998. He commanded the *Soyuz* mission to *Mir* in April–June 2000, being part of the final resident crew on *Mir*. He was selected as an astronaut in 1990.

***Zarya* (Russian: 'sunrise')** The control module of the *International Space Station* (*ISS*), and the first part to be launched. It was put into orbit by a Russian Proton rocket in November 1998. The Zarya Control Module, also known as the Functional Cargo Block, is 12.6 m long and 4.1 m across at its widest, with two solar wings 10.7 m long. Its side docking ports accommodate Russian *Soyuz* crew ferries and unpiloted *Progress* resupply craft. The module's sixteen fuel tanks can hold more than 5.4 tonnes of propellant. Zarya provided the initial propulsion, communications, and power for the *ISS* but after the arrival of the Zvezda Service Module and the US Destiny Laboratory it became used primarily for storage, docking, and as a fuel tank. Although built and launched by Russia, it is owned by NASA.

Z axis An imaginary line that runs through a spacecraft vertically, from top to bottom.

zenith The uppermost point of the celestial horizon, immediately above the observer; the *nadir is below, diametrically opposite. *See* CELESTIAL SPHERE.

Zenit rocket A two-stage Soviet rocket. It was first launched in 1985 and was the USSR's first new rocket for 20 years. The liquid-fuelled Zenit is able to launch a satellite weighing more than 13 500 kg into *low Earth orbit. On 21 October 2000, the rocket (operated by Sea Launch and equipped with a third stage) launched the world's heaviest commercial communication satellite, Boeing's *Thuraya-1*, weighing 5 063 kg. The Zenit may also be used for transportation, with Sea Launch offering the rockets to NASA to deliver cargo to the *International Space Station*. A Zenit-3SL rocket was used by Sea Launch to place a Brazilian satellite into orbit on 11 January 2004, launching it from a platform in the Pacific Ocean.

Zentrum für Angewandte Raumfahrttechnologie und Mikrogravitation (ZARM; Centre of Applied Space Technology and Microgravity) A scientific institution at the University of Bremen, Germany. ZARM researches the guidance, navigation, and control of spacecraft, and has conducted several experiments aboard NASA's space shuttles. It has conducted extensive research on fluid mechanics under *microgravity conditions. Its best-known facility is the Drop Tower Bremen, a 146-m concrete shaft that provides 4.74 seconds of weightlessness to objects dropped down it.

zero g(ravity) An absence of gravity or its apparent force. This phenomenon produces the state of free fall or *weightlessness. Earth-orbiting spacecraft such as the space shuttle remain within the influence of gravity, demonstrated by their adherence to an orbital route. The term 'zero gravity' is sometimes also used for weightlessness in such an orbit, but 'microgravity' is the more correct term.

Zond A series of seven Soviet interplanetary and lunar spacecraft launched 1964–70. *Zond 1* and *Zond 2* were sent towards Venus and Mars in 1964 but contact with the ships was lost. *Zond 3* flew past the Moon in July 1965. *Zond 5* (September 1968), *Zond 6* (November 1968), *Zond 7* (August 1969), and *Zond 8* (October 1970) all went around the Moon and returned to Earth; they were later revealed to be modified *Soyuz* spacecraft being tested for a crewed flight to the Moon.

Zvezda (Russian: 'star') A Russian-owned and -built module of the *International Space Station* (*ISS*), the third component to arrive. It was launched by a Russian Proton rocket in July 2000. Zvezda is similar in layout to the core section of Russia's earlier *Mir* space station. It is a stepped cylinder 13.1 m long and has two solar arrays with a wingspan of 29.7 m. Zvezda contains three pressurized compartments: a small, spherical Transfer Compartment at the forward end; the long, cylindrical main Work Compartment; and the small, cylindrical Transfer Compartment at the aft end. Spacewalks can be performed by using the Transfer Compartment as an airlock. Zvezda provided the early station living quarters, and will become the centre of the Russian segment of the *ISS* as further US and international sections are added.

Appendix

Table 1. First astronauts to orbit earth

Name and country	Spacecraft	Date of orbit	Number of orbits
Yuri Gagarin (USSR)	*Vostok 1*	12 April 1961	1
Gherman Titov (USSR)	*Vostok 2*	6–7 August 1961	17
John Glenn (USA)	*Friendship 7*	20 February 1962	3
M Scott Carpenter (USA)	*Aurora 7*	24 May 1962	3
Andriyan Nikolayev (USSR)	*Vostok 3*	11–15 August 1962	64
Pavel Popovich (USSR)	*Vostok 4*	12–15 August 1962	48
Walter M Schirra (USA)	*Sigma 7*	3 October 1962	6
L Gordon Cooper (USA)	*Faith 7*	15–16 May 1963	22
Valery Bykovsky (USSR)	*Vostok 5*	14–19 June 1963	81
Valentina Tereshkova (USSR)	*Vostok 6*	16–19 June 1963	48

Table 2. First astronauts to walk on the Moon

All 12 astronauts who visited the Moon are US citizens.

Spacecraft	Landing date	Name
Apollo 11	20 July 1969	Neil Armstrong Buzz Aldrin
Apollo 12	18 November 1969	Pete Conrad Alan Bean
Apollo 14	3 February 1971	Alan B Shepard Edgar Mitchell
Apollo 15	30 July 1971	David R Scott James B Irwin
Apollo 16	20 April 1972	John Young Charles M Duke Jr
Apollo 17	11 December 1972	Eugene Cernan Harrison H Schmitt

Table 3. First women in space

Name and country	Dates of first flight	Spacecraft or mission	Comments
Valentina Tereshkova (USSR)	16–19 June 1963	*Vostok 6*	first woman in space
Svetlana Savitskaya (USSR)	19–27 August 1982	*Soyuz/Salyut 7*	first woman to walk in space
Sally Ride (USA)	18–24 June 1983	*Challenger*	first US woman in space
Judith Resnik (USA)	30 August–5 September 1984	*Discovery*	
Kathryn Sullivan (USA)	5–13 October 1984	*Challenger*	first US woman to walk in space
Anna Fisher (USA)	8–16 November 1984	*Discovery*	
Rhea Seddon (USA)	12–19 April 1985	*Discovery*	
Shannon Lucid (USA)	17–24 June 1985	*Discovery*	in 1996 spent 188 days in space as member of joint Russian–US crew aboard *Mir*, setting US space endurance record
Bonnie Dunbar (USA)	30 October–6 November 1985	*Challenger*	
Mary Cleave (USA)	26 November–3 December 1985	*Atlantis*	

Table 4. Temperatures, surface gravities, and escape velocities of the planets

Planet	Average surface temperature °C	°F	Gravity (m/s^2)	Escape velocity (km/s)
Mercury	480	896	3.70	4.25
Venus	167	333	8.87	10.36
Earth	22	72	9.78	11.18
Mars	−23	−9	3.72	5.02
Jupiter	−150	−238	22.88	59.56
Saturn	−180	−292	9.05	35.49
Uranus	−210	−346	7.77	21.30
Neptune	−220	−364	11.00	23.50
Pluto	−230	−382	0.40	1.22

Table 5. First artificial satellites to orbit earth

Satellite and country	Date of launch
Sputnik 1 (USSR)	4 October 1957
Sputnik 2 (USSR)	3 November 1957
Explorer 1 (USA)	31 January 1958
Vanguard 1 (USA)	17 March 1958
Explorer 3 (USA)	26 March 1958
Sputnik 3 (USSR)	15 May 1958
Explorer 4 (USA)	26 July 1958
SCORE (USA)	18 December 1958
Vanguard 2 (USA)	17 February 1959
Discoverer 1 (USA)	28 February 1959

Table 6. Crewed spacecraft accidents

Date	Spacecraft	Event
16 March 1966	*Gemini 8*	Command pilot Neil Armstrong and pilot Dave Scott had just completed the first space docking with an Agena target rocket when the combination went into a spin. One of the spacecraft's thrusters was firing continuously due to a short circuit. Armstrong undocked from the Agena and the spinning got worse. An emergency set of thrusters on *Gemini 8* was fired to bring the capsule under control and an immediate return to Earth was ordered
27 January 1967	*Apollo 1*	A fire in the spacecraft, during pre-launch testing in preparation for an Earth orbital test flight on 21 February, killed three astronauts at Cape Canaveral's Pad 34. The *Apollo* programme was delayed by 21 months. The astronauts killed were Gus Grissom, one of the 'Original Seven' crew, who made the USA's second crewed space flight in 1961; Edward White, the first US astronaut to undertake a space walk (from *Gemini 4* in June 1965); and newcomer Roger Chaffee
24 April 1967	*Soyuz 1*	The first crewed flight of this new Soviet spaceship, piloted by lone cosmonaut Vladimir Komarov, hit serious technical problems soon after reaching orbit on 23 April. The launch of *Soyuz 2* to dock with *Soyuz 1* was cancelled and attempts to return Komarov to Earth were made. With the spacecraft tumbling in orbit, Komarov re-entered the Earth's atmosphere on the 18th orbit. The craft's main parachute failed to deploy properly and *Soyuz 1* plummeted to Earth, catching fire on impact. Komarov was killed instantly. The accident delayed crewed *Soyuz* flights for 17 months
13 April 1970	*Apollo 13*	A routine flight to the Moon to complete the third crewed landing was dramatically interrupted by an explosion in a fuel cell oxygen tank in the service module. With power dwindling from the command module, the crew—James Lovell, Jack Swigert, and Fred Haise—were extremely fortunate that they were still attached to the lunar module, which provided just enough power and propulsion to allow the spacecraft to return to Earth
30 June 1971	*Soyuz 11*	Having completed a record breaking 23 days in space, most of it aboard the first Soviet space station, *Salyut 1*, Georgi Dobrovolsky, Vladimir Volkov, and Viktor Patsayev were killed in space just before re-entry when the cabin of *Soyuz 11* sprang a leak (causing depressurization and loss of oxygen). The cosmonauts were not equipped with spacesuits, which could have saved them. The craft made an automatic re-entry and landing and the crew was found dead inside
5 April 1975	*Soyuz* 18–1	A long-duration visit to the *Salyut 4* space station was on the schedule for cosmonauts Vasili Lazarev and Oleg Makarov as they lifted off aboard their *Soyuz* spacecraft atop a booster of the same name. The second stage failed to separate properly from the first-stage booster section and the vehicle went off course. A computer-controlled emergency landing was made close to the border with China after a flight lasting 21 minutes 27 seconds. It could have been worse: a parachute line that snarled on a branch of a tree saved the capsule from plunging over a ravine

Table 6. (Contd.)

Date	Spacecraft	Event
27 September 1983	*Soyuz* T-10-1	Cosmonauts Vladimir Titov and Gennady Strekalov were on board their spacecraft awaiting launch to *Salyut 7*, for a long duration shift aboard the *Salyut* space station, when the *Soyuz* launcher caught fire before ignition of the rocket engines. Moments before the rocket exploded, the *Soyuz* craft's emergency escape rockets fired, hauling the spacecraft away towards a safe landing far from the blazing launch pad. During the five-minute abort flight, Titov and Strekalov endured a record acceleration force of more than 20 g
28 January 1986	*Challenger* STS-51L	The 25th space shuttle mission was launched from Pad 39B at the Kennedy Space Center, carrying six astronauts, Dick Scobee, Mike Smith, Judith Resnick, Ellison Onizuka, Ron McNair, and Gregory Jarvis, plus a teacher, Christa McAuliffe, who was to become the first US 'citizen' in space. The right-hand solid rocket booster suffered a severe mechanical fault at lift-off and 73 seconds later the shuttle broke apart and all the crew were killed instantly. The accident led to a major redesign of the solid rocket boosters and other shuttle systems and delayed the programme until September 1988
25 June 1997	*Mir*	The Russian space station *Mir* was hit by the uncrewed Progress M34 supply ship when it went out of control during a docking practice manually controlled by cosmonauts inside the space station. The Progress broke one of the solar panels on the *Spektr* module of the space station and its impact depressurized *Spektr*. The *Spektr* hatch was open and had it not been closed rapidly, the whole space station could have lost pressure and air, killing the three-person crew—Russians Vasili Tsiblyev and Alexander Lazutkin, with visiting US astronaut Michael Foale—or at least resulted in an emergency evacuation to the *Soyuz* ferry vehicle for return to Earth
1 February 2003	*Columbia* STS-107	The space shuttle, carrying the US astronauts Rick Husband, William McCool, Michael Anderson, David Brown, Kalpana Chawla, and Laurel Clark, and the first Israeli astronaut, Ilan Ramon, launched on 16 January 2003 for a 16-day microgravity research mission. The shuttle broke up and burned on re-entry, 16 minutes before landing, killing all the crew. The apparent cause was a hole in the heat shield of a wing, caused just after lift-off by foam insulation that had broken off

Table 7. Space flight: key dates

1903	Russian scientist Konstantin Tsiolkovsky publishes the first practical paper on astronautics
1926	US engineer Robert Goddard launches the first liquid-fuel rocket
1937–45	In Germany, Wernher von Braun develops the V2 rocket
4 October 1957	The USSR launches the first space satellite, *Sputnik 1*, into Earth orbit
3 November 1957	The USSR launches *Sputnik 2*, which carries a dog called Laika; she dies on board after seven days
1958	*Explorer 1*, the first US satellite, discovers the Van Allen radiation belts
12 April 1961	The USSR launches the first crewed spaceship, *Vostok 1*, with Soviet cosmonaut Yuri Gagarin on board. Before landing he completes a single orbit of 89.1 minutes at an altitude of 142–175 km/88–109 mi
5 May 1961	Alan Shepard, of the USA, makes a 15-minute sub-orbital flight reaching an altitude of 185 km/115 mi aboard the *Mercury* capsule *Freedom 7*
20 February 1962	John Glenn in *Friendship 7* becomes the first US astronaut to orbit the Earth
July 1962	*Telstar*, a US communications satellite, sends the first live television transmission between the USA and Europe
16–19 June 1963	Soviet cosmonaut Valentina Tereshkova becomes the first woman in space, making 48 orbits of the Earth in *Vostok 6*
18 March 1965	Soviet cosmonaut Alexei Leonov performs the first space walk, outside the spacecraft *Voskhod 2*
27 January 1967	US astronauts Gus Grissom, Edward White, and Roger Chaffee die during a simulated countdown when a flash fire sweeps through the cabin of *Apollo 1*
24 April 1967	Vladimir Komarov is the first person to be killed on a space mission, when his ship, *Soyuz 1* (USSR), crash-lands on the Earth
20 July 1969	US astronaut Neil Armstrong of *Apollo 11* becomes the first person to walk on the Moon
April 1970	The *Apollo 13* mission to the Moon is cut short after an on-board explosion; all three astronauts survive the return to Earth
November 1970	*Luna 17* (USSR) is launched; its space probe, *Lunokhod*, takes photographs and makes soil analyses of the Moon's surface
1971	Cosmonauts Georgi Dobrovolsky, Viktor Patsayev, and Vladislav Volkov, aboard the *Soyuz 11* mission, are the first people to die in space when their cabin depressurizes during re-entry
April 1971	The Soviet *Salyut 1*, the first orbital space station, is established; it is later visited by the *Soyuz 11* crewed spacecraft
1973	*Skylab 2*, the first US orbital space station, is established
1975	The crafts *Apollo 18* of the USA and *Soyuz 19* of the USSR make a joint flight and link up in space
1979	The European Space Agency's satellite launcher, Ariane 1, is launched
1981	The first reusable crewed spacecraft, the US space shuttle *Columbia*, is launched
18 June 1983	The US space shuttle *Challenger* carries the first five-person crew into orbit, including Sally Ride, the first US woman in space
28 November 1983	The space shuttle *Columbia*, with a crew of six, launches the European experimental platform *Spacelab*
1984	Soviet cosmonaut Svetlana Savitskaya, one of the crew of the *Soyuz* T-12, becomes the first woman to walk in space
28 January 1986	The US space shuttle *Challenger* explodes shortly after take-off, killing all seven crew members

Table 7. (Contd.)

February 1986	The Russian space station *Mir* is launched into orbit and is visited by its first crew.
1988	The US shuttle programme resumes with the launch of *Discovery*
15 November 1988	The Soviet shuttle *Buran* is launched from the rocket Energiya
December 1988	Soviet cosmonauts Musa Manarov and Vladimir Titov in the space station *Mir* complete a whole year—actually 365 days 59 minutes—in space
April 1990	The US *Hubble Space Telescope* is launched from Cape Canaveral
August 1990	The US space probe *Magellan*, launched in 1989, reaches Venus and begins mapping the planet in detail
5 April 1991	The *Gamma-Ray Observatory* is launched from the space shuttle *Atlantis* to survey the sky at gamma-ray wavelengths
18 May 1991	Helen Sharman, the first Briton in space, is launched with Anatoli Artsebarsky and Sergei Krikalev to the *Mir* space station, returning to Earth on 26 May in *Soyuz* flight TM-11 with Viktor Afanasyev and Musa Manarov
1992	European satellite *Hipparcos*, launched in 1989 to measure the position of 120,000 stars, fails to reach geostationary orbit and goes into a highly elliptical orbit, swooping to within 500 km/308 mi of the Earth every ten hours. The satellite is later retrieved
	Astronauts aboard the space shuttle *Endeavour* successfully carry out a mission to replace the *Hubble Space Telescope's* solar panels and repair its mirror
October 1992	*LAGEOS-2* (Laser Geodynamics Satellite) is released from the space shuttle *Columbia* into an orbit so stable that it will still be circling the Earth in billions of years
1994	Japan's heavy-lifting H-II rocket is launched successfully, carrying an uncrewed shuttle craft
1995	The US space shuttle *Atlantis* docks with *Mir*, exchanging crew members
4 June 1996	The Ariane 5 rocket disintegrates almost immediately after take-off, destroying four *Cluster* satellites
1997	Astronauts aboard the US space shuttle *Discovery* carry out a second service mission to improve the performance of the *Hubble Space Telescope*
	Mir undergoes increasing difficulties, following a collision with a cargo ship that depressurizes one of its modules, *Spektr*
October 1998	NASA launches the space probe *Deep Space 1* to explore interplanetary space
November 1998	A Russian Proton rocket lifts *Zarya*, the first module of the *International Space Station* (*ISS*) into Earth orbit. The *ISS* is an international project involving 15 countries, including the USA and Russia
1999	The Russian space agency abandons the problem-ridden *Mir* space station NASA loses two important Mars probes, *Mars Climate Orbiter* and *Mars Polar Lander*
July 1999	The US space shuttle *Columbia* launches the *Chandra X-Ray Observatory*
November 1999	A space shuttle crew aboard *Discovery* carry out another servicing mission on the *Hubble Space Telescope* to fix faulty gyroscopes
December 1999	The European Space Agency (ESA) launches its *X-Ray Multi-Mirror Mission Observatory* atop an Ariane 5 rocket; it is later renamed *XMM-Newton*
2000	A Dutch-based company, MirCorp, takes over the *Mir* space station with support from the Russian government
November 2000	The first crew of scientists move into the *ISS*
2001	The second and third crews take up residency in the *ISS* NASA terminates its *X-33* experimental spaceship programme

Table 7. (Contd.)

23 March 2001	The *Mir* space station is finally decommissioned and deliberately brought down into the Pacific Ocean
April 2001	Dennis Tito, 60, visits the *ISS* as the first space tourist Canadarm 2 is installed on the *ISS*
1 February 2003	The space shuttle *Columbia* breaks up on re-entry, killing all seven crew, including the first Israeli astronaut Ilan Ramon
15 October 2003	Yang Liwei becomes the first Chinese astronaut with the launch of *Shenzhou-5*

Table 8. Longest space flights

Duration (days)	Name	Country	Spacecraft
Men			
437.7	Valeri Poliakov	USSR/Russia/CIS	*Soyuz* TM-18/*Mir*/TM-20
379.6	Sergei Avdeyev	USSR/Russia/CIS	*Soyuz* TM-28/*Mir*/TM-29
365.9	Musa Manarov	USSR/Russia/CIS	*Soyuz* TM-4/*Mir*/TM-6
365.9	Vladimir Titov	USSR/Russia/CIS	*Soyuz* TM-4/*Mir*/TM-6
326.5	Yuri Romanenko	USSR	*Soyuz* TM-2/*Mir*/TM-3
311.8	Sergei Krikalyov	USSR/Russia/CIS	*Soyuz* TM-12/*Mir*/TM-13
240.9	Valeri Poliakov	USSR/Russia/CIS	*Soyuz* TM-6/*Mir*/TM-7
237.0	Leonid Kizim	USSR	*Soyuz* T-10b/*Salyut 7*/T-11
237.0	Vladimir Solovyov	USSR	*Soyuz* T-10b/*Salyut 7*/T-11
237.0	Oleg Atkov	USSR	*Soyuz* T-10b/*Salyut 7*/T-11
211.4	Valentin Lebedev	USSR	*Soyuz* T5/*Salyut 7*/T7
211.4	Anatoli Berezovoi	USSR	*Soyuz* T5/*Salyut 7*/T7
207.5	Nikolai Budarin	Russia/CIS	*Soyuz* TM-27/*Mir*/TM-29
207.5	Talgat Musabayev	Russia/CIS	*Soyuz* TM-27/*Mir*/TM-29
Women			
188.2	Shannon Lucid	USA	Space Shuttle *Atlantis* STS-76/*Mir*/STS-79
169.2	Yelena Kondakova	Russia/CIS	*Soyuz* TM 20/*Mir*

Table 9. Most-experienced space travellers

Duration (days in space)	Name	Country	Number of flights
Men			
747.6	Sergei Avdeyev	Russia/CIS	3
678.7	Valeri Poliakov	USSR/Russia/CIS	2
651.0	Anatoli Solovyov	USSR/Russia/CIS	5
624.4	Sergei Krikalev	USSR/Russia/CIS	5
609.9	Alexander Kaleri	Russia/CIS	4
555.8	Viktor Afansyev	USSR/Russia/CIS	4
552.9	Yuri Usachyov	USSR/Russia/CIS	4
541.0	Musa Manarov	USSR/Russia/CIS	2
489.1	Alexander Viktorenko	USSR/Russia/CIS	4
444.1	Nikolai Budarin	USSR	3
Women			
223.1	Shannon Lucid	USA	5
211.0	Susan Helms	USA	5
184.9	Peggy Whitson	USA	1
178.5	Yelena Kondakova	Russia/CIS	2
63.1	Tamara Jernigan	USA	5
55.9	Marsha Ivins	USA	5
50.4	Bonnie Dunbar	USA	5
49.1	Janice Voss	USA	5
41.7	Nancy Currie	USA	4
40.8	Ellen Ochoa	USA	4

Table 10. Most space flights

Record	Name	Country	Number
Men	Franklin Chang-Diaz	USA	7
	Jerry Ross	USA	
Women	Bonnie Dunbar	USA	5
	Susan Helms	USA	
	Marsha Ivins	USA	
	Tamara Jernigan	USA	
	Shannon Lucid	USA	
	Janice Voss	USA	

Table 11. Most space walks

Number of space walks	Name	Country
Men		
16	Anatoli Solovyov	USSR/Russia/CIS
10	Alexander Serebrov	USSR/Russia/CIS
	Sergei Avdeyev	Russia/CIS
9	Nikolai Budarin	Russia/CIS
	Jerry Ross	USA
	Vladimir Dezhurov	Russia/CIS
8	Talgat Musabayev	Russia/CIS
	Yuri Onufrienko	Russia/CIS
	Leonid Kizim	USSR
	Vladimir Solovyov	USSR
Women		
3	Kathryn Thornton	USA
2	Linda Godwin	USA
	Tamara Jernigan	USA
1	Susan Helms	USA
	Peggy Whitson	USA
	Svetlana Savitskaya	USSR
	Kathryn Sullivan	USA

Table 12. Oldest space travellers

Age	Name	Country
Men		
77	John Glenn	USA
61	Story Musgrave	USA
60	Dennis Tito	USA
59	Vance Brand	USA
	Jean-Loup Chrétien	France
58	Valery Ryumin	Russia
	Karl Henize	Germany
56	Roger Crouch	USA
	William Thornton	USA
55	Claude Nicollier	Switzerland
Women		
53	Shannon Lucid	USA
49	Marsha Ivins	USA
46	Roberta Bondar	Canada
40	Yelena Kondakova	USSR

Table 13. Space probes: key dates

13 September 1959	*Luna 2* hits the Moon, the first craft to do so
10 October 1959	*Luna 3* photographs the far side of the Moon
14 December 1962	*Mariner 2* flies past Venus; launch date 26 August 1962
31 July 1964	*Ranger 7* hits the Moon, having sent back 4 316 pictures before impact
14 July 1965	*Mariner 4* flies past Mars; launch date 28 November 1964
3 February 1966	*Luna 9* achieves the first soft landing on the Moon, having transmitted 27 close-up panoramic photographs; launch date 31 January 1966
2 June 1966	*Surveyor 1* lands softly on the Moon and returns 11 150 pictures; launch date 30 May 1965
13 November 1971	*Mariner 9* enters orbit of Mars; launch date 30 May 1971
3 December 1973	*Pioneer 10* flies past Jupiter; launch date 3 March 1972
29 March 1974	*Mariner 10* flies past Mercury; launch date 3 November 1973
22 October 1975	*Venera 9* lands softly on Venus and returns its first pictures; launch date 8 June 1975
20 July 1976	*Viking 1* first lands on Mars; launch date 20 August 1975
3 September 1976	*Viking 2* transmits data from the surface of Mars
20 August 1977	*Voyager 2* is launched
5 September 1977	*Voyager 1* is launched. With a faster trajectory, it will reach Jupiter sooner than *Voyager 2*
4 December 1978	*Pioneer-Venus 1* orbits Venus; launch date 20 May 1978
5 March 1979	*Voyager 1* encounters Jupiter
9 July 1979	*Voyager 2* encounters Jupiter
12 November 1980	*Voyager 1* reaches Saturn
25 August 1981	*Voyager 2* flies past Saturn
1 March 1982	*Venera 13* transmits its first colour pictures of the surface of Venus; launch date 30 October 1981
10 October 1983	*Venera 15* maps the surface of Venus from orbit; launch date 2 June 1983
2 July 1985	The European Space Agency (ESA) probe *Giotto* is launched to Halley's Comet
24 January 1986	*Voyager 2* encounters Uranus
13–14 March 1986	*Giotto* meets Halley's Comet, closest approach 596 km, at a speed 50 times faster than that of a bullet
4 May 1989	*Magellan* launches from space shuttle *Atlantis* on a 15-month cruise to Venus across 15 million km of space
25 August 1989	*Voyager 2* reaches Neptune (4 400 million km from Earth), approaching it to within 4 850 km
18 October 1989	*Galileo* is launched from space shuttle *Atlantis* for a six-year journey to Jupiter
10 August 1990	*Magellan* arrives at Venus and transmits its first pictures on 16 August 1990
6 October 1990	The ESA probe *Ulysses* is launched from space shuttle *Discovery*, to study the Sun
29 October 1991	*Galileo* makes the closest-ever approach to an asteroid, Gaspra, flying within 1 600 km
8 February 1992	*Ulysses* flies past Jupiter at a distance of 380 000 km from the surface, just inside the orbit of Io and closer than 11 of Jupiter's 16 moons
10 July 1992	*Giotto* flies at a speed of 14 kps to within 200 km of comet Grigg-Skjellerup, 12 light years (240 million km) away from Earth

Table 13. (Contd.)

25 September 1992	*Mars Observer* is launched from Cape Canaveral, the first US mission to Mars for 17 years
10 October 1992	*Pioneer Venus 1* burns up in the atmosphere of Venus
21 August 1993	Contact is lost with *Mars Observer* three days before it is due to drop into orbit around Mars
28 August 1993	*Galileo* flies past the asteroid Ida
December 1995	*Galileo*'s probe enters the atmosphere of Jupiter. It radios information back to the orbiter for 57 minutes before it is destroyed by atmospheric pressure
17 February 1996	NASA's *Near-Earth Asteroid Rendezvous* (*NEAR*) is launched to study Eros
1997	*Galileo* begins orbiting Jupiter's moons. It takes photographs of Europa for a potential future landing site, and detects molecules containing carbon and nitrogen on Callisto, suggesting that life once existed there
17 June 1997	The US *NEAR* spacecraft flies within 1 200 km of the asteroid Mathilde,taking high-resolution photographs and revealing a 25-km crater covering the 53-km asteroid
4 July 1997	The NASA *Mars Pathfinder* becomes the third spacecraft to soft land on Mars. It deploys a micro-roving vehicle called Sojourner. The *Mars Pathfinder* images are posted on Internet Web sites and accessed by millions of people all over the world
11 September 1997	NASA's *Mars Global Surveyor* enters orbit around the Red Planet to begin a highly successful mission which continues into 2000, taking high-resolution images of almost the entire Martian surface, as well as gathering vital climate data
15 October 1997	The NASA–ESA *Cassini–Huygens* mission gets underway with the successful launch of a Titan IV Centaur booster from Cape Canaveral. *Cassini* entered orbit around Saturn in July 2004 and deployed the ESA *Huygens* spacecraft to land on Titan, the largest moon of the ringed planet
1998	Analysis of high-resolution images from the *Galileo* spacecraft suggests that the icy crust of Europa, Jupiter's fourth-largest moon, may hide a vast ocean warm enough to support life
6 January 1998	NASA's *Lunar Prospector* is launched by an Athena 2 rocket from Cape Canaveral and enters orbit around the Moon to conduct a comprehensive survey of the Moon's chemical composition, including a discovery, announced in March 1998, that there could be frozen water beneath the surface of the polar regions. *Lunar Prospector* later impacts the south pole of the Moon on 31 July 1999 but no trace of water is found in the resulting impact plume
April 1998	Daniel Goldin, the administrator of NASA, gives the go-ahead for the development of the fourth and last Great Observatory series spacecraft, the *Space Infrared Telescope Facility*, to be launched in 2001. The Great Observatory series includes the *Hubble Space Telescope* which was launched in 1990
4 July 1998	Japan launches its first probe to explore Mars using an M5 booster from Kagoshima. Called *Nozomi*, the spacecraft flew past the Red Planet in 2003 but due to a propulsion failure did not go into orbit as intended
24 October 1998	The first spacecraft in NASA's New Millennium programme, called *Deep Space 1*, is launched aboard a Delta II booster from Cape Canaveral, to demonstrate a range of new technologies needed for future Solar System exploration, including ion propulsion and 'intelligent spacecraft' systems during a flight that featured a fly-by of an asteroid in 1999

Table 13. (Contd.)

11 December 1998	The *Mars Climate Orbiter* is launched by a Delta II from Cape Canaveral as part of NASA's long-term exploration of Mars. However, due to a metric–imperial value data error—made during the manufacture of the spacecraft—the craft makes a wrong approach to Mars and crashes into the Martian surface during the planned orbital insertion burn on 23 September 1999
3 January 1999	The second Mars '99 programme spacecraft, the *Mars Polar Lander*, is launched by another Delta II from Cape Canaveral but is also lost (on 6 December 1999) as it attempts to make a soft landing in a region close to the Red Planet's south pole. The dual loss of the Mars missions is a major blow to NASA and to the future exploration of Mars
7 April 1999	ESA's *Mars Express* mission is given formal approval with a contract to build the Mars orbiting spacecraft. *Mars Express* was launched in June 2003 and entered orbit around Mars in December that year, although the *Beagle 2* lander failed
June 1999	NASA gives formal approval to the development of the Next-Generation Space Telescope (NGST) to succeed the *Hubble Space Telescope* and to be launched in about 2008. NASA also says that *Hubble* will remain operational until 2010 and may be brought back to Earth aboard a space shuttle
23 July 1999	The space shuttle *Columbia* STS-93 is launched and deploys the *Chandra X-Ray Observatory*, NASA's third Great Observatory series spacecraft, into orbit
July 1999	NASA gives approval for two new *Discovery* programme spacecraft: *Messenger*, which will make two close fly-bys of the planet Mercury in 2008, the first such mission since *Mariner 10* in 1974; and *Deep Impact*, a mission to penetrate the comet P/Tempel 1 in July 2005
10 December 1999	ESA's *X-Ray Multi-Mirror telescope* is launched by an Arianespace Ariane 5 booster from Kourou, French Guiana
7 February 2000	NASA's *Stardust* spacecraft, en route to the comet Wild 2 in 2004, collects its first samples of interstellar dust, which are due to be returned to Earth in 2006, along with samples of comet dust from Wild 2
14 February 2000	NASA's *NEAR* spacecraft becomes the first to orbit an asteroid—Eros. *NEAR* takes thousands of high-resolution images and makes detailed scientific observations during a year-long mission
March 2000	NASA scientists announce that they may have to crash the *Galileo* probe in order to prevent it from contaminating Europa, one of Jupiter's moons, with microbes from Earth. They believe their best chance of locating extraterrestrial life within our Solar System is on Europa
4 June 2000	NASA's *Gamma-Ray Observatory* is deliberately de-orbited by the space agency as a precaution after gyroscope control problems
22 June 2000	Images from the *Mars Global Surveyor* revealing 'water gullies' on the planet are released by NASA
July 2000	NASA announces that twin Mars Exploration Rover missions will be launched in 2003 to land at separate locations on the Red Planet in 2004
September 2000	ESA approves the BepiColombo mission to explore the planet Mercury, to be launched in 2009
December 2000	As NASA's *Cassini* spacecraft flies past Jupiter en route to Saturn, making observations of Jupiter, the first dual-space exploration of a planet is conducted, with the Galileo Jupiter orbiter
January 2001	Contact with the first deep-space interplanetary explorer, *Pioneer 10*, is lost. The spacecraft, which explored Jupiter in 1973, was over 11 billion km/7 billion mi from Earth

Table 13. (Contd.)

12 February 2001	NASA's *NEAR Shoemaker* makes a spectacular finale to its mission at the asteroid Eros by making a carefully thruster-controlled soft landing on the surface. The originally unplanned manoeuvre makes the craft the first to land on an asteroid
March 2001	NASA budget cuts threaten to cancel the only planned mission to explore the final planet in the solar system, Pluto. The *Pluto Kuiper Express*, to fly past Pluto and its moon, Charon, and to explore the region beyond Pluto in which other objects orbit the Sun, was to have reached Pluto by 2020
7 April 2001	NASA's *Mars Odyssey* is successfully launched by a Delta rocket from Cape Canaveral. The craft reached Mars on 24 October 2001, and attained its working orbit around the planet in January 2002. The probe carries instruments designed to examine from orbit the chemical make-up of the planet's surface, and it is hoped that the information gathered may help to determine whether life has ever existed on Mars
2 June 2003	ESA launches its *Mars Express* Orbiter carrying the British-built lander *Beagle 2*. It entered orbit in December of that year but the lander failed
July 2003	NASA launches two Mars Exploration rovers, Spirit and Opportunity
25 December 2003	ESA's *Beagle 2* lander descends from the *Mars Express* Orbiter towards the Isidis Planitia region on the Martian surface. However, transmission is lost and further contact proves impossible
January 2004	NASA's rovers Spirit and Opportunity successfully descend by parachute to the Martian surface
March 2004	NASA's Opportunity rover discovers evidence that its landing site on the flat Meridiani Planum region of Mars was previously covered in water. The accompanying Spirit rover on the opposite side of the planet finds that water had once existed in the Gusev Crater. The continuing success of the mission leads NASA to extend original three-month schedule by a further five months

Table 14. Space shuttle missions

Date 1981–90	Shuttle	Crew	Duration	Remarks
12 April 1981	*Columbia* STS-1	John Young, Bob Crippen	2 days 6 hr 20 min	first flight of space shuttle
12 November 1981	*Columbia* STS-2	Joe Engle, Dick Truly	2 days 6 hr 13 min	first crewed flight of used vehicle
22 March 1982	*Columbia* STS-3	Jack Lousma, Gordon Fullerton	8 days 4 min	third test flight
27 June 1982	*Columbia* STS-4	Ken Mattingly, Hank Hartsfield	7 days 1 hr 9 min	military flight; final test flight
11 November 1982	*Columbia* STS-5	Vance Brand, Robert Overmyer, Joe Allen, William Lenoir	5 days 2 hr 14 min	first commercial mission of shuttle; deployed two communications satellites; first four-person flight
4 April 1983	*Challenger* STS-6	Paul Weitz, Karol Bobko, Don Peterson, Story Musgrave	5 days 23 min	deployed *TDRS 1*; performed first shuttle space walk
18 June 1983	*Challenger* STS-7	Bob Crippen, Rick Hauck, John Fabian, Sally Ride, Norman Thagard	6 days 2 hr 24 min	satellite deployment mission is first by five people and includes first US woman in space
30 August 1983	*Challenger* STS-8	Richard Truly, Dan Brandenstein, Dale Gardner, Guion Bluford, William Thornton	6 days 1 hr 8 min	night launch and landing
28 November 1983	*Columbia* STS-9	John Young, Brewster Shaw, Owen Garriott, Robert Parker, Byron Lichtenberg, Ulf Merbold	10 days 7 hr 47 min	flight of European *Spacelab 1*; Merbold from West Germany; first six-person flight
3 February 1984	*Challenger* STS-41B	Vance Brand, Robert Gibson, Bruce McCandless, Robert Stewart, Ronald McNair	7 days 23 hr 15 min	first independent space walk using manned manoeuvring unit by McCandless; first space mission to end at launch site
6 April 1984	*Challenger* STS-41C	Bob Crippen, Dick Scobee, George Nelson, Terry Hart, James van Hoften	6 days 23 hr 40 min	captured, repaired, and redeployed *Solar Max* satellite
30 August 1984	*Discovery* STS-41D	Hank Hartsfield, Michael Coats, Judith Resnik, Steven Hawley, Michael Mullane, Charles Walker	6 days 56 min	launch pad abort on 27 June; three satellites deployed; Walker first astronaut to represent private company

5 October 1984	*Challenger* STS-41G	Bob Crippen, Jon McBride, Sally Ride, Kathy Sullivan, David Leestma, Marc Garneau, Paul Scully Power	8 days 5 hr 23 min	first seven-person flight; first carrying two women; Ride first woman in space twice; Sullivan first US woman to walk in space; Garneau from Canada
8 November 1984	*Discovery* STS-51A	Rick Hauck, Dave Walker, Joe Allen, Dale Gardner, Anna Fisher	7 days 23 hr 45 min	two space walks to retrieve lost communications satellites and return them to Earth
24 January 1985	*Discovery* STS-51C	Ken Mattingly, Loren Shriver, Ellison Onizuka, James Buchli, Gary Payton	3 days 1 hr 33 min	military mission; Payton first USAF crewed space flight engineer
12 April 1985	*Discovery* STS-51D	Karol Bobko, Don Williams, Rhea Seddon, Jeff Hoffman, David Griggs, Charlie Walker, Jake Garn	6 days 23 hr 55 min	deployed three communications satellites; unscheduled extravehicular activity (EVA); Senator Jake Garn first passenger observer in space
29 April 1985	*Challenger* STS-51B	Bob Overmyer, Fred Gregory, Don Lind, William Thornton, Norman Thagard, Lodewijk van den Berg, Taylor Wang	7 days 8 min	*Spacelab* science research mission
17 June 1985	*Discovery* STS-51G	Dan Brandenstein, John Creighton, Shannon Lucid, Steve Nagel, John Fabian, Patrick Baudry, Abdul Aziz Al-Saud	7 days 1 hr 38 min	satellite deployment and research mission; first with three nations represented; Baudry from France, Al-Saud from Saudi Arabia
20 July 1985	*Challenger* STS-51F	Gordon Fullerton, Roy Bridges, Karl Henize, Anthony England, Story Musgrave, John-David Bartoe, Loren Acton	7 days 22 hr 45 min	launch pad abort on 12 July; one engine shut down during launch causing abort-to-orbit *Spacelab 2* research mission
27 August 1985	*Discovery* STS-51I	Joe Engle, Dick Covey, William Fisher, James van Hoften, Mike Lounge	7 days 2 hr 1 min	three satellites deployed; *Leasat 3* captured, repaired, and redeployed
23 October 1985	*Atlantis* STS-51J	Karol Bobko, Ron Grabe, Dale Hilmers, Bob Stewart, William Pailes	4 days 1 hr 45 min	military mission; Pailes is second USAF crewed space flight engineer

Table 14. (Contd.)

Date 1981–90	Shuttle	Crew	Duration	Remarks
30 October 1985	*Challenger* STS-61A	Hank Hartsfield, Steve Nagel, Bonnie Dunbar, Guion Gluford, James Buchli, Ernst Messerschmitt, Reinhard Furrer, Wubbo Ockels	7 days 44 min	West German-funded *Spacelab* D1 mission; Messerschmitt and Furrer from West Germany; Ockels from the Netherlands; first eight-person mission
27 November 1985	*Atlantis* STS-61B	Brewster Shaw, Bryan O'Connor, Mary Cleave, Jerry Ross, Sherwood Spring, Rudolpho Neri Vela, Charlie Walker	6 days 21 hr 4 min	structures assembled during EVAs; Neri Vela from Mexico
12 January 1986	*Columbia* STS-61C	Robert Gibson, Charles Bolden, Franklin Chang-Diaz, George Nelson, Steve Hawley, Robert Cenker, Bill Nelson	6 days 2 hr 4 min	much-delayed flight; Bill Nelson, a member of Congress, second 'political' passenger
28 January 1986	*Challenger* STS-51	Dick Scobee, Mike Smith, Judith Resnik, Ronald McNair, Ellison Onizuka, Christa McAuliffe, Gregory Jarvis	1 min 13 sec	broke apart at 14 325 m crew killed; first flight to take off but not to reach space; first US in-flight fatalities
29 September 1988	*Discovery* STS-26	Rick Hauck, Dick Covey, Mike Lounge, David Hilmers, George Nelson	4 days 1 hr	USA's return to space 32 months after *Challenger* disaster
2 December 1988	*Atlantis* STS-27	Robert Gibson, Guy Gardner, Jerry Ross, Mike Mullane, William Shepherd	4 days 9 hr 5 min	military mission to deploy *Lacrosse* spy satellite
13 March 1989	*Discovery STS-29*	Michael Coats, John Blaha, James Buchli, James Bagian, Robert Springer	4 days 23 hr 38 min	deployed *TDRS* satellite
4 May 1989	*Atlantis* STS-30	David Walker, Ron Grabe, Norman Thagard, Mary Cleave, Mark Lee	4 days 57 min	deployed *Magellan* for its journey to orbit planet Venus; the first deployment of a planetary spacecraft from a crewed spacecraft
8 August 1989	*Columbia* STS-28	Brewster Shaw, Richard Richards, David Leestma, James Adamson, Mark Brown	5 days 1 hr	military mission to deploy *KH-12* reconnaissance satellite
18 October 1989	*Atlantis* STS-34	Donald Williams, Michael McCulley, Shannon Lucid, Franklin Chang-Diaz, Ellen Baker	4 days 23 hr 39 min	deployed Jupiter orbiter *Galileo*
22 November 1989	*Discovery* STS-33	Frederick Gregory, John Blaha, Story Musgrave, Manley Carter, Kathryn Thornton	5 days 6 min	military mission to deploy *Magnum* elite spacecraft

9 January 1990	*Columbia* STS-32	Dan Brandenstein, James Wetherbee, Bonnie Dunbar, Marsha Ivins, David Low	10 days 21 hr	retrieved *LDEF* from orbit
28 February 1990	*Atlantis* STS-36	John Creighton, John Casper, Mike Mullane, David Hilmers, Pierre Thuot	4 days 10 hr 18 min	military mission to deploy *KH-12* reconnaissance satellite
24 April 1990	*Discovery* STS-31	Loren Shriver, Charles Bolden, Steven Hawley, Bruce McCandless, Kathryn Sullivan	5 days 1 hr 16 min	deployed *Hubble Space Telescope*
6 October 1990	*Discovery* STS-41	Richard Richards, Robert Canbana, Thomas Akers, Bruce Melnick, William Shepherd	4 days 2 hr 10 min	deployed *Ulysses* solar polar orbiter
15 November 1990	*Atlantis* STS-38	Richard Covey, Frank Culbertson, Robert Springer, Carl Meade, Sam Gemar	4 days 21 hr 54 min	military mission to deploy *Magnum* elite satellite
2 December 1990	*Columbia* STS-35	Vance Brand, Guy Gardner, Jeff Hoffman, Bob Parker, Mike Lounge, Ronald Parise, Samuel Durrance	8 days 23 hr 5 min	flight of *Astro 1* observatory
1991–2005				
5 April 1991	*Atlantis* STS-37	Steven Nagel, Kenneth Cameron, Linda Godwin, Jerry Ross, Jay Apt	5 days 23 hr 32 min	launched *Gamma-Ray Observatory*
28 April 1991	*Discovery* STS-39	Michael Coats, Blaine Hammond, Gregory Harbaugh, Donald McMonagle, Guion Blurford, Charles Veach, Richard Hieb	8 days 7 hr 22 min	developed sensors for identifying missiles in space
5 June 1991	*Columbia* STS-40	Bryan O'Connor, Sidney Gutierrez, James Bagian, Tamara Jernigan, Rhea Seddon, Drew Gaffney, Millie Hughes-Fulford	9 days 2 hr 14 min	carried *Life Sciences Laboratory*
2 August 1991	*Atlantis* STS-43	John Blaha, Michael Baker, Shannon Lucid, David Low, James Anderson	8 days 21 hr 21 min	launched *TDRS-E* communications satellite
13 September 1991	*Discovery* STS-48	John Creighton, Kenneth Reightler, James Buchli, Mark Brown, Sam Gemar	5 days 8 hr 27 min	launched *Upper-Atmosphere Research Satellite*
24 November 1991	*Atlantis* STS-44	Fred Gregory, Tom Henricks, James Voss, Story Musgrave, Mario Runco, Tom Hennen	6 days 22 hr 50 min	launched Defense Support Program early-warning satellite

Table 14. (Contd.)

Date 1991–2005	Shuttle	Crew	Duration	Remarks
22 January 1992	*Discovery* STS-42	Ronald Grabe, Stephen Oswald, Norman Thagard, William Readdy, David Hilmers, Roberta Bondar, Ulf Merbold	8 days 1 hr 14 min	carried *International Microgravity Laboratory 1*
24 March 1992	*Atlantis* STS-45	Charles Bolden, Brian Duffy, Kathryn Sullivan, David Leestma, Michael Foale, Dirk Frimout, Byron Lichtenberg	8 days 22 hr 9 min	carried Atmospheric Laboratory or ATLAS-1
7 May 1992	*Endeavour* STS-49	Daniel Brandenstein, Kevin Chilton, Bruce Melnick, Pierre Thuot, Richard Hieb, Kathryn Thornton, Thomas Akers	8 days 21 hr	rescue of *Intelsat V1* satellite
25 June 1992	*Columbia* STS-50	Richard Richards, Kenneth Bowersox, Carl Meade, Ellen Baker, Bonnie Dunbar, Lawrence Delucas, Eugene Trinh	13 days 19 hr 30 min	carried *US Microgravity Laboratory*
31 July 1992	*Atlantis* STS-46	Loren Shriver, Andrew Allen, Franklin Chang-Diaz, Jeffrey Hoffman, Claude Nicollier, Marsha Ivins, Franco Malerba	7 days 23 hr 15 min	launched *European Retrievable Carrier (EURECA)* and *Tethered Satellite*
12 September 1992	*Endeavour* STS-47	Robert Gibson, Curtis Brown, Jan Davis, Jerome Apt, Mae Jemison, Mark Lee, Mamoru Mohri	7 days 22 hr 31 min	launched Japanese *Spacelab* mission
22 October 1992	*Columbia* STS-52	James Wetherbee, Michael Baker, William Shepherd, Tamara Jernigan, Charles Veach, Steven MacLean	9 days 20 hr 56 min	launched *Laser Geodynamics Satellite 2*; carried US processing experiments and Canadian experiments
2 December 1992	*Discovery* STS-53	David Walker, Robert Cabana, Guion Bluford, James Voss, Michael Clifford	7 days 7 hr 19 min	launched Department of Defense satellite
13 January 1993	*Endeavour* STS-54	John Casper, Donald McMonagle, Gregory Harbaugh, Mario Runco, Susan Helms	5 days 23 hr 38 min	launched *TDRS-F* communications satellite
8 April 1993	*Discovery* STS-56	Kenneth Cameron, Stephen Oswald, Michael Foale, Kenneth Cockrell, Ellen Ochoa	9 days 6 hr 8 min	carried ATLAS-2 atmospheric laboratory
26 April 1993	*Columbia* STS-55	Steven Nagel, Tom Henricks, Jerry Ross, Charles Precourt, Bernard Harris, Ulrich Walter, Hans Schlegel	9 days 23 hr 39 min	carried second German *Spacelab* mission

21 June 1993	*Endeavour* STS-57	Ronald Grabe, Brian Duffy, David Low, Nancy Sherlock, Jeff Wisoff, Janice Voss	9 days 23 hr 46 min	carried *Spacelab 1* pressurized module; retrieved *EURECA* satellite
12 September 1993	*Discovery* STS-51	Frank Culbertson, William Readdy, James Newman, Daniel Bursch, Carl Walz	9 days 20 hr 11 min	launched *Advanced Communications Technology Satellite*
18 October 1993	*Columbia* STS-58	John Blaha, Richard Searfoss, Rhea Seddon, William McArthur, David Wolf, Shannon Lucid, Martin Fettman	14 days 13 min	carried second *Spacelab* life-sciences mission
2 December 1993	*Endeavour* STS-61	Richard Covey, Kenneth Bowersox, Story Musgrave, Jeffrey Hoffman, Tom Akers, Kathryn Thornton, Claude Nicollier	10 days 19 hr 58 min	mission to repair *Hubble Space Telescope*
3 February 1994	*Discovery* STS-60	Charles Bolden, Kenneth Reightler, Jan Davis, Ronald Sega, Franklin Chang-Diaz, Sergei Krikalev	8 days 7 hr 9 min	carried *Spacelab* pressurized module
4 March 1994	*Columbia* STS-62	John Casper, Andrew Allen, Pierre Thuot, Sam Gemaar, Marsha Ivins	13 days 23 hr 16 min	carried microgravity experiments
9 April 1994	*Endeavour* STS-59	Sidney Gutierrez, Kevin Chilton, Jay Apt, Rich Clifford, Linda Godwin, Thomas Jones	11 days 5 hr 49 min	carried *Space Radar Laboratory*
8 July 1994	*Columbia* STS-65	Robert Cabana, James Halsell, Richard Hieb, Carl Walz, Leroy Chiao, Donald Thomas, Chiaki Mukai	14 days 17 hr 55 min	carried *Spacelab* pressurized module for experiments in weightlessness
9 September 1994	*Columbia* STS-64	Richard Richards, Blaine Hammond, Jerry Linenger, Susan Helms, Carl Meade, Mark Lee	10 days 22 hr 49 min	carried out atmospheric studies and experiments on processing materials in space
30 September 1994	*Endeavour* STS-68	Michae Baker, Terry Wilcutt, Tom Jones, Steve Smith, Dan Bursch, Jeff Wisoff	11 days 5 hr 46 min	carried radar to study the Earth's surface and oceans
3 November 1994	*Atlantis* STS-66	Donald McMonagle, Curtis Brown, Ellen Ochoa, Joseph Tanner, Jean-François Clervoy (ESA), Scott Parazynski	10 days 22 hr 34 min	carried instruments to study the Earth's atmosphere
3 February 1995	*Discovery* STS-63	James Wetherbee, Eileen Collins, Bernard Harris, Michael Foale, Janice Voss, Vladimir Titov	8 days 6 hr 28 min	Eileen Collins first woman shuttle pilot; conducted science experiments in orbit

Table 14. (Contd.)

Date 1991–2005	Shuttle	Crew	Duration	Remarks
2 March 1995	*Endeavour* STS-67	Stephen Oswald, William Gregory, Tamara Jernigan, John Grunsfeld, Wendy Lawrence, Samuel Durrance, Ronald Parise	16 days 15 hr 8 min	carried *Astro 2* observatory
27 June 1995	*Atlantis* STS-71	Robert Gibson, Charles Precourt, Ellen Baker, Gregory Harbaugh, Anatoli Solovyov, Nikolai Budarin	9 days 19 hr 23 min	first shuttle–*Mir* mission, landed with eight people
13 July 1995	*Discovery* STS-70	Tom Henricks, Kevin Kregal, Don Thomas, Nancy Sherlock, Mary Ellen Weber	8 days 23 hr 20 min	deployed *TDRS* satellite
7 September 1995	*Endeavour* STS-69	David Walker, Ken Cockrell, James Voss, James Newman, Michael Hernhardt	10 days 20 hr 28 min	deployed and retrieved spacecraft and conducted space walks
20 October 1995	*Columbia* STS-73	Ken Bowersox, Kent Rominger, Kathryn Thornton, Michael Lopez-Algeria, Catherine Coleman, Fred Leslie, Albert Sacco	15 days 21 hr 52 min	*US Microgravity Laboratory 2* mission
12 November 1995	*Atlantis* STS-74	Ken Cameron, James Halsell, Jerry Ross, Bill McArthur, Chris Hadfield	8 days 4 hr 30 min	second shuttle–*Mir* mission
11 January 1996	*Endeavour* STS-72	Brian Duffy, Brent Jett, Leroy Chiao, Daniel Barry, Winston Scott, Kiochi Wakata	8 days 22 hr	deployments, retrievals, space walks
22 February 1996	*Columbia* STS-75	Andrew Allen, Scott Horowitz, Maurizio Cheli, Claude Nicollier, Jeff Hoffman, Franklin Chang-Diaz, Umberto Guidoni	15 days 5 hr 15 min	deployed Italian tethered satellite which broke loose
22 March 1996	*Atlantis* STS-76	Kevin Chilton, Richard Searfoss, Ronald Sega, Rich Clifford, Linda Godwin, Shannon Lucid	9 days 5 hr 15 min	third shuttle–*Mir* mission; delivered Lucid
19 May 1996	*Endeavour* STS-77	John Caspar, Curtis Brown, Dan Bursch, Mario Runco Andrew Thomas, Marc Garneau	10 days 39 min	science, deployments, and retrievals
20 June 1996	*Columbia* STS-78	Tom Henricks, Kevin Kregal, Susan Helms, Charles Brady, Richard Linnehan, Jean-Jaques Favier, Robert Thirsk	16 days 21 hr 47 min	*Spacelab* Life and Microgravity mission
16 September 1996	*Atlantis* STS-79	William Readdy, Terrance W lcutt, Tom Akers, Jerome Apt, Carl Walz, John Blaha	10 days 13 hr 18 min	fourth shuttle–*Mir* mission; returned Lucid, delivered Blaha

19 November 1996	*Columbia* STS-80	Ken Cockrell, Kent Rominger, Tamara Jernigen, Thomas Jones, Story Musgrave	17 days 15 hr 53 min	science, deployments, retrievals; longest shuttle mission
12 January 1997	*Atlantis* STS-81	Mike Baker, Brent Jett, John Grunsfield, Jeff Wisoff, Marsha Ivins, Jerry Linenger	10 days 4 hr 55 min	fifth shuttle–*Mir* mission; delivered Linenger, returned Blaha
11 February 1997	*Discovery* STS-82	Ken Bowersox, Scott Horowitz, Steven Hawley, Mark Lee, Joe Tanner, Steve Smith, Greg Harbaugh	9 days 23 hr 37 min	second mission to service the *Hubble Space Telescope*; five space walks
4 April 1997	*Columbia* STS-83	Jim Halsell, Susan Still, Janice Voss, Don Thomas, Mike Gernhardt, Roger Crouch, Greg Linteris	3 days 23 hr 12 min	*Microgravity Science Laboratory* mission; aborted after fuel cell fault
15 May 1997	*Atlantis* STS-84	Charles Precourt, Eileen Collins, Carlos Noriega, Jean Francois Clervoy, Edward Lu, Michael Foale, Yelena Kondakova	9 days 5 hr 15 min	sixth shuttle–*Mir* mission; delivered Foale, returned Linenger
1 July 1997	*Columbia* STS-94	James Halsell, Susan Still, Janice Voss, Don Thomas, Mike Gernhardt, Roger Crouch, Greg Linteris	15 days 16 hr 44 min	reflight of the STS-83 mission; fastest return to space for any space travellers and first complete reflight of an entire crew
7 August 1997	*Discovery* STS-85	Curtis Brown, Kent Rominger, Jan Davis, Robert Curbeam, Steven Robinson, Bjarni Tryggvason	11 days 20 hr 26 min	Earth observation, science mission
26 September 1997	*Atlantis* STS-86	James Wetherbee, Mike Bloomfield, Scott Parazinsky, Vladimir Titov, Jean Loup Chretien, Wendy Lawrence, David Wolf	10 days 19 hr 29 min	seventh shuttle–*Mir* mission; delivered Wolf, returned Foale
19 November 1997	*Columbia* STS-87	Hevin Kregel, Steven Lindsey, Winston Scott, Kalpana Chawla, Takeo Doi, Leonid Kadenyuk	15 days 16 hr 34 min 4 sec	science flight, two space walks; Doi from Japan, Chawla born in India (resident in USA), Kadenyuk from the Ukraine
22 January 1998	*Endeavour* STS-89	Terrence Willcutt, Joe Frank Edwards, Bonnie Dunbar, Michael Anderson, James Reilly, Andrew Thomas	8 days 19 hr 46 min 54 sec	eighth shuttle–*Mir* mission; Thomas replaces Wolf who returns with a flight time of 127 days
2 June 1998	*Discovery* STS-91	Charles Precourt, Dominic Gorie, Wendy Lawrence, Franklin Chang-Diaz, Janet Kavandi, Valery Ryumin, Andrew Thomas	8 days 19 hr 53 min 36 sec	final shuttle–*Mir* mission; Thomas returns with a flight time of 4.5 months; Chang-Diaz sets a new record for shuttle flight times with 1 211 hours

Table 14. (Contd.)

Date 1991–2005	Shuttle	Crew	Duration	Remarks
29 October 1998	*Discovery* STS-95	Curtis Brown, Steven Lindsey, Scott Parazynski, Stephen Robinson, Pedro Duque, Chiaki Mukai, John Glenn	8 days 21 hr 45 min	*Spacelab* flight with microgravity and life science experiments; Glenn becomes the oldest astronaut at 77; Duque from European Space Agency, Mukai from Japan
4 December 1998	*Endeavour* STS-88	Robert Cabana, Frederick Sturckow, Nancy Currie, Jerry Ross, James Newman, Sergei Krikalev	11 days 19 hr 18 min	first assembly of the *International Space Station*, connecting the Unity module to the Zarya module; Krikalev from Russia
27 May 1999	*Discovery* STS-96	Kent Rominger, Rick Husbard, Ellen Ochoa, Tamara Jernigan, Daniel Barry, Julie Payette, Valery Tokarev	9 days 19 hr 13 min	logistics and resupply mission to *International Space Station*; Payette from Canada, Tokarev from Russia
23 July 1999	*Columbia* STS-93	Eileen Collins, Jeffrey Ashby Steven Hawley, Catherine Coleman, Michel Tognini	4 days 22 hr 49 min	deploys *Chandra X-Ray Observatory*; Tognini from France
19 December 1999	*Columbia* STS-103	Curtis Brown, Scott Kelly, Steven Smith, Michael Foale, John Grunsfeld, Claude Nicollier, Jean-François Clervoy	7 days 23 hr 10 min 47 sec	third servicing mission for *Hubble Space Telescope*; second-, third-, and fourth-longest space walks ever
11 February 2000	*Columbia* STS-99	Kevin Kregel, Dominic Gorie, Janet Kavandi, Janice Voss, Mamoru Mohri, Gerhard Thiele	11 days 5 hr 39 min	Shuttle Radar Topography Mission to map the Earth; Mohri from Japan, Thiele from Germany
19 May 2000	*Atlantis* STS-101	James Halsell, Scott Horowitz, Mary Ellen Weber, Jeffrey Williams, James Voss, Susan Helms, Yuri Usachev	9 days 20 hr 10 min 10 sec	refurbishment and resupply mission to the *International Space Station*; Usachev from Russia
8 September 2000	*Atlantis* STS-106	Terence Wilcutt, Scott Altman, Daniel Burbank, Edward Lu, Richard Mastracchio, Yuri Malenchenko, Brosi Morukov	11 days 18 hr 40 min	preparation of the living quarters of the *International Space Station* for its first residents; Malenchenko and Morukov from Russia
11 October 2000	*Discovery* STS-92	Brian Duffy, Pamela Melroy, Leroy Chiao, Peter Wisoff, Michael Lopez-Alegria, William McArthur, Koichi Wakata	12 days 21 hr 42 min	equipment and supplies taken to the *International Space Station*; rescue backpack demonstrated; Wakata from Japan
30 November 2000	*Endeavour* STS-97	Brent Jett, Michael Bloomfield, Joseph Tanner, Carlos Noriega, Marc Garneau	11 days 19 hr 57 min	delivery of new solar arrays to the *International Space Station*; Garneau from Canada
7 February 2001	*Atlantis* STS-98	Kenneth Cockrell, Mark Polansky, Robert Curbeam, Thomas Jones, Marsha Ivins	12 days 21 hr 20 min	delivery of the US laboratory Destiny to the *International Space Station*

8 March 2001	*Discovery* STS-102	James Wetherbee, James Kelly, Andrew Thomas, Paul Richards, James Voss, Susan Helms	12 days 19 hr 49 min	delivery of Voss and Helms to the *International Space Station* to replace William Shepherd, Yuri Gidzenko, and Sergei Krikalev
19 April 2001	*Endeavour* STS-100	Kent Rominger, Jeffrey Ashby, Chris Hadfield, John Phillips, Scott Parazynski, Umberto Guidoni, Yuri Lonchakov	10 days 19 hr 19 min	delivery of the Canadarm 2 and the *Raffaello* Multi-Purpose Logistics Module (MPLM) to the *International Space Station*; Guidoni from Italy, Lonchakov from Russia
12 July 2001	*Atlantis* STS-104	Steven Lindsey, Charles Hobaugh, Michael Gernhardt, Janet Kavandi, James Reilly	12 days 18 hr 35 min	delivery of the Quest joint airlock to the *International Space Station*
10 August 2001	*Discovery* STS-105	Dominic Gorie, Mark Kelly, Linda Godwin Daniel Tani, Yuri Onufrienko, Carl Walz, Daniel Bursch Frank Culbertson, Mikhail Turin, Vladimir Dezhurov	11 days 21 hr 12 min	logistics and crew exchange flight to the *International Space Station*
5 December 2001	*Endeavour* STS-108	Scott Horowitz, Rick Sturchow, Dan Barry, Patrick Forrester, Frank Culberston, Vladimir Dezhurov, Mikhail Tyurin	10 days 19 hr 57 min	logistics flight to the *International Space Station*
1 March 2002	*Columbia* STS-109	Scott Altman, Duane Carey, John Grunsfeld, Nancy Currie, James Newman, Richard Linnehan, Michael Massimino	10 days 22 hr 11 min	visit and service the *Hubble Space Telescope*
8 April 2002	*Atlantis* STS-110	Michael Bloomfield, Stephen Frick, Jerry Ross, Steven Smith, Ellen Ochoa, Lee Morin, Rex Walheim	10 days 19 hr 43 min	Deliver the *International Space Station* Center Integrated Truss Assembly (ITS)
5 June 2002	*Endeavour* STS-111	Kenneth Cockrell, Paul Lockhart, Franklin Chang-Diaz, Philippe Perrin, Valeri Korzun, Peggy Whitson, Sergei Treschev, Yuri Onufriyenko, Carl Walz, Daniel Bursch	13 days 25 hr 13 min	logistics flight to the *International Space Station*
7 October 2002	*Atlantis* STS-112	Jeffrey Ashby, Pamela Melroy, David Wolf, Piers Sellers, Sandra Magnus, Fyodor Yurchikhin	10 days 19 hr 58 min	*International Space Station* Assembly Mission 9A
23 November 2002	*Endeavour* STS-113	James Wetherbee, Paul Lockhart, Michael Lopez-Alegria, John Herrington, Kenneth Bowersox, Nikolai Budarin, Donald Pettit, Valeri Korzun, Peggy Whitson, Sergei Treschev	13 days 18 hr 48 min	*International Space Station* Assembly Mission 11A Integrated Truss Assembly P1
16 January 2003	*Columbia* STS-107	Rick Husband, William McCool, Michael Anderson, Kalpana Chawla, David Brown, Laurel Clark, Ilan Ramon	15 days 22 hr 20 min	research mission, Freestar. Broke apart on return to Earth while travelling at 20 100 kph (Mach 18.3) at an altitude of over 63 000 m, resulting in the loss of both vehicle and crew
26 July 2005	*Discovery* STS-114	Eileen Collins, James Kelly, Soichi Noguchi, Stephen Robinson, Andrew Thomas, Wendy Lawrence, Charles Camarda	13 days 21 hr 33 min	Shuttle's 'return to flight' after modifications following loss of *Columbia* and crew in February 2003. Evaluated new safety procedures, serviced *International Space Station*, and made unscheduled repairs to the shuttle's heat shield

Table 15. Most-experienced space walkers

Total duration of space walks	Name(s)	Country	Remarks
Men			
77 hr 41 min	Anatoli Solovyov	Russia	16 space walks
58 hr 27 min	Jerry Ross	USA	9 space walks
49 hr 34 min	Steven Smith	USA	7 space walks
46 hr 14 min	Nikolai Budarin	Russia	9 space walks
43 hr 13 min	James Newman	USA	6 space walks
Women			
21 hr 15 min	Kathryn Thornton	USA	3 space walks
10 hr 14 min	Linda Godwin	USA	2 space walks
8 hr 56 min	Susan Helms	USA	1 space walk
8 hr 41 min	Tamara Jernigan	USA	2 space walks
4 hr 23 min	Peggy Whitson	USA	1 space walk
Longest Time on Moon			
3 days 2 hr 59 min	Eugene Cernan, Jack Schmitt	USA	including 22 hr 5 min in three Moon walks

Table 16. Longest space walks

Duration	Name(s)	Country	Mission	Spacecraft	Date
8 hr 56 min	James Voss, Susan Helms	USA	Earth orbit	space shuttle *Discovery* STS-105	11 March 2001
8 hr 29 min	Tom Akers, Rick Hieb, Pierre Thuot	USA	Earth orbit	space shuttle *Endeavour* STS-49	13 May 1992
8 hr 15 min	Steven Smith, John Grunsfeld	USA	Earth orbit	space shuttle *Discovery* STS-103	22 December 1999
8 hr 10 min	Michael Foale, Claude Nicollier	USA	Earth orbit	space shuttle *Discovery* STS-103	23 December 1999
8 hr 8 min	Steven Smith, John Grunsfeld	USA	Earth orbit	space shuttle *Discovery* STS-103	24 December 1999

Table 17. Youngest space travellers

Age on date of first launch	Name	Country
Men		
25	Gherman Titov	Russia
26	Boris Yegorov	Russia
27	Yuri Gagarin	Russia
28	Dumitriu Prunariu	Romania
	Valery Bykovsky	Russia
	Mark Shuttleworth	South Africa
29	Salman Al-Saud	Saudi Arabia
Women		
26	Valentina Tereshkova	Russia
28	Helen Sharman	UK
32	Sally Ride	USA
	Tamara Jernigan	USA

Oxford Paperback Reference

A Dictionary of Psychology
Andrew M. Colman

Over 10,500 authoritative entries make up the most wide-ranging dictionary of psychology available.

'impressive ... certainly to be recommended'

Times Higher Educational Supplement

'Comprehensive, sound, readable, and up-to-date, this is probably the best single-volume dictionary of its kind.'

Library Journal

A Dictionary of Economics
John Black

Fully up-to-date and jargon-free coverage of economics. Over 2,500 terms on all aspects of economic theory and practice.

A Dictionary of Law

An ideal source of legal terminology for systems based on English law. Over 4,000 clear and concise entries.

'The entries are clearly drafted and succinctly written ... Precision for the professional is combined with a layman's enlightenment.'

Times Literary Supplement

OXFORD